# Weather Prediction by Numerical Process

Second edition

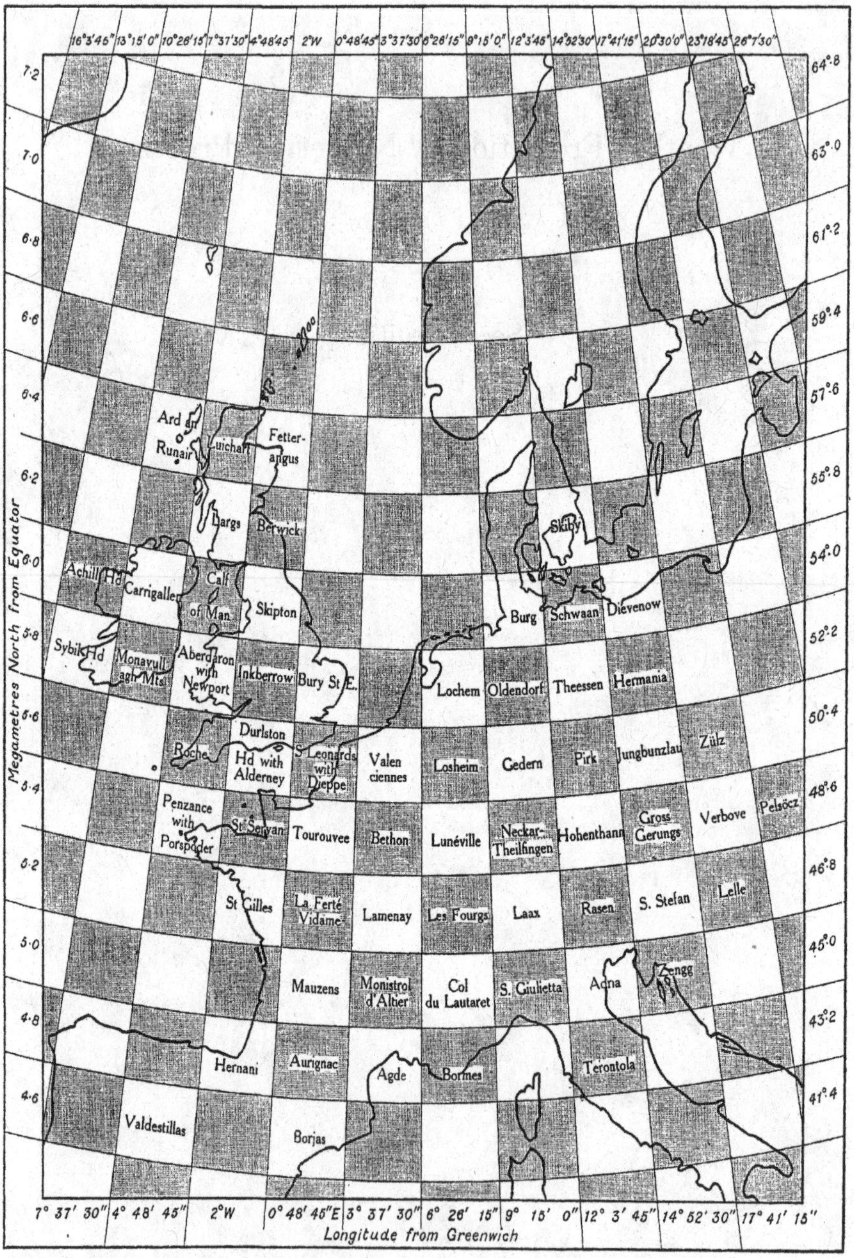


An arrangement of meteorological stations designed to fit with the chief mechanical properties of the atmosphere. Other considerations have been here disregarded. Pressure to be observed at the centre of each shaded chequer, velocity at the centre of each white chequer. The numerical coordinates refer to these centres as also do the names, although as to the latter there may be errors of ·5 or 10 km. The word "with" in "St Leonards with Dieppe" etc. is intended to suggest an interpolation between observations made at the two places. See page 9, and Chapters 3 and 7. Contrast the existing arrangement shown on p. 184.

# WEATHER PREDICTION

BY

# NUMERICAL PROCESS

Second edition

BY

LEWIS F. RICHARDSON, B.A., F.R.MET.SOC., F.INST.P.

FORMERLY SUPERINTENDENT OF ESKDALEMUIR OBSERVATORY

LECTURER ON PHYSICS AT WESTMINSTER TRAINING COLLEGE

with a Foreword by Peter Lynch
University College, Dublin

CAMBRIDGE
UNIVERSITY PRESS

CAMBRIDGE
UNIVERSITY PRESS

Shaftesbury Road, Cambridge CB2 8EA, United Kingdom

One Liberty Plaza, 20th Floor, New York, NY 10006, USA

477 Williamstown Road, Port Melbourne, VIC 3207, Australia

314–321, 3rd Floor, Plot 3, Splendor Forum, Jasola District Centre, New Delhi – 110025, India

103 Penang Road, #05–06/07, Visioncrest Commercial, Singapore 238467

Cambridge University Press is part of Cambridge University Press & Assessment, a department of the University of Cambridge.

We share the University's mission to contribute to society through the pursuit of education, learning and research at the highest international levels of excellence.

www.cambridge.org
Information on this title: www.cambridge.org/9780521680448



First edition published 1922
Second edition published 2007

*A catalogue record for this publication is available from the British Library*

ISBN 978-0-521-68044-8 Paperback

## FOREWORD

Accurate weather forecasts based on computer simulation are now produced as a routine, and have reached such a level of reliability that the rare forecast failures evoke a strong reaction in the media and amongst users. Numerical simulation of an ever-increasing range of geophysical phenomena is adding enormously to our understanding of complex processes in the Earth system. The consequences for mankind of ongoing climate change will be far-reaching. Earth System Models are capable of replicating climate regimes of past millennia and are the best means we have of predicting the future of our climate.

The basic ideas of numerical forecasting and climate modelling date from long before the first electronic computer was constructed. These techniques were first developed by Lewis Fry Richardson about a century ago, and set down in this book. Richardson was concerned with establishing a scientific method of predicting the weather. Since he was not aware of the dominant role of dynamics in the short term, he gave as much weight to small-scale physical processes as to large-scale dynamics. As a result, the algorithm he produced amounts, in essence, to a general circulation model of the atmosphere, capable of describing both weather and climate.

The first explicit analysis of the weather prediction problem from a scientific viewpoint was undertaken by the Norwegian scientist Vilhelm Bjerknes. Richardson's forecasting scheme amounts to a precise and detailed iniplementation of Bjerknes' programme. Richardson had developed a versatile technique for calculating approximate solutions of nonlinear partial differential equations by numerical approximation. Realizing that it could be applied to the evolution of atmospheric flows, he laid out the principles of scientific weather prediction in this book. He constructed a systematic algorithm for generating the numerical solution of the governing equations, and he applied it to a real-life case, calculating the initial changes in pressure and wind.

Although mathematically correct, Richardson's prediction was physically unrealistic. The essence of the problem is that a delicate dynamical balance between the fields of mass and motion prevails in the atmosphere. This was absent from the initial data used by Richardson; only later did he come to understand this problem, The consequence of the imbalance was the contamination of the forecast by spurious noise. As a result, his 'forecast' was a failure. The significance of Richardson's work was not therefore immediately evident, and his book had little influence in the initial decades after its appearance. The computational complexity of the process and the disastrous results of the single trial forecast tended to deter others from following the trail mapped out by him.

* * *

Richardson's life and work are discussed in a biography by Oliver Ashford and his *Collected Papers* have been published by Cambridge University Press. Lewis Fry Richardson was born in 1881, the youngest of seven children of David Richardson and Catherine Fry. He was educated at Bootham, the Quaker school in York, entered King's College, Cambridge, in 1900 and graduated in 1903. In 1909 he married Dorothy Garnett. Richardson began serious work on weather prediction in 1913 when he joined the Meteorological Office and was appointed Superintendent of Eskdalemuir Observatory. In May 1916 he resigned from the Met. Office in order to work with the Friends Ambulance Unit in France. During his two years as an ambulance driver, he carried out the computations for the trial forecast that he describes in this book. After the war, he rejoined the Met. Office to work at Benson with W. H. Dines. However, when the Office came under the authority of the Air Ministry, he felt obliged, as a committed pacifist, to resign once more. His meteorological research now focussed primarily on atmospheric turbulence. Several of his publications during this period are still cited by scientists. In one of these he derived a criterion for the onset of turbulence, introducing what we now call the Richardson Number.

Around 1926, Richardson made a deliberate break with meteorological research: he was distressed that his turbulence research was being exploited for military purposes. From about 1935 until his death in 1953, Richardson thrust himself energetically into peace studies, developing mathematical theories of human conflict and the causes of war. He pioneered the application of quantitative methods in this extraordinarily difficult area. In the course of these studies, he digressed to consider the lengths of geographical borders and coastlines, discovering the scaling properties that later resulted in the theory of fractals.

Richardson's genius was to apply mathematical methods to problems that had traditionally been regarded as beyond quantitative assault. The continuing relevance and usefulness of his work confirms the value of his ideas. The approximate methods that he developed for the solution of differential equations are extensively used today in the numerical treatment of physical problems.

* * *

Ch. 1 of the book is a summary of its contents. Richardson's plan is to apply his finite difference method to the problem of weather forecasting. The fundamental idea is that the numerical values of atmospheric pressures, velocities, etc., are tabulated at certain latitudes, longitudes and heights so as to give a general description of the synoptic state of the atmosphere. The physical laws determine how these quantities change with time. The laws are used to formulate an arithmetical procedure which, when applied to the numerical tables, yields the corresponding values after a brief interval of time, $\Delta t$. The process can be repeated so as to yield the state of the atmosphere after $2\Delta t$, $3\Delta t$, and so on, until the desired forecast length is reached.

In Ch. 2 the method of numerical integration is illustrated by application to a simple linear 'shallow-water' model. Richardson's step-by-step description of his method and calculations is clear and explicit and still serves as a good introduction to the process of numerical weather prediction. It is a remarkable coincidence that the initial state that Richardson chose (illustrated on page 6) corresponds closely to a natural oscillation of the atmosphere, the gravest symmetric rotational Hough mode or 'five-day wave' (the pressure and meridional winds are identical; only the zonal winds differ). This mode progresses westward, with a period of about five days. Richardson was unaware of this and, observing that 'actual cyclones move eastward', rejected geostrophic initial winds as unsuitable.

Ch, 3 describes the choice of coordinates and the discrete grid to be used. The following three chapters, comprising half the book, are devoted to assembling a set of equations suitable for Richardson's purposes. The complete system of fundamental equations was, for the first time, set down in a systematic way in Ch. 4. Richardson formulated a description of atmospheric phenomena in terms of seven differential equations. To solve them, he divided the atmosphere into discrete columns of extent 3° east–west and 200 km north–south, giving 12 000 columns to cover the globe. Each of these columns was divided vertically into five boxes. The values of the variables were given at the centre of each box, and the differential equations were approximated by expressing them in finite difference form – the computational grid is illustrated on the frontispiece of the book. The rates of change of the variables could then be calculated by arithmetical means.

Hidden away on page 66 is Richardson's famous rhyme, 'Big whirls have little whirls that feed on their velocity, and little whirls have lesser whirls and so on to viscosity', which beautifully encapsulates the turbulent energy cascade in the atmosphere. Scientists are still debating the nature of this multi-stage energy transfer. As Richardson assumed an atmosphere in hydrostatic balance, there was no prognostic equation for the vertical velocity. Ch. 5 is devoted to the derivation of a diagnostic equation for this quantity, a major contribution to dynamic meteorology. Ch. 6 considers the special measures that must be taken for the uppermost layer, the stratosphere, a region later described by Richardson as 'a happy hunting-ground for meteorological theorists'. Ch. 7 gives details of the finite difference scheme, explaining the rationale for the choice of a staggered grid.

In Ch. 8 the forecasting 'algorithm' is presented in detail. The description of the method is sufficiently detailed and precise to enable a computer program based on it to be written. Ch. 9 describes the celebrated trial forecast and its unfortunate results. The preparation of the initial data is outlined – the data are tabulated on page 185. The calculations themselves are presented on a set of 23 computer forms, rather like in a modern spread-sheet program. But the forms were completed manually: 'multiplications were mostly worked by a 25 centim slide rule'. The rate of rise of surface pressure, found on Form $P_{XIII}$, was 145 millibars in 6 hours, a totally unrealistic value. Richardson described his forecast as 'a fairly correct deduction from a somewhat unnatural initial distribution'.

In Ch. 10 Richardson discusses five smoothing techniques. Such methods are crucial for the success of modern computer forecasting models. In a sense, this chapter contains the key to solving the difficulties with Richardson's forecast. He certainly appreciated its importance for he stated, at the beginning of the following chapter, 'The scheme of numerical forecasting has developed so far that it is reasonable to expect that when the smoothing . . . has been arranged, it may give forecasts agreeing with the actual smoothed weather.' Ch. 11 considers 'Some Remaining Problems' relating to observations and to eddy diffusion, and also contains the oft-quoted passage depicting the Forecast Factory (page 291). Finally, Ch. 12 deals with units and notation and contains a full list of symbols, giving their meanings in English and in Ido, a then-popular international language.

* * *

In the aftermath of the First World War, there was considerable delay in printing the book. It was thoroughly revised in 1920–1 and was finally published by Cambridge University Press in 1922 at a price of 30 shillings (£1.50), the print run being 750 copies. The book was certainly not a commercial success. The impracticality of the method and the abysmal failure of the solitary sample forecast inevitably attracted adverse criticism. Napier Shaw, reviewing the book for *Nature,* wrote that Richardson 'presents to us a *magnum opus* on weather prediction'. However, in regard to the forecast, he observed that the wildest guess at the pressure change would not have been wider of the mark. The book was re-issued in 1965 as a Dover paperback and the 3000 copies, priced at $2, about the same as the original hardback edition, were sold out within a decade. The Dover edition was identical to the original except for a six-page introduction by Sydney Chapman.

In his Preface, Richardson wrote that the investigation of numerical prediction 'grew out of a study of finite differences and first took shape in 1911 as the fantasy which is now relegated to Ch. 11/2'. Richardson's forecast was confined to the calculation of the initial changes in two columns over central Europe, one for mass variables and one for winds. The computation of these *twenty numbers* (which appear on page 211) took him some two years. Recognizing that a practical implementation of his method would involve a phenomenal amount of numerical calculation, he imagined a fantastic Forecast Factory with a huge staff of human computers busily calculating the terms in the fundamental equations and combining their results in an ingeniously organized way to produce a weather forecast. This may be the earliest example of massively parallel processing. Richardson estimated that 64000 people would be required to compute the atmospheric changes at the speed that they were taking place. Coincidentally, the fastest computer in the TOP500 list as of June 2005 was the IBM BlueGene/L with 65536 (64K) processors!

Richardson expressed a dream that, 'some day in the dim future', numerical weather prediction would become a practical reality, However, there were several major practical obstacles to be overcome before numerical prediction could be put into practice. A fuller understanding of atmospheric dynamics allowed the development of simplified systems of equations; regular radiosonde observations of the free atmosphere and, later, satellite data, provided the initial conditions; stable finite difference schemes were developed; and powerful electronic computers provided a practical means of carrying out the prodigious calculations required to predict the changes in the weather.

Progress in weather forecasting and in climate modelling over the past fifty years has been dramatic. The useful range of deterministic prediction is increasing by about one day each decade, and seasonal forecasting skill is expected to increase significantly in the near future, As our knowledge of the atmosphere grows, so does our understanding of Richardson's remarkable vision and audacity. While his book had little effect in the short term, his methods are at the core of atmospheric simulation and it may be reasonably claimed that his work is the basis of modern weather and climate forecasting.

Peter Lynch
*UCD Dublin*

# CONTENTS

# PREFACE

THE process of forecasting, which has been carried on in London for many years, may be typified by one of its latest developments, namely Col. E. Gold's *Index of Weather Maps**. It would be difficult to imagine anything more immediately practical. The observing stations telegraph the elements of present weather. At the head office these particulars are set in their places upon a large-scale map. The index then enables the forecaster to find a number of previous maps which resemble the present one. The forecast is based on the supposition that what the atmosphere did then, it will do again now. There is no troublesome calculation, with its possibilities of theoretical or arithmetical error. The past history of the atmosphere is used, so to speak, as a full-scale working model of its present self.

But—one may reflect—the *Nautical Almanac*, that marvel of accurate forecasting, is not based on the principle that astronomical history repeats itself in the aggregate. It would be safe to say that a particular disposition of stars, planets and satellites never occurs twice. Why then should we expect a present weather map to be exactly represented in a catalogue of past weather? Obviously the approximate repetition does not hold good for many days at a time, for at present three days ahead is about the limit for forecasts in the British Isles. This alone is sufficient reason for presenting, in this book, a scheme of weather prediction, which resembles the process by which the *Nautical Almanac* is produced, in so far as it is founded upon the differential equations, and not upon the partial recurrence of phenomena in their ensemble.

The scheme is complicated because the atmosphere is complicated. But it has been reduced to a set of computing forms. These are ready† to assist anyone who wishes to make partial experimental forecasts from such incomplete observational data as are now available. In such a way it is thought that our knowledge of meteorology might be tested and widened, and concurrently the set of forms might be revised and simplified. Perhaps some day in the dim future it will be possible to advance the computations faster than the weather advances and at a cost less than the saving to mankind due to the information gained. But that is a dream.

The present distribution of meteorological stations on the map has been governed by various considerations: the stations have been outgrowths of existing astronomical or magnetic observatories; they have adjoined the residence of some independent enthusiast, or of the only skilled observer available in the district; they have been set out upon the confines of the British Isles so as to include between them as much weather as possible; or they have been connected with aerodromes in order to

* *Meteor. Office Geophysical Memoir*, No. 16, deals mainly with types of pressure distribution but foreshadows a more general indexing.

† Printed blank forms may be obtained from the Cambridge University Press, Fetter Lane, E.C. 4.

exchange information with airmen. On the map the dots representing the positions of the stations look as if they had fallen from a pepperpot. The nature of the atmosphere, as summarized in its chief differential equations, appears to have been without influence upon the distribution. We shall examine in Ch. VII what would happen if these differential equations were the sole consideration. The result is represented in the frontispiece.

The extensive researches of V. Bjerknes and his School are pervaded by the idea of using the differential equations for all that they are worth. I read his volumes on *Statics and Kinematics** soon after beginning the present study, and they have exercised a considerable influence throughout it; especially, for example, in the adoption of conventional strata, in the preference for momentum-per-volume rather than of velocity, in the statical treatment of the vertical column, and in the forced vertical motion at the ground. But whereas Prof. Bjerknes mostly employs graphs, I have thought it better to proceed by way of numerical tables. The reason for this is that a previous comparison† of the two methods, in dealing with differential equations, had convinced me that the arithmetical procedure is the more exact and the more powerful in coping with otherwise awkward equations. Graphical methods are sometimes elegant when the problem involves irregularly curved boundaries. But the atmospheric boundary, at the earth, nearly coincides with one of the coordinate surfaces, so that graphs would have no advantage over arithmetic in that respect.

It has been customary to regard line-squalls and other marked discontinuities as curious exceptions to the otherwise smoothly gradated distribution of the atmosphere. But in the last two years Prof. V. Bjerknes and his collaborators J. Bjerknes, H. Solberg and T. Bergeron at Bergen have enunciated the view, based on detailed observation, that discontinuities are the vital organs supplying the energy to cyclones‡. The question then arises: how are we to deal with discontinuities by finite differences? For such purposes graphs have a special facility which numerical tables lack. But it is not to be expected that a knowledge of the position and motion of surfaces of discontinuity will prove to be sufficient for forecasting, any more than "vital" organs alone would suffice to keep an animal alive. So probably the most thorough treatment will be reached by tabulating quantities numerically, where they vary continuously, and by drawing a line on the table where there is a discontinuity. The line will be a notification to the computer that one may interpolate up to it from either side, but not across it.

This investigation grew out of a study of finite differences and first took shape in 1911 as the fantasy which is now relegated to Ch. 11/2. Serious attention to the problem was begun in 1913 at Eskdalemuir Observatory with the permission and encouragement of Sir Napier Shaw, then Director of the Meteorological Office, to whom I am greatly indebted for facilities, information and ideas. I wish to thank

* Carnegie Institution, Washington, 1910, 1911.

† L. F. Richardson, *Phil. Mag.* Feb. 1908; *Proc. Roy. Soc. Dublin*, May 1908; *Phil. Trans.* A, Vol. 210, p. 307 (1910); *Proc. Phys. Soc. London*, Feb. 1911.

‡ *Q. J. R. Met. Soc.* 1920 April, and *Nature*, 1920, June 24.

Mr W. H. Dines, F.R.S., for his interest in some early arithmetical experiments, and Dr A. Crichton Mitchell, F.R.S.E., for some criticisms of the first draft. The arithmetical reduction of the balloon, and other observations, was done with much help from my wife. In May 1916 the manuscript was communicated by Sir Napier Shaw to the Royal Society, which generously voted £100 towards the cost of its publication. The manuscript was revised and the detailed example of Ch. IX was worked out in France in the intervals of transporting wounded in 1916—1918. During the battle of Champagne in April 1917 the working copy was sent to the rear, where it became lost, to be re-discovered some months later under a heap of coal. In 1919, as printing was delayed by the legacy of the war, various excrescences were removed for separate publication, and an introductory example was added. This was done at Benson, where I had again the good fortune to be able to discuss the hypotheses with Mr W. H. Dines. The whole work has been thoroughly revised in 1920, 1921. As the cost of printing had by this time much increased, an application was made to Dr G. C. Simpson, F.R.S., for a further grant in aid, and the sum of fifty pounds was provided by the Meteorological Office. For the construction of the index we are indebted to Mr M. A. Giblett, M.Sc. The discernment and accuracy with which the Cambridge Press have set the type have been constant sources of satisfaction.

L. F. R.

LONDON
1921 *Oct.* 10

## GUIDING SIGNS

For finding one's way about this book it is most helpful to realise that:—

(i) A complete list of symbols and notation is given in Ch. XII at the end of the book.

(ii) An impression such as (Ch. 9/12/3) is intended to refer the reader to the 3rd subdivision of the 12th division of Chapter IX.

(iii) The mark # is used to refer to equations, expressions, or statements which have had numbers assigned to them in the right-hand margin. Thus (Ch. 9/12/3 # 16) means the equation, expression or statement numbered 16 in the aforesaid subdivision. The mark # is often omitted where the meaning is plain without it.

(iv) The physical units are those of the centimetre-gram-second system, unless a different unit is expressly mentioned. Temperatures are in degrees centigrade absolute. Energy, whether by itself or as involved in entropy or specific heats, is expressed in ergs not in calories.

# CHAPTER I

## SUMMARY

FINITE arithmetical differences have proved remarkably successful in dealing with differential equations; for instance, approximate particular solutions of the equation for the diffusion of heat $\partial^2\theta/\partial x^2 = \partial\theta/\partial t$ can be obtained quite simply and without any need to bring in Fourier analysis. An example is worked out in a paper published in *Phil. Trans.* A, Vol. 210*. In this book it is shown that similar methods can be extended to the very complicated system of differential equations, which expresses the changes in the weather. The fundamental idea is that atmospheric pressures, velocities, etc. should be expressed as numbers, and should be tabulated at certain latitudes, longitudes and heights, so as to give a general account of the state of the atmosphere at any instant, over an extended region, up to a height of say 20 kilometres. The numbers in this table are supposed to be given, at a certain initial instant, by means of observations.

It is shown that there is an arithmetical method of operating upon these tabulated numbers, so as to obtain a new table representing approximately the subsequent state of the atmosphere after a brief interval of time, $\delta t$ say. The process can be repeated so as to yield the state of the atmosphere after $2\delta t$, $3\delta t$ and so on. There is a limit however to the possible number of repetitions, because each table is found to be smaller than its predecessor, in longitude and latitude, having lost a strip round its edge. Only if the table included the whole globe could the repetitions be endless. Also the errors increase with the number of steps.

In Ch. 2 the working of the method is shown by its application to a specially simplified case. In Ch. 3 the coordinate differences are considered in relation to the average size of European cyclones, and the following differences are provisionally selected: in time 6 hours, in longitude the distances between 128 equally spaced meridians, in latitude 200 kilometres of the earth's circumference, and in height the intervals between fixed heights nearly corresponding to the normal pressures of 8, 6, 4, 2 decibars. Thus small-scale phenomena, such as local thunderstorms, have to be smoothed out.

In Ch. 4 the fundamental equations are collected from various sources, set in order and completed where necessary. Those for the atmosphere are then integrated with respect to height so as to make them apply to the mean values of the pressure, density, velocity, etc., in the several conventional strata. Incidentally certain constants relating to friction and to radiation are collected from observational data. It is found to be necessary to eliminate the vertical velocity from all the equations, and in Ch. 5 it is shown how this can be done. Special difficulties arise in connection with the uppermost stratum on account of its great thickness and the enormous ratio of density

* pp. 312, 313.

between its upper and lower surfaces. These difficulties are removed in Ch. 6, as far as high latitudes are concerned. In particular it is shown how the total mass transport above any level may be deduced from a pilot balloon observation which extends well into the stratosphere.

In Ch. 7 the arrangement of the tabular numbers in space and time is discussed with a view to securing the best representation of differential coefficients by difference ratios.

In Ch. 8 the whole system of arithmetical operations is reviewed in order. With regard to the horizontal differential coefficients the general method may be briefly described in the following four sentences: Take the differential equations and replace everywhere the infinitesimal operator $\partial$ by the finite difference operator $\delta$. Use arithmetic instead of symbols. Attend carefully to the centering of the differences. Leave the errors due to the finiteness of the differences over for consideration at the end of the process. With regard to the vertical differential coefficients, on the contrary, it is often possible to effect an exact transformation to differences, by means of a vertical integration. In arranging the computing, it has constantly to be borne in mind that the rate of change with time of every one of the discrete values of the dependent variables must be calculable from their instantaneous distribution in time and space, excepting only those values near the edge of the horizontal area represented in the table. We may refer to this necessary property by saying, for brevity, that the system must be "lattice-reproducing."

In Ch. 9 will be found an arithmetical table showing the state of the atmosphere observed over middle Europe at 1910 May 20 d. 7 h. G.M.T. This region and instant were chosen because the observations form the most complete set known to me at the time of writing, and also because V. Bjerknes has published large scale charts of the isobaric surfaces, together with collated data for wind, cloud and precipitation. Starting from the table of the initially observed state of the atmosphere at this instant, the method described in the preceding paragraphs is applied, and so the rates of change of the pressures, winds, temperatures, etc. are obtained. Unfortunately this "forecast" is spoilt by errors in the initial data for winds. These errors appear to arise mainly from the irregular distribution of pilot balloon stations, and from their too small number.

In Ch. 10 the smoothing of initial observations is discussed.

In Ch. 11 is a collection of problems still waiting to be solved, with some suggestions for their treatment.

Ch. 12 is a list of Notation.

Pressures fictitiously "reduced to sea-level" are not used in the present method. Instead, the varying height of the land is dealt with by the variation of the lower limit of an integral with respect to height. See Ch. 4/2, Ch. 4/4.

The problem of weather prediction is of the "marching" variety. To explain this statement it should be pointed out that the ease or difficulty with which physical problems involving differential equations can be solved, depends on very different things according to whether symbolic methods or arithmetical differences are to be employed. In the former case the main facts in the situation are the "order" and "degree" of the equations and whether they are ordinary or partial. In the latter case what usually matters most is the relation of the "body equation" to the boundary conditions. By "body equation" is here meant the differential equation which holds throughout the region of space and the interval of time with which we have to deal. The "boundary" must be understood to be the limits of either this time or this space. According to the relation between the body equation and the boundary conditions, problems are divided into:

(i) "Jury" problems in which the integral must be determined with reference to the boundary as a whole: for instance the problem of a stone thrown from a given point to hit another given point; or that of the stresses inside a loaded dam. Cases like these frequently require troublesome successive approximations, before a statement is obtained with which the "jurymen," seated round the boundary, will all agree.

(ii) "Marching" problems in which the integral can be stepped out from a part of the boundary: for instance the problem of a stone thrown with given initial vector-velocity, or that of the cooling of a body with given initial and superficial temperatures. Other things being equal, these problems are much more easily solved than those in division (i) above. Weather prediction falls into the "marching" category.

Whilst dealing with the general subject of finite differences it may be well to mention two important properties brought to notice by Mr W. F. Sheppard.

(*a*) The great gain in accuracy, in the representation of a differential coefficient, when the differences are centered instead of progressive; a gain secured by a slight increase of work.

(*b*) That the errors due to centered differences, when small enough, are proportional to the square of the coordinate difference. This fact provides a universal means of checking and correcting the errors.

For further information about centered differences the reader is referred to "Central-Difference Formulae," by W. F. Sheppard, *Proc. Lond. Math. Soc.* Vol. XXXI. (1899) and to "The Approximate Arithmetical Solution by Finite Differences of Physical Problems," by L. F. Richardson, *Phil. Trans.* A, Vol. 210, p. 307 (1910).

# CHAPTER II

## INTRODUCTORY EXAMPLE

BEFORE attending to the complexities of the actual atmosphere and their treatment by this numerical method, it may be well to exhibit the working of a much simplified case. Lest the reader, catching sight of numbers of 7 digits, should suppose that these are necessary, let me at once point out that they have been introduced in order to measure the errors due to finite differences, which in this example are very small. An intelligible picture of the sequence of phenomena would remain after the last 4 places of digits had been cut off everywhere.

Suppose now that there is no precipitation, clouds or water vapour, neither solar nor terrestrial radiation, no eddies, and no mountains or land, but an atmosphere in which we can ignore or summarize variations with height moving upon a globe covered by sea. Further to simplify the problem, let us neglect all the quadratic terms in the dynamical equations. Then, in order to summarize the vertical velocity and the density, let us perform an integration with respect to height upon the horizontal dynamical equations and upon the equation of continuity of mass. If the limits of integration are sea-level, and a height so great that the density there is negligible, we thus arrive at a set of equations similar to those used by Laplace in his discussion of Tides on a Rotating Globe (*vide* Lamb, *Hydrodynamics*, 4th ed. § 214):

$$\frac{\partial M_E}{\partial t} = -H'\frac{\partial p_G}{\partial e} + 2\omega\sin\phi \,.\, M_N, \quad \text{.............................(1)}$$

$$\frac{\partial M_N}{\partial t} = -H'\frac{\partial p_G}{\partial n} - 2\omega\sin\phi \,.\, M_E, \quad \text{.............................(2)}$$

$$\frac{\partial p_G}{\partial t} = -g\left\{\frac{\partial M_E}{\partial e} + \frac{\partial M_N}{\partial n} - \frac{M_N\tan\phi}{a}\right\} \quad \text{.......................(3)}$$

$$= -g\left\{\frac{\partial M_E}{\partial e} + \frac{1}{\cos\phi}\frac{\partial}{\partial n}(M_N\cos\phi)\right\}. \quad \text{.................(3a)}$$

Here $M_E$, $M_N$ are the components of the whole momentum of the column of atmosphere standing upon a horizontal square centimetre at sea-level, $p_G$ is the pressure at sea-level, $g$ is gravity, $t$ is time, $\phi$ is latitude, $a$ is the radius of the earth, $\omega$ its angular velocity, $\partial e$ and $\partial n$ are distances to east and to north, and $H'$ is an empirical height, used to convert $\int_{h=0}^{h=\infty}\frac{\partial p}{\partial e}\,dh$ into $H'\frac{\partial p_G}{\partial e}$. The data given by Mr W. H. Dines for the difference of pressure between cyclones and anticyclones up to 14 km, and at 20 km, when combined with the extrapolation into the stratosphere according to the method to be described in Ch. 6 below, indicate that $H' = 9{\cdot}2$ km $= 0{\cdot}92 \times 10^6$ cm, on the average.

Proof of (1), (2), (3) is not here given, because at this stage we are more concerned with the procedure of solving these equations, than with their naturalness. Let it suffice to point out that this is a set of three simultaneous equations, involving three dependent variables $M_E$, $M_N$ and $p_G$ and three independent variables, time, latitude and longitude.

The atmospheric pressure at sea-level, $p_G$, plays the same rôle in these equations as does the elevation of the sea above its equilibrium level in the tidal equations. Much of tidal theory* is directly applicable, but its interest has centered mainly in forced and free oscillations, whereas now we are concerned with unsteady circulations.

The free periods of oscillation of the hypothetical atmosphere governed by equations (1), (2), (3) are the same as those of an ocean of uniform depth $H'$ in which the particles of water do not attract one another to an appreciable extent.

To proceed to the numerical solution of (1), (2) and (3) we take a piece of paper ruled in large squares, like a chessboard, and let it represent a map. The lines forming the squares are taken as meridians and parallels of latitude—an unusual thing in map projection. Next, we lay down the convention that all numbers written inside a square relate to the latitude and longitude of its centre. The longitude of the centres of the squares is written down at the top of the table, and the latitude along the left-hand margin. The difference $\delta n$ is constant and equal to 400 km between the points where like quantities such as $p$ and $p$, or $M$ and $M$, are tabulated, although the squares are only 200 km in the side. The distance $\delta e$, between the points where like quantities are tabulated, is that between meridians equally spaced at the rate of 64 to the equator; it is nearly 400 km in latitude 50°.

The symbols $p_G$, $M_E$, $M_N$ are written in the squares in the places shown in the table. It is seen that pressure and momentum alternate in a pattern, which is such that, if a chessboard had been used, the pressures would all appear on the red squares, and the momenta all on the white ones, or vice versa. The reason for this pattern will appear as we go on.

Now to represent the initial observations of pressure we are at liberty to write down any arbitrary set of numbers, at the points of the map where $p_G$ is required, only with this qualification: that if the assumed pressure gradients be unnaturally steep, the consequent changes will be perplexingly violent. Since $p_G$ enters the equations only by way of its differential, we may dispense with superfluous digits by tabulating differences of pressure from any standard value such as the general mean. These differences are here denoted by $\Delta p_G$.

When the pressure distribution has been chosen, we have next to represent the initial observations of momentum-per-area by writing down numbers in the alternate chequers. These numbers might have been chosen independently of the pressure, and in fact quite arbitrarily, with a qualification similar to that mentioned above. But it has been thought to be more interesting to sacrifice the arbitrariness in order to test our familiar idea, the geostrophic wind, by assuming it initially and watching the ensuing changes.

* E.g. Lamb, *Hydrodynamics*, 4th ed. §§ 213 to 223 and 314 to 316. *Vide* also Gold in M. O. publication 203 on diurnal variation of the trade winds of the Atlantic Ocean.

In order to convince the reader of the reliability of the numerical method, a problem has been selected which has been solved analytically. This has imposed a further great restriction on the arbitrariness. One of the most powerful of analytic methods is that of expansion in a series, by Maclaurin's Theorem. (*Vide* Forsyth's *Differential Equations.*) But in order that this method may be applicable, the initial distribution must be free from discontinuities even in remote regions. To get isobars running partly north and south, we may try $p_G = \sin\lambda$, where $\lambda$ is longitude eastward. To avoid a discontinuity of pressure at the pole we can multiply by $\cos\phi$. To avoid an infinite geostrophic wind at the equator we can multiply by $(\sin\phi)^2$.

So the selected form of the initial pressure distribution has been

$$\Delta p_G = \sin\lambda \cos\phi\,(\sin\phi)^2 \times 10^5 \text{ dynes cm}^{-2}, \quad \ldots\ldots\ldots\ldots\ldots\ldots(4)$$

where $\Delta p_G$ signifies the deviation from the general mean.

The isobars are shown in Fig. 1 on the map of the globe. The equator and the meridian of Greenwich are isobars. There is high pressure over Asia, and over the

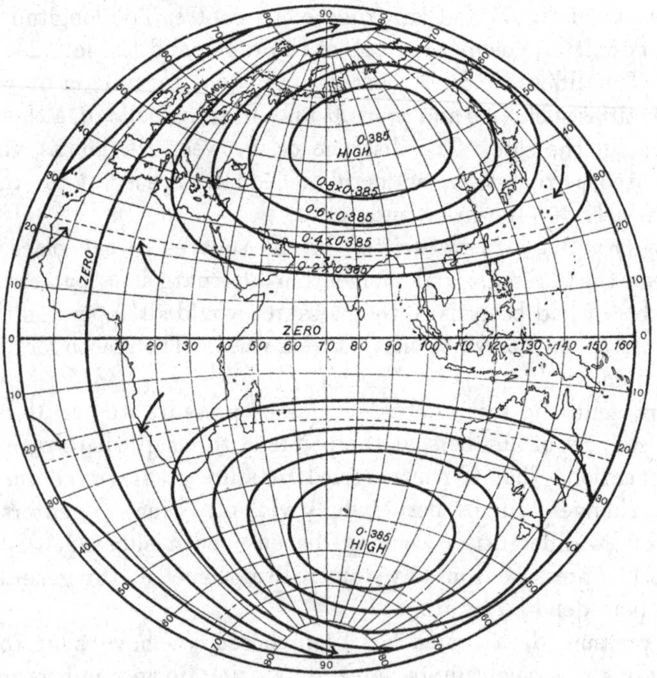


Fig. 1. Isopleths of $\sin l \cos\phi\,(\sin\phi)^2$ corresponding to the isobars of the initial distribution studied in Chapter II.

south Indian Ocean. Low pressure is at the antipodes of each high. The maxima and minima of pressure occur in latitude 54°·7, that is $6{\cdot}08 \times 10^8$ cm from the equator, and amount to 38·5 millibars above or below the mean.

A small portion of this distribution of pressure near England is entered numerically on the table.

To find the geostrophic momenta-per-area, we next insert the above value of $\Delta p_G$

in equations (1) and (2), at the same time putting $\partial M_E/\partial t = 0$; $\partial M_N/\partial t = 0$. Then it follows that initially

$$M_E = -\frac{H'}{2\omega a} \sin\lambda \{\tfrac{1}{2} + \tfrac{3}{2}\cos 2\phi\} \times 10^5 \text{ grm sec}^{-1}\text{ cm}^{-1}, \quad \ldots\ldots\ldots\ldots(5)$$

$$M_N = +\frac{H'}{2\omega a} \cos\lambda \sin\phi \times 10^5 \text{ grm sec}^{-1}\text{ cm}^{-1}. \quad \ldots\ldots\ldots\ldots(6)$$

These initial momenta-per-area, expressed in numbers, are entered at the appropriate points of the numerical table.

Next, inserting the numbers from the table into equations (1), (2), (3), and treating difference-ratios as if they were differential coefficients, we get the rates of increase of $M_E$, $M_N$, $p_G$. These rates have been multiplied by a $\delta t$ equal to 2700 seconds, that is by $\frac{3}{4}$ hour, in order to get the increases in that time. An example of the calculation of the increase of $M_E$ will now be given at the point having longitude $-3\delta\lambda$, latitude $6{\cdot}4 \times 10^8$ cm from equator, which will be found in the top left-hand corner of the table of initial data:

$$M_N = 827578{\cdot}6 \text{ grm cm}^{-1}\text{ sec}^{-1},$$

$$2\omega \sin\phi\,.\,\delta t = 0{\cdot}3324746, \text{ a pure number};$$

multiplying these together we get

$$2\omega \sin\phi\,.\,M_N = +275148{\cdot}9 \text{ grm cm}^{-1}\text{ sec}^{-1},$$

Nearest $\Delta p_G$ to east $= -3744{\cdot}11$ grm cm$^{-1}$ sec$^{-2}$

Nearest $\Delta p_G$ to west $= -7452{\cdot}17$ ,, ,,

difference $+3708{\cdot}06$ ,, ,,

but $$\frac{\delta t\,.\,H'}{\delta e} = +74{\cdot}17322 \text{ seconds};$$

multiplying, we get

$$-\frac{\delta t\,.\,H'\,.\,\delta p_G}{\delta e} = -275038{\cdot}8 \text{ grm cm}^{-1}\text{ sec}^{-1};$$

adding, we get from equation (1),

$$\frac{\partial M_E}{\partial t}\delta t = +110{\cdot}1 \text{ grm cm}^{-1}\text{ sec}^{-1}.$$

This change of 110·1 is entered on the initial table in parenthesis under the value of $M_E$. Its value is, in this case, a measure of the error due to the finite difference, for we took $M_N$ such as to make $\partial M_E/\partial t$ vanish, when the calculation was performed exactly by analysis. It is seen that the error is in this example quite small, being only 1/7000 of the resultant of $M_E$ and $M_N$. That is why such large numbers of digits have been taken. The changes in $M_E$ at other points are all worked in the same way, but of course the coefficients vary with latitude. The computation of $\delta t\,.\,\partial M_N/\partial t$ from equation (2) is so similar that it need not be illustrated.

Next, as to the pressure changes: it happens conveniently that $\cos\phi\,.\,\delta n$ is a fixed

| Auxiliary functions of latitude | | | | Distance north of equator centim $10^8 \times$ | Initial distribution, with, in parentheses, increases in $\frac{3}{4}$ hour | | | | | | | | |
|---|---|---|---|---|---|---|---|---|---|---|---|---|---|
| $\frac{2700\,H'}{\delta e}$ | $2700 \times 2\omega \sin \phi$ | $\frac{2700\,g}{\delta e}$ | $\frac{\delta e}{\delta n}$ | | $-4\delta\lambda$ | $-3\delta\lambda$ | $-2\delta\lambda$ | $-\delta\lambda$ | Longitude 0 | $\delta\lambda$ | $2\delta\lambda$ | $3\delta\lambda$ | $4\delta\lambda$ |
| 78·07616 | 0·3389380 | $8·308322 \times 10^{-2}$ | ·7953772 | 6·6 | | $\Delta p_G - 5533·74$ | $M_N$ 848791·2 | $\Delta p_G - 1850·52$ | $M_N$ 852898·2 | $\Delta p_G$ 1850·52 | $M_N$ 848791·2 | $\Delta p_G$ 5533·74 | |
| 74·17322 | ·3324746 | $7·892998 \times 10^{-2}$ | ·8372294 | 6·4 | $\Delta p_G - 7452·17$ | $M_E - 20161·5$<br>(+110·1)<br>$M_N$ 827578·6<br>(−6·7) | $\Delta p_G - 3744·11$<br>(+2192·21) | $M_E - 6742·2$<br>(+111·7)<br>$M_N$ 835626·0<br>(−2·1) | $\Delta p_G$ 0·00<br>(+2202·79) | $M_E$ 6742·2<br>(+111·7)<br>$M_N$ 835626·0<br>(+2·1) | $\Delta p_G$ 3744·11<br>(+2192·21) | $M_E$ 20161·5<br>(+110·1)<br>$M_N$ 827578·6<br>(+6·7) | $\Delta p_G$ 7452·17 |
| 70·70837 | ·3256830 | $7·524293 \times 10^{-2}$ | ·8782553 | 6·2 | $M_E - 10088·4$ | $\Delta p_G - 5641·79$<br>(2285·89) | $M_E - 5068·6$<br>(+106·3)<br>$M_N$ 815597·2<br>(−7·9) | $\Delta p_G - 1886·65$<br>(+2308·14) | $M_E$ 0·0<br>(+107·5)<br>$M_N$ 819543·4<br>(0·0) | $\Delta p_G$ 1886·65<br>(2308·14) | $M_E$ 5068·6<br>(+106·3)<br>$M_N$ 815597·2<br>(+7·9) | $\Delta p_G$ 5641·79<br>(+2285·89) | $M_E$ 10088·4 |
| 67·61653 | ·3185700 | $7·195281 \times 10^{-2}$ | ·9184145 | 6·0 | $\Delta p_G - 7505·32$ | $M_E + 5303·1$<br>(+101·8)<br>$M_N$ 792967·7<br>(−17·7) | $\Delta p_G - 3770·82$<br>(+2405·14) | $M_E + 1773·2$<br>(+102·5)<br>$M_N$ 800679·0<br>(−6·0) | $\Delta p_G$ 0·00<br>(+2416·75) | $M_E - 1773·2$<br>(+102·5)<br>$M_N$ 800679·0<br>(+6·0) | $\Delta p_G$ 3770·82<br>(+2405·14) | $M_E - 5303·1$<br>(+101·8)<br>$M_N$ 792967·7<br>(+17·7) | $\Delta p_G$ 7505·32 |
| 64·84507 | ·3111427 | $6·900361 \times 10^{-2}$ | ·9576673 | 5·8 | $M_E + 24543·9$ | $\Delta p_G - 5614·87$<br>(+2492·95) | $M_E$ 12331·3<br>(+97·2)<br>$M_N$ 779184·4<br>(−15·8) | $\Delta p_G - 1877·65$<br>(+2517·16) | $M_E$ 0·0<br>(+97·9)<br>$M_N$ 782954·5<br>(0·0) | $\Delta p_G$ 1877·65<br>(+2517·16) | $M_E - 12331·3$<br>(+97·2)<br>$M_N$ 779184·4<br>(+15·8) | $\Delta p_G$ 5614·87<br>(+2492·95) | $M_E - 24543·9$ |
| 62·35096 | ·3034082 | $6·634956 \times 10^{-2}$ | ·9959750 | 5·6 | $\Delta p_G - 7382·85$ | $M_E + 31830·7$<br>(+92·4)<br>$M_N$ 755228·0<br>(−28·5) | $\Delta p_G - 3709·29$<br>(+2608·54) | $M_E$ 10644·4<br>(+92·8)<br>$M_N$ 762572·0<br>(−9·7) | $\Delta p_G$ 0·00<br>(+2621·15) | $M_E - 10644·4$<br>(+92·8)<br>$M_N$ 762572·0<br>(+9·7) | $\Delta p_G$ 3709·29<br>(+2608·54) | $M_E - 31830·7$<br>(+92·4)<br>$M_N$ 755228·0<br>(+28·5) | $\Delta p_G$ 7382·85 |
| 60·09873 | ·2953744 | $6·395288 \times 10^{-2}$ | 1·0332998 | 5·4 | $M_E + 60313·5$ | $\Delta p_G - 5459·81$<br>(+2690·07) | $M_E$ 30302·7<br>(+88·0)<br>$M_N$ 739696·5<br>(−22·5) | $\Delta p_G - 1825·80$<br>(+2716·23) | $M_E$ 0·0<br>(+88·0)<br>$M_N$ 743275·3<br>(0·0) | $\Delta p_G$ 1825·80<br>(+2716·23) | $M_E - 30302·7$<br>(+88·0)<br>$M_N$ 739696·5<br>(+22·5) | $\Delta p_G$ 5459·81<br>(+2690·07) | $M_E - 60313·5$ |
| 58·05882 | ·2870490 | $6·178216 \times 10^{-2}$ | 1·0696048 | 5·2 | $\Delta p_G - 7096·70$ | $M_E + 59004·1$<br>(+82·5)<br>$M_N$ 714507·5<br>(−39·0) | $\Delta p_G - 3565·52$<br>(+2801·67) | $M_E$ 19731·4<br>(+83·2)<br>$M_N$ 721455·5<br>(−12·8) | $\Delta p_G$ 0·00<br>(2815·24) | $M_E - 19731·4$<br>(+83·2)<br>$M_N$ 721455·5<br>(+12·8) | $\Delta p_G$ 3565·52<br>(+2801·67) | $M_E - 59004·1$<br>(+82·5)<br>$M_N$ 714507·5<br>(+39·0) | $\Delta p_G$ 7096·70 |
| 56·20651 | ·2784404 | $5·981105 \times 10^{-2}$ | 1·1048544 | 5·0 | | $\Delta p_G - 5187·70$ | $M_N$ 697289·2 | $\Delta p_G - 1734·80$ | $M_N$ 700663·0 | $\Delta p_G$ 1734·80 | $M_N$ 697289·2 | $\Delta p_G$ 5187·70 | |

Note: $\frac{2700\,H'}{\delta n}$ and $\frac{2700\,g}{\delta n}$ are constant and equal to 62·1 exactly and 0·06608250 respectively. For $H'$, $g$ and $\omega$ see page 13.

multiple of $\delta e$, so that by combining the multiplier with $M_N \cos\phi$, the arithmetic is considerably shortened. Equation (3a) may then be written

$$\delta_t p_G = -\frac{g\delta t}{\delta e}\left\{\delta_E M_E + \delta_N\left(M_N\frac{\delta e}{\delta n}\right)\right\}, \qquad (7)$$

where the suffixes in $\delta_t$, $\delta_E$, $\delta_N$ indicate the variable which alone is varied. As an example of the computation of these pressure-changes we may take the following, which refers to the point at longitude $-3\delta\lambda$, latitude $6{\cdot}2\times10^8$ cm north.

| | | | |
|---|---|---|---|
| Cm from equator | $10^8\times6{\cdot}4$ | $10^8\times6{\cdot}2$ | $10^8\times6{\cdot}0$ |
| $M_N$ | 827578·6 | | 792967·7 grm cm$^{-1}$ sec$^{-1}$ |
| $M_N\frac{\delta e}{\delta n}$ | 692873·1 | | 728273·0 ,, ,, |
| $\delta_N\left(M_N\frac{\delta e}{\delta n}\right)$ | | $-$35399·9 grm cm$^{-1}$ sec$^{-1}$ | |
| $\delta_E M_E$ | | + 5019·8 ,, ,, | |
| Sum | | $-$30380·1 ,, ,, | |

Sum $\times -g\,.\,\delta t/\delta e = +2285{\cdot}89 = \delta_t p$ expressed in dynes cm$^{-2}$ per 2700 secs.

This time-change is written in parenthesis under the initial pressure. The pressure-changes at all the points are worked in the same way, except that the coefficients vary with latitude. The advantage of the chessboard pattern is now seen to be that the time-rates are given at the points at which the variables are initially tabulated. By adding the changes in $\frac{3}{4}$ hour we obtain a new table which has the same pattern as the initial one, so that it can in turn be taken as a starting-point. In this way the process can be continued with no limit; except that set by the loss of a strip, round the edge of the map, at every step.

If we had begun with $p_G$, $M_E$, $M_N$ all three tabulated in every square, the distribution might have been regarded as two interpenetrating chessboard patterns. In the subsequent steps these interpenetrating systems would have been propagated quite independently of each other.

Let us now compare with observation the result so far obtained. In this deduction the distribution distorts and moves west. Actual cyclones move eastward. It is natural to expect a slip of a sign in the process, but that expectation may be very simply disposed of. For, substitute the geostrophic momenta from (1) and (2) into (3), then there results

$$\frac{\partial p_G}{\partial t} = \frac{gH'}{2\omega\sin\phi}\cdot\frac{\partial p_G}{\partial e}\cdot\frac{\cot\phi}{a} \qquad (8)$$

The deduction of this equation has been abundantly verified. It means that where pressure increases towards the east, there the pressure is rising if the wind be momentarily geostrophic. This statement may be taken as a criticism of the geostrophic wind as an adequate idea when pressure-changes have to be deduced.

The same point was brought to notice by Sir Napier Shaw in his *Principia Atmospherica*. But he was able to escape from the dilemma by supposing that the congestion of air in a northward wind was somehow relieved in an upper level. Here we are dealing with the effect of all levels combined by integration, and the dilemma remains.

I am indebted to Dr E. H. Chapman for a value of the correlation between $\partial p_G/\partial t$ and $\partial p_G/\partial e$. Taking $\delta p_G/\delta t$ for 12 hours at Pembroke and representing $\delta p_G/\delta e$ by the excess of pressure at Bath over that at Roche's Point, an interval of space which has Pembroke nearly at its centre, he finds for observations at $7^h$ and $18^h$ on 1915 July 1st to 26th a correlation of $-0{\cdot}5_7$, the negative sign of which is in direct conflict with the equation above.

These facts bring one into sympathy with the view emphasized by Dr Harold Jeffreys (*Phil. Mag.* Jan. 1919) that, to understand cyclones, it is essential that the small terms in the equations be taken into account. But owing to treating latitude as constant in deducing his equation (9) from his (6), Dr Jeffreys concludes that a geostrophic wind implies no change of pressure. The success of Sir Napier Shaw's comparison between theory and the weather map for the case of a rigid portion of air, moving like a wheel with its hub rolling along a parallel of latitude (*Manual of Meteorology*, IV, Ch. 9), again turns on the fact that he has taken exact account of the small terms in the dynamical equations.

$(\delta t)^2 \dfrac{\partial^2}{\partial t^2}$ *(of initial distribution) with, written below,* $(\delta t)^3 \dfrac{\partial^3}{\partial t^3}$ *of the same*

| | $-3\delta\lambda$ | $-2\delta\lambda$ | $-\delta\lambda$ | 0 | $\delta\lambda$ | $2\delta\lambda$ | $3\delta\lambda$ |
|---|---|---|---|---|---|---|---|
| 6·4 | | | $M_E''$ − 785·5 | | $M_E''$ 785·5 | | |
| 6·2 | | $M_E''$ − 1575·9<br>$M_N''$ 13188·4 | $p_G''$ − 0·43 | $M_E''$ 0<br>(4255·1)<br>$M_N''$ 13251·9<br>(0·0) | $p_G''$ 0·43 | $M_E''$ 1575·9<br>$M_N''$ 13188·4 | |
| 6·0 | $M_N''$ 12826·0 | $p_G''$ − 0·64 | $M_E''$ − 786·9<br>(4081·4)<br>$M_N''$ 12947·4<br>(256·3) | $p_G''$ 0·00<br>(− 78·12) | $M_E''$ 786·9<br>(4081·4)<br>$M_N''$ 12947·4<br>(− 256·3) | $p_G''$ 0·64 | $M_N''$ 12826·0 |
| 5·8 | | $M_E''$ − 1574·8<br>$M_N''$ 12600·9<br>(+ 494·3) | $p_G''$ − 0·34<br>(− 81·53) | $M_E''$ 0<br>(3895·8)<br>$M_N''$ + 12662·7<br>(0·0) | $p_G''$ + 0·34<br>(− 81·53) | $M_E''$ 1574·8<br>$M_N''$ 12600·9<br>(− 494·3) | |
| 5·6 | $M_N''$ 12213·2 | $p_G''$ − 0·57 | $M_E''$ − 789·1<br>(3706·7)<br>$M_N''$ 12334·0<br>(244·4) | $p_G''$ 0·00<br>(− 84·75) | $M_E''$ 789·1<br>(3706·7)<br>$M_N''$ 12334·0<br>(− 244·4) | $p_G''$ 0·57 | $M_N''$ 12213·2 |
| 5·4 | | $M_E''$ − 1579·2<br>$M_N''$ 11967·4 | $p_G''$ − 0·26 | $M_E''$ 0<br>(3521·2)<br>$M_N''$ 12027·0 | $p_G''$ 0·26 | $M_E''$ 1579·2<br>$M_N''$ 11967·4 | |
| 5·2 | | | $M_E''$ − 791·6 | | $M_E''$ 791·6 | | |

| Coeff. 0 | In expansion of $M_E \times 10^{-5}$ | In expansion of $M_N \times 10^{-5}$ | In expansion of $p_G \times 10^{-5}$ |
|---|---|---|---|
| 1 | $-\frac{H'}{2\omega a}\sin\lambda\left\{\cos 2\phi + (\cos\phi)^2\right\}$ | $+\frac{H'}{2\omega a}\cos\lambda . \sin\phi$ | $\sin\lambda . \cos\phi . (\sin\phi)^2$ |
| $t$ | zero | zero | $+\cos\lambda . \cos\phi\,\frac{gH'}{2\omega a^2}$ |
| $\frac{t^2}{2!}$ | $+\frac{gH'^2}{2\omega a^3}\sin\lambda$ | $+\frac{gH'^2}{2\omega a^3}\cos\lambda . \sin\phi$ | zero |
| $\frac{t^3}{3!}$ | $+\frac{gH'^2}{a^3}\cos\lambda . (\sin\phi)^2$ | $-\frac{gH'^2}{a^3}\sin\lambda . \sin\phi$ | $-\frac{g^2H'^2}{\omega a^4}\cos\lambda . \cos\phi$ |
| $\frac{t^4}{4!}$ | $-\frac{g^2H'^3}{\omega a^5}\sin\lambda\left\{1+\frac{2\omega^2a^2}{gH'}(\sin\phi)^2\right\}$ | $-\frac{g^2H'^3}{\omega a^5}\cos\lambda . \sin\phi\left\{1+\frac{2\omega^2a^2}{gH'}(\sin\phi)^2\right\}$ | $+\frac{g^2H'^2}{a^4}\sin\lambda . \cos\phi$ |
| $\frac{t^5}{5!}$ | $-\frac{g^2H'^3}{a^5}\cos\lambda\left[1+2(\sin\phi)^2\left\{1+\frac{2\omega^2a^2}{gH'}(\sin\phi)^2\right\}\right]$ | $\frac{g^2H'^3}{a^5}\sin\lambda . \sin\phi\left[3+4\frac{\omega^2a^2}{gH'}(\sin\phi)^2\right]$ | $+\frac{2g^3H'^3}{\omega a^6}\cos\lambda . \cos\phi\left\{1+\frac{4\omega^2a^2}{gH'}(\sin\phi)^2\right\}$ |
| $\frac{t^6}{6!}$ | $+\frac{2g^3H'^4}{\omega a^7}\sin\lambda\left[1+\frac{\omega^2a^2}{gH'}(\sin\phi)^2\left\{7+4\frac{\omega^2a^2}{gH'}(\sin\phi)^2\right\}\right]$ | $\frac{2g^3H'^4}{\omega a^7}\cos\lambda . \sin\phi\left[1+\frac{\omega^2a^2}{gH'}\left\{-7\cos 2\phi+4\frac{\omega^2a^2}{gH'}(\sin\phi)^4\right\}\right]$ | $-4\frac{g^3H'^3}{a^6}\sin\lambda . \cos\phi\left\{1+3\frac{\omega^2a^2}{gH'}(\sin\phi)^2\right\}$ |

But it is quite possible that the present discrepancy may arise from treating $H'$ as a constant height, while in reality it may be different in different portions of the pressure distribution.

Nevertheless it is generally agreed that the terms on the left-hand sides of our equations (1), (2), (3) are far larger than the terms which have been neglected. On this account it will be of interest, in spite of the disagreement with observation, to make several further time-steps in order to observe the errors which arise from finite differences and from the curtailment of digits. The ordinary procedure would be to add to the initial distribution the changes in the first interval of time, so as to obtain a complete new distribution; and then to find its rate of change. If the non-linear terms had been included in the dynamical equations, that would probably have been the best way. But here, as equations (1), (2), (3) are linear in $M_E$, $M_N$, $p_G$ we can simplify the process, because the second time-rates $\partial^2 M_E/\partial t^2$, $\partial^2 M_N/\partial t^2$, $\partial^2 p_G/\partial t^2$ are given by inserting, in the equations, the first time-rates $\partial M_E/\partial t$, $\partial M_N/\partial t$, $\partial p_G/\partial t$ in place of $M_E$, $M_N$, $p_G$ respectively. This would not have been the case if the squares and products of $M_E$, $M_N$ had been retained. By taking advantage of this fact we save arithmetic. The result is a set of tables of successive time-rates, all referring to the initial instant. The unit for $t$ is taken as 1 second throughout. But as one second is an uncomfortably small unit of time, for this purpose, the $n$th time-rates have been multiplied by $(\delta t)^n$ where $\delta t = \frac{3}{4}$ hour $= 2700$ seconds. The actual progress of the variables in time is easily obtained from Maclaurin's expansion.

Now let us put the foregoing solution by finite differences to the test, by comparing it with the analytical solution. To obtain the latter we expand $M_E$, $M_N$, $p_G$ separately in powers of the time by Maclaurin's Theorem. For instance, if suffix zero denote values at the initial instant, then

$$M_E = (M_E)_0 + t\left(\frac{\partial M_E}{\partial t}\right)_0 + \frac{t^2}{2!}\left(\frac{\partial^2 M_E}{\partial t^2}\right)_0 + \frac{t^3}{3!}\left(\frac{\partial^3 M_E}{\partial t^3}\right)_0 + \text{etc.} \quad \ldots\ldots\ldots\ldots(9)$$

And there are similar series for $M_N$ and $p_G$. Now $(\partial M_E/\partial t)_0$ is given by the left-hand side of (1) when the initial values are assigned to $p_G$ and $M_N$. Next by differentiating (1)

$$\left(\frac{\partial^2 M_E}{\partial t^2}\right)_0 = -H'\frac{\partial}{\partial e}\left(\frac{\partial p_G}{\partial t}\right)_0 + 2\omega \sin\phi \left(\frac{\partial M_N}{\partial t}\right)_0, \quad \ldots\ldots\ldots\ldots\ldots\ldots(10)$$

and the differential coefficients on the left of (10) are supplied by equations (2) and (3). If the expansions of $M_E$, $M_N$, $p_G$ be worked simultaneously, all the differential coefficients come to hand as they are required, those of any order being derived from those of the next lower order, which in turn have already been expressed in terms of the initial distribution. For the special initial distribution defined by (4), (5), and (6) the expansions are set out in the adjoining table (p. 11), which goes as far as the sixth power.

Looking at these coefficients in the Maclaurin expansion we see that, broadly speaking, they describe the following succession of atmospheric changes in the neighbourhood of England. The isobars initially run north and south, and the high pressure is to the east. The winds at first do not change because they are geostrophic and because non-linear terms are omitted from the dynamical equations. But the pressure rises, especially in the south, so that the isobars swing round in the sense of "veering."

The greater increase of pressure to the south subsequently produces an increase in the northward component of the wind*, while there is still no eastward component across the meridian of Greenwich. The pressure changes continue at first at a uniform rate, because at first the winds have not changed. The wind begins to veer perceptibly only when the effect of the coefficient of $t^3/3!$ begins to show itself; and there is a corresponding slowing down of the rate of rise of pressure.

The numerical constants which occur in the coefficients of these expansions have been given precisely the same values as those used in the finite difference solution, namely: the earth is taken as a sphere having a circumference of $4 \times 10^9$ centim,

$$H' = 0{\cdot}92 \times 10^6 \text{ cm}, \quad g = 979 \text{ cm sec}^{-2}, \quad 2\omega = 1{\cdot}458423 \times 10^{-4} \text{ sec}^{-1}.$$

The numbers obtained by finite and infinitesimal differences are contrasted in the tables below. Bearing in mind that the resultant momentum is of the order of a million, it is seen that the discrepancies are relatively very small. The finite-difference equivalent for $\partial^2 p_G/\partial t^2$ is small, where it should be zero. The higher differential coefficients change signs at the central meridian in the proper way. The errors are small chiefly because the initial distribution varies slowly with latitude and longitude. If an ordinary cyclone were analysed in this way, the errors would be much greater; though still, I think, within the useful limit.

The error due to the curtailment of digits can be recognised in the finite difference tables as the cause of the irregularities in the last digit or two digits. The calculations were performed by a Mercedes-Euklid multiplying machine lent by the Statistical Division of the Meteorological Office.

## TABLES OF EXPANSIONS IN POWERS OF $t/2700$

WHERE, AS USUAL, $t$ IS IN SECONDS, SO THAT $t/2700$ BECOMES UNITY AFTER $\frac{3}{4}$ HOUR

For $M_E$, $M_N$, at the centre of the region, in longitude 0°, latitude $5{\cdot}8 \times 10^8$ cm north from the equator.

| Coefficient of | Expansion of $M_E$ | | Expansion of $M_N$ | |
|---|---|---|---|---|
| | by finite differences | by infinitesimals | by finite differences | by infinitesimals |
| $t/2700$ | 97·9 | zero | zero | zero |
| $\frac{1}{2!}(t/2700)^2$ | zero | zero | 12662·7 | 12684·53 |
| $\frac{1}{3!}(t/2700)^3$ | 3895·8 | 3946·7 | zero | zero |
| $\frac{1}{4!}(t/2700)^4$ | zero | zero | −1623·8 | −1639·0 |
| $\frac{1}{5!}(t/2700)^5$ | unfindable | −612·4 | unfindable | zero |
| $\frac{1}{6!}(t/2700)^6$ | unfindable | zero | unfindable | 219·53 |

*Note: The resultant initial momentum is here as large as 800000· roughly. The higher coefficients are unfindable because of the limited extent of the initial table.*

* Since the geostrophic condition is only initial.

To show up the terms in $\sin\lambda$ the following comparison has been made at a point away from the origin of longitude.

*At longitude* $-\delta\lambda$; *Latitude* $6{\cdot}0\times10^8$ *cm north of equator*

| Coefficient of | Expansion of $M_E$ | | Expansion of $M_N$ | |
|---|---|---|---|---|
| | by differences | by differentials | by differences | by differentials |
| $t/2700$ | $+102{\cdot}5$ | zero | $-6{\cdot}0$ | zero |
| $\frac{1}{2!}(t/2700)^2$ | $-786{\cdot}9$ | $-787{\cdot}7$ | $+12947{\cdot}4$ | $12971{\cdot}7$ |
| $\frac{1}{3!}(t/2700)^3$ | $4081{\cdot}4$ | $4132{\cdot}4$ | $256{\cdot}3$ | $250{\cdot}9$ |

*Note: The resultant initial momentum is here about* 800000.

To test the progress of $p_G$, a comparison has been made as follows:

*At longitude* $-\delta\lambda$; *Latitude* $5{\cdot}8\times10^8$ *cm north of equator*

| Coefficient of | Expansion of $\Delta p_G$ | |
|---|---|---|
| | Finite | Infinitesimal |
| $t/2700$ | $2517{\cdot}16$ | $2518{\cdot}62$ |
| $\frac{1}{2!}(t/2700)^2$ | $-0{\cdot}34$ | zero |
| $\frac{1}{3!}(t/2700)^3$ | $-81{\cdot}53$ | $-81{\cdot}61$ |

*Note: The maximum initial* $\Delta p_G$ *is as large as* 38500, *but occurs in a distant part of the globe.*

There are many alternative ways of utilizing these successive differential coefficients in order to obtain a chart of the progress of the variables. The discussion as to which way is the neatest may be left over to Ch. 7.

Enough has been done to show the power and the limitations of arithmetical finite differences in dealing with a set of simultaneous differential equations known to resemble those of the weather.

It has been made abundantly clear that a geostrophic wind behaving in accordance with the linear equations (1), (2), (3) cannot serve as an illustration of a cyclone. These equations are of the same form as those which successfully describe free Tidal oscillations. We may question whether the inadequacy resides in the equations (1), (2), (3) or in the initial geostrophic wind, or in both. In any case we receive a hint that it would be well to analyse weather maps more closely, in order to find out what are the real relationships.

Before leaving equations (1), (2), (3), it may be noted that on eliminating $M_E$, $M_N$ they yield what purports to be the general equation for an isobaric map at sea-level. It runs

$$0 = -\frac{a}{g}\left[\frac{\partial^2}{\partial t^2} + \Omega^2\right]^2 \frac{\partial p_G}{\partial t} + H'a\left(\frac{\partial^2}{\partial t^2} + \Omega^2\right)\nabla^2 \frac{\partial p_G}{\partial t} - \frac{H'2\omega}{a}\left(\frac{\partial^2}{\partial t^2} + \Omega^2\right)\frac{\partial p_G}{\partial \lambda}$$

$$-\frac{H'}{a}\, 8\omega^2 \sin\phi \,.\, \cos\phi \left(\frac{\partial^2 p_G}{\partial t\, \partial\phi} - \frac{2\omega \sin\phi}{\cos\phi}\frac{\partial p_G}{\partial\lambda}\right),$$

where $\Omega = 2\omega \sin\phi$ and $\nabla^2$ stands for

$$\frac{1}{a^2}\left(\frac{\partial^2}{\partial\phi^2} - \tan\phi\,\frac{\partial}{\partial\phi} - \frac{1}{(\cos\phi)^2}\frac{\partial^2}{\partial\lambda^2}\right).$$

No wonder that isobaric maps look complicated, if this be their differential equation, derived from most stringently restricted hypotheses.

I am indebted to Mr W. H. Dines, F.R.S., for having read and criticised the manuscript of this chapter.

# CHAPTER III

## THE CHOICE OF COORDINATE DIFFERENCES

THE choice would have to be guided by four considerations: (1) the scale of variation of atmospheric disturbances, (2) the errors due to replacing infinitesimals by finite differences, (3) the accuracy which is necessary in order to satisfy public requirements, (4) the cost, which increases with the number of points in space and time that have to be dealt with.

### CH. 3/1. EXISTING PRACTICE

In general the distance apart of telegraphic stations in the existing distribution is the safest guide to the required scale of proceedings. In a "Map showing the position of the Meteorological Stations, the observations from which are used in the preparation of the daily weather report Jan.—June 1918," the mean distance between a station and its immediate neighbours appears to be about 130 kilometres, if we confine attention to the British Isles, to which the map principally relates. Or to put it another way: the number of stations marked on the British Isles is 32, and the area of the polygon formed on the map by stretching a string round the outermost stations is about $56 \times (100 \text{ km})^2$. So that if the stations were, in imagination, rearranged in rectangular order, there would be enough of them to put one at the centre of each square of 132 kilometres in the side.

From the open sea there are indeed valuable wireless reports of observations on ships (see International Section of *Daily Weather Report of Meteor. Office*). But, in comparison with reports coming from land stations, they are scarce and irregular, and refer only to surface conditions.

With regard to time intervals, the existing practice of making observations for telegraphic purposes every 12 hours, or sometimes every 6 hours, is again our safest guide.

### CH. 3/2. THE DIVISION INTO HORIZONTAL LAYERS

In making a conventional division of the atmosphere into horizontal layers the following considerations have to be borne in mind. It is desirable to have one conventional dividing surface at or near the natural boundary between the stratosphere and troposphere, at an average height of 10·5 km * over Europe. Secondly, that to represent the convergence of currents at the bottom of a cyclone and the divergence at the top †, the troposphere must be divided into at least two layers. Thirdly, that the lowest kilometre is distinguished from all the others by the disturbance due to the

* E. Gold, *Geophys. Mem.* v.; W. H. Dines, *Geophys. Mem.* II.

† W. H. Dines, *Q. J. R. Met. Soc.* 38, pp. 41—50.

ground. Thus it appears desirable to divide the atmosphere into not less than 4 layers. If the layers are of equal or approximately equal mass the treatment of many parts of the subject is greatly simplified, e.g. radiation, atmospheric mixing, etc. To facilitate comparison with V. Bjerknes' charts and tables, I have chosen 5 strata divided at approximately 2, 4, 6, and 8 decibars.

This being granted, there are various ways in which the divisions may be taken:

(1) Divisions at the instantaneous pressures of 2, 4, 6, 8 decibars. This is Bjerknes' system, except that he takes 10 sheets, not 5. The heights of the isobaric surfaces become the dependent variables in place of the pressures. This system readily yields elegant approximations. But it entails the inconveniences of deformable coordinates, for it is equivalent to taking $p$ as an independent variable in place of $h$. The corresponding alterations in the equations can be carried out by means of the following set of substitutions.

Let $A$ be any variable and let the suffixes denote the quantities which remain constant during the differentiations.

Then when $\phi$ and $t$ are constant,

$$\left(\frac{\partial A}{\partial \lambda}\right)_h = \left(\frac{\partial A}{\partial \lambda}\right)_p - \left(\frac{\partial A}{\partial h}\right)_\lambda \left(\frac{\partial h}{\partial \lambda}\right)_p.$$

There is a similar equation in $\partial\phi$ when $\lambda$ and $t$ are constant.

Also when $\lambda$, $\phi$ and $t$ are constant,

$$\frac{\partial A}{\partial h} = -g\rho \frac{\partial A}{\partial p}.$$

Also when $\lambda$ and $\phi$ are constant,

$$\left(\frac{\partial A}{\partial t}\right)_h = \left(\frac{\partial A}{\partial t}\right)_p - \left(\frac{\partial A}{\partial h}\right)_t \left(\frac{\partial h}{\partial t}\right)_p.$$

The result of these substitutions is to produce a large number of terms. The additional terms are small, but they are not always negligible in comparison with the errors of observations. As observations improve they are likely to become more significant. On this account I have preferred to use instead the following system:

(2) The divisions between the five conventional strata are taken at fixed heights above mean sea-level, so chosen as to correspond to the mean heights of the 2, 4, 6 and 8 decibar surfaces. These mean positions* are at about 2·0, 4·2, 7·2, 11·8 kilometres over Europe.

Gravity has a small north-south component parallel to a surface at a fixed height above sea-level, which is taken into account. The "horizontal" components of wind are taken parallel to the sea surface in the theory, which in this particular is much too good for the observations.

(3) Another alternative, and a rather tempting one, would have been to have taken the divisions between the strata at fixed values of the gravity potential corresponding to mean pressures of 2, 4, 6, 8 decibars. This system would be particularly

* W. H. Dines, *Geophys. Mem.* 2. Also Bjerknes' synoptic charts.

elegant in the stratosphere. The horizontal components of wind would be defined as at right angles to the force of gravity. The variation of the thickness of a stratum with latitude could be taken into account in the equation of continuity of mass. Analytically $\psi$ would be an independent variable in place of $h$.

### CH. 3/3. EFFECT OF VARYING THE SIZE OF THE FINITE DIFFERENCES

The reduction of the differences of latitude and longitude in the ratio $n$ would multiply by $n^2$ the number of stations falling on any territory. The cost of maintaining observing stations is but little affected by a change in the number of points in a vertical line at which observations are required, and would only be affected by a reduction of the interval of time, if this necessitated an extra shift of observers, as for instance a night-shift. Together we may reckon the effect of reducing both $\delta h$ and $\delta t$ in the ratio $n$, as increasing the cost perhaps in the ratio $n$ instead of $n^2$. We have then

$$\text{Cost of stations} \quad = An^3,$$
$$\text{Cost of computing} = Bn^4,$$

where $A$ and $B$ are constants.

Administrative expenses would contain an element $C$ independent of $n$, as well as other parts which may be debited to the stations and to the computing, by changing the constants $A$ and $B$ to $A'$ and $B'$. The whole cost would then be of the form

$$A'n^3 + B'n^4 + C.$$

Next, as to the accuracy obtained: it has been shown* that in the representation of a smooth continuous function by finite differences, which are in any case sufficiently small, the finite size of the differences produces errors proportional to $1/n^2$.

Thus as $n$ varied, *the errors would be inversely as the square root of the cost of computing alone.* The relation to the whole cost

$$A'n^3 + B'n^4 + C$$

cannot be simply expressed until we know the values of $A'$, $B'$, $C$.

### CH. 3/4. THE PATTERN ON THE MAP

As has already been illustrated in the introductory example, it is very desirable to arrange pressure and momentum in a special pattern like that of a chessboard. The consideration of this question is deferred to Ch. 7 as it cannot be properly discussed until we have the full differential equations before us. Suffice it to point out here, that the pattern is such that the least difference in latitude or longitude between similar quantities, such as pressure and pressure, or wind and wind, is twice the least difference between dissimilar quantities, such as pressure and wind. To what then are we to compare the distance of 130 km, which we have seen is the mean distance between existing British telegraphic stations? If we take 130 km as more than one alternative and as less than the other, we arrive at approximately 100 km between dissimilar and 200 km between similar stations, for a densely populated area.

* L. F. Richardson, *Phil. Trans.* A, Vol. 210, 1910, p. 310.

## CH. 3/5. DEVICES FOR MAINTAINING A NEARLY SQUARE CHEQUER

There is some advantage in a chequer which is nearly a square; for the degree of detail with which it is desirable to study the weather, is roughly the same in all directions at any fixed point. A square-shaped chequer might be maintained in all latitudes by making $\delta n$ decrease towards the poles in proportion to $\cos\phi$. But then the treatment would be the more detailed in the high uninhabited latitudes where detail matters less. A more economical plan would be to maintain $\delta n$ at the same value in all inhabited latitudes and to make it increase in the uninhabited polar zones; and, when the chequers became too elongated, on approaching the poles, to omit alternate meridians. To make this possible right up to the poles, it would be necessary that the number of meridians used on the equator should be divisible by 2 many times over. For instance, instead of 120 meridians 3° apart it would be better to have

$$2^7 = 128 \text{ meridians } 2^\circ\, 48'\, 45''\cdot 0 = 0\cdot 0490874 \text{ radian apart.}$$

Unfortunately this was not thought of, until after the 3° difference of longitude had been used in the example of Ch. 9.

## CH. 3/6. SUMMARY ON COORDINATE DIFFERENCES

The adopted size of chequer might be a compromise between that indicated by the existing practices on land and on sea; for abundance of observations on the former does not compensate for scarcity of observations on the latter.

A satisfactory arrangement would appear to be to divide the surface into chequers by parallels of latitude separated by 200 km and by meridians spaced uniformly at the rate of 128 to the whole equator. The chequer is then nearly a square of 200×200 km in latitude 50°. At the equator even, it is not too elongated, as it measures there 313·09 km east×200 km north. In latitude 63° the chequer is about 142 km from east to west. As $\frac{142}{200} = \sqrt{1/2}$ this would be the latitude for the first omission of alternate meridians. The discussion of the polar caps is deferred to Ch. 7. In the example of Ch. 9 the width of the chequers is taken as 3° of longitude; but that must be considered as an inferior choice. The centres of the chequers are supposed to be the points for observing and recording the meteorological elements. If we imagine the chequers to be coloured alternately red and white, so as to produce a pattern resembling a chessboard, then it will be shown in Ch. 7 below, that the red chequers should bear the pressure, temperature and humidity, and the white chequers should bear the two components of momentum; or vice versa.

Horizontally the atmosphere may conveniently be divided into five conventional strata at the fixed heights of 2·0, 4·2, 7·2, 11·8 km above mean sea-level, heights which correspond to mean pressures of 8, 6, 4, and 2 decibars. Ten layers would give four times the accuracy obtained with these five layers. A technically more perfect division would be to take the separating surfaces at constant gravity potential instead of, as in this book, at constant height.

The interval of time $\delta t$ has usually been taken in what follows, as 6 hours.

## Ch. 3/7. THE ORIGIN OF LONGITUDE

This might be determined so as to get the greatest number of stations on land, or on steamship routes. If the whole world were considered, the origin would be a matter of indifference, as the coast lines are so irregular. For a small group of islands like the British Isles, if they could be treated as independent (as they certainly cannot), choice of origin would be important. Thus taking Greenwich as origin and meridians spaced equally at the rate of 128 to the equator, we find two stations about 50 kilometres west of the Irish coast, and three others at a like distance west of the Hebrides, of the Isle of Man and of Cornwall. By shifting the origin to longitude 2° W. of Greenwich these five stations come on land, or within about 10 kilometres of it; while only one other, which moves from Flamborough Head into the North Sea, is lost. The arrangement with the origin at 2° W. of Greenwich is shown in the frontispiece. The stations are supposed to be at the centres of the chequers.

**Postscript:** A study of turbulence, in Ch. 4/8, has disclosed several reasons for taking thinner strata near the ground. The divisions might for example be at heights of 50, 200 and 800 metres. The dividing surfaces would need to follow, more or less, the slope of the land.

Again difficulties connected with the stratosphere in the tropics may make it necessary to divide the atmosphere above 11·8 km into two conventional layers.

# CHAPTER IV

## THE FUNDAMENTAL EQUATIONS

### CH. 4/0. GENERAL

THERE are four independent variables:

$t$ time.
$h$ height above mean sea-level.
$\lambda$ longitude, reckoned eastward.
$\phi$ latitude, reckoned negative in the southern hemisphere.

Seven dependent variables have been taken, namely:

$v_E$ velocity horizontally towards the east.
$v_N$ ,, ,, ,, ,, north.
$v_H$ ,, vertically upwards.
$\rho$ density.
$\mu$ joint mass of solid, liquid and gaseous water per mass of atmosphere.
$\theta$ temperature absolute centigrade.
$p$ pressure, expressed in dynes $\text{cm}^{-2}$.

If an eighth dependent variable had been taken, it might perhaps have specified the amount of dust in the air.

The rates of change of the seven dependent variables are given by the following seven main equations. The "tributaries" in the table supply the values of certain terms occurring in the main equations.

| | Main equations | Tributaries |
|---|---|---|
| $\partial v_E/\partial t$ | eastward dynamical equation ... ... | eddy-viscosity |
| $\partial v_N/\partial t$ | northward ,, ,, ... ... | ,, ,, |
| $\partial v_H/\partial t$ | upward ,, ,, ... ... | eddy-conduction of heat |
| $\partial \rho/\partial t$ | indestructibility of mass ... ... ... | precipitation |
| $\partial \mu/\partial t$ | conveyance of water ... ... ... | precipitation, stirring |
| $\partial \theta/\partial t$ | conveyance of heat ... ... ... | precipitation, stirring, radiation and clouds |
| $\partial p/\partial t$ | characteristic gas equation, $\partial\theta/\partial t$ and $\partial\rho/\partial t$ | |

One of the first questions which had to be decided was whether to eliminate any of the dependent variables before proceeding to the numerical process. Now if a variable be eliminated between two differential equations the resulting equation is usually more complicated, so that the saving in arithmetical toil due to the absence of a variable, is partly or entirely compensated by the increase in toil due to the complication. There is also a clear advantage in keeping to the familiar variables which are observed, to the avoidance of stream-functions and other quantities which cannot be observed. Therefore I decided to do without analytical preparation of this kind except in two cases:

(i) To eliminate temperature between the characteristic equation $p = b\rho\theta$ and any other equations in which $\theta$ occurs. This introduces no complications, because the characteristic equation is not differential.

(ii) To solve for the vertical velocity. This is necessary, because sufficient observations of the vertical velocity are not available, nor likely to become so. The solution can be obtained because the vertical equilibrium can be treated as a static one, that is to say $Dv_H/Dt$ can be neglected. It is done in Ch. 5.

After this change $v_H$ is given in terms of mixing, precipitation, and radiation, and of the instantaneous values of the five remaining variables $v_E$, $v_N$, $\rho$, $\mu$, $p$. The time rates of the remaining variables are then given by the following five equations:

| | Main equations | Tributaries |
|---|---|---|
| $\partial(\rho v_E)/\partial t$ | two horizontal dynamical equations, | $v_H$, eddy-viscosity |
| $\partial(\rho v_N)/\partial t$ | modified by equation continuity | $v_H$, eddy-viscosity |
| $\partial\rho/\partial t$ | indestructibility of mass ... ... | $v_H$, precipitation |
| $\partial\mu/\partial t$ | conveyance of water ... ... | $v_H$, precipitation, mixing |
| $\partial p/\partial t$ | vertical static ... ... ... | $v_H$ and $\partial p/\partial t$ |

The equations are further modified so as to make them apply to conventional strata. The way in which they then fit and lock together will be explained in detail in Ch. 8. As the arithmetical method allows us to take account of the terms which are usually neglected, many of these terms have been included. But as all terms cannot be treated with equally small time- and space-differences (see Ch. 7), they cannot all be treated with equal accuracy, and so it is necessary to know beforehand which terms are likely to be the more important. Some figures, representing the extreme values ordinarily attained by the various terms, are set out underneath certain of the equations. These figures have been obtained by a casual inspection of observational data and they may be uncertain except as to the power of ten. They are expressed in C.G.S. units. They relate only to the large-scale phenomena which can be represented by the chosen coordinate differences of 200 kilometres horizontally, one-fifth of the pressure vertically and by the time step of six hours*.

* Very much fuller data concerning the size of terms has been collected by Hesselberg and Friedmann *Geoph. Inst.*, Leipsig, Spezialarb. Ser. 2, Heft 6.

## CH. 4/1. CHARACTERISTIC EQUATIONS OF DRY AND MOIST AIR

The units in the following are C.G.S. centigrade throughout. The equations are taken from Hertz's* paper. Dry air

$$p = 2{\cdot}870 \times 10^6 \rho\theta, \quad \ldots\ldots (1)$$

moist unsaturated air

$$p = (2{\cdot}870 + 1{\cdot}746\mu) \times 10^6 \rho\theta, \quad \ldots\ldots (2)$$

saturated with liquid water or with ice

$$(p - p_w) = 2{\cdot}870 \times 10^6 (1 - \mu)\rho\theta, \quad \ldots\ldots (3)$$

here $p_w$ is the saturation pressure of vapour in equilibrium with water or with ice. A table of $p_w$, in $10^3$ dynes/cm² as unit, is given in Sir Napier Shaw's *Principia Atmospherica*.

For our purposes it is convenient to employ $w_s$, the density of the saturated vapour, instead of $p_w$. On taking account of the relative densities of air and water-vapour equation (3) yields the following equation in $w_s$,

$$p = \theta\{\rho(1-\mu) \times 2{\cdot}87 \times 10^6 + w_s \times 4{\cdot}616 \times 10^6\}. \quad \ldots\ldots (4)$$

We shall frequently require to find $\theta$ from the characteristic equations, being given $p$, $\rho$, $\mu$. Now in order to find $\theta$ we must know whether to use equation (4) or equation (2), that is to say we must know whether the air is saturated or not. For this purpose we require the relation between $p$, $\rho$ and $\mu$ when the air is just saturated. This relation is to be found by treating (2) and (4) as simultaneously true, and then eliminating $\theta$ between them. Thus solving (2) for $\theta$, and substituting in (4), there results

$$w_s = \mu\rho, \quad \ldots\ldots (5)$$

in which it is important to notice that $w_s$ corresponds to the temperature given by (2).

If the numerical values of $p$, $\rho$, $\mu$ are such as to make (5) an identity, then the air is just saturated. If $w_s$ is too large to fit in (5), then the air is unsaturated and the temperature already found from (2) is correct. If on the other hand $w_s$ is too small to fit in (5), then the air is saturated, and both $w_s$ and $\theta$ must be recalculated by solving (4).

## CH. 4/2. THE INDESTRUCTIBILITY OF MASS

This principle leads to the well-known "equation of continuity (of mass)."

Let **v** be the vector velocity, $\rho$ be the density, so that $\mathbf{v}\rho$ is the momentum-per-volume, which for brevity we denote by **m**. Then the equation runs as follows, in vector notation

$$-\frac{\partial\rho}{\partial t} = \operatorname{div}(\rho\mathbf{v}) = \operatorname{div}\mathbf{m} = \mathbf{v}\nabla\rho + \rho\operatorname{div}\mathbf{v}, \quad \ldots\ldots (1)$$

or when expanded in spherical coordinates on a globe of radius $a$:

$$\underset{10^{-9}}{-\frac{\partial\rho}{\partial t}} = \underset{10^{-7}}{\frac{\partial m_E}{\partial e}} + \underset{10^{-7}}{\frac{\partial m_N}{\partial n}} - \underset{10^{-8}\tan\phi}{\frac{m_N\tan\phi}{a}} + \underset{<10^{-7}}{\frac{\partial m_H}{\partial h}} + \underset{10^{-10}}{\frac{2m_H}{a}}. \quad \ldots\ldots (2)$$

* *Miscellaneous Papers*, No. XIX. (Macmillan & Co.).

The term $\frac{2m_H}{a}$ is small, but the other term involving the vertical velocity is by no means negligible. The term $\frac{\partial\rho}{\partial t}$ is usually small in comparison with either $\frac{\partial m_E}{\partial e}$ or $\frac{\partial m_N}{\partial n}$; that is to say, the atmosphere, although compressible, flows almost as if the density were constant at a fixed point. This is a much better approximation as V. Bjerknes* and later Hesselberg and Friedmann† have pointed out, than to suppose that the atmosphere moves like a liquid. As cyclones and anticyclones sweep past an observatory, W. H. Dines has shown that the increases of pressure are compensated by increases of temperature in the troposphere, so that the density there has a standard deviation, at a fixed point, as small as 1·5 °/ₒ of its mean value. In the stratosphere, up to 13 km of height, the variation of density at a fixed point is about the same in absolute magnitude as in the troposphere, but therefore much greater relative to its mean‡.

The fact that the momentum-per-volume **m** is nearly a non-divergent vector, has been the reason for using it in preference to the velocity. However, although $\partial\rho/\partial t$ is small, it will not be neglected.

To adapt the equation of continuity of mass to our conventional strata let us integrate it across a stratum with respect to height, and apply the rule for differentiating a definite integral, remembering that all the limits of integration are constant during $\frac{\partial}{\partial e}$ and $\frac{\partial}{\partial n}$ except only the limit at the ground; and that all, without exception, are independent of time. The limits are denoted by $G$ for the ground and 8, 6, 4, 2 for the conventionally fixed heights which correspond approximately to mean pressures of 8, 6, 4, 2 decibars. Taking the lowest stratum as example, equation (2) becomes

$$-\frac{\partial}{\partial t}\int_G^8 \rho\, dh = \frac{\partial}{\partial e}\int_G^8 m_E\, dh + \frac{\partial}{\partial n}\int_G^8 m_N\, dh + [m_H]_8$$
$$+\left(m_E\frac{\partial h}{\partial e} + m_N\frac{\partial h}{\partial n} - m_H\right)_G$$
$$-\frac{\tan\phi}{a}\int_G^8 m_N\, dh + \frac{2}{a}\int_G^8 m_H\, dh. \quad \text{.......................(3)}$$

Now $\left(m_E\frac{\partial h}{\partial e} + m_N\frac{\partial h}{\partial n} - m_H\right)_G$ is the scalar product of the vector **m** and of a vector at right angles to the surface of the ground, and therefore it vanishes since the ground is impervious to the wind. .................................................................(4)

Now let capital letters denote the integrals, with respect to $h$, of the corresponding small letters across the thickness of the stratum. Here $R$ is taken as corresponding to $\rho$. Then the following equation is precisely true (for a spherical sea-surface). In it the differentiations $\frac{\partial}{\partial e}, \frac{\partial}{\partial n}$ now follow the stratum as a whole,

$$-\frac{\partial R}{\partial t} = \frac{\partial M_E}{\partial e} + \frac{\partial M_N}{\partial n} + [m_H]_8 - \frac{\tan\phi\, M_N}{a} + \frac{2M_H}{a}. \quad \text{..............(5)}$$

* *Dynamic Meteorology and Hydrography*, Art. 117.

† *Geoph. Inst.*, Leipsig, Ser. 2, Heft 6, p. 164.

‡ W. H. Dines, *Characteristics of the Free Atmosphere*, Meteor. Office, London, 1919, p. 76.

For any layer, except the lowest, there is a term $m_H$ at the lower limit to be subtracted from the right side of this equation. Because they fit in this equation, the quantities $R$, $M_E$, $M_N$ have been chosen as dependent variables throughout this book. $R$ is the mass per unit horizontal area of the stratum. $M_E$, $M_N$, $M_H$ are the components of the total momentum of the stratum per unit horizontal area. Since $dh = -\frac{dp}{g\rho}$ we have

$$R = \frac{p_G - p_S}{g}, \quad \text{.....................................(6)}$$

$$M_E = -\frac{1}{g}\int_G^S v_E \, dp. \quad \text{..............................(7)}$$

This formula for $M_E$ has been used in deriving $M_E$ from observations in which a registering balloon had been followed by a theodolite.

The velocity $v_E$ was plotted against $p$ and $M_E$ measured as an area on the diagram. It is rather difficult to say what is the best way to take account of precipitation in the continuity of mass. It depends on how one defines $m_H$ the vertical momentum per unit volume, when rain is falling. If $m_H$ is defined to be the total momentum including that of the rain, as well as that of the air, then the equation of continuity of mass is correct as it stands. If on the other hand $m_H$ refers solely to the moist air, then a correcting term is necessary. This is discussed in Ch. 4/6.

## CH. 4/3. THE CONVEYANCE OF WATER, WHEN MOLECULAR DIFFUSION IS NEGLECTED, AND EDDIES ARE NOT AVERAGED-OUT

It is a familiar observation that winds coming from over a warm sea bring water with them. This large-scale conveyance of water will now be put into equations, which will be equally applicable to small-scale motions provided that the equations are applied to the actual detailed distribution of the velocities, humidities and the like, and provided also that the detail is not so small as to bring molecular diffusion into the question. The modifications introduced, when smoothed velocities replace the actual velocities, will be considered in the section on eddy-motion. Precipitation will also be reserved for separate treatment.

Apart from these complications, the air carries the water along with it, so that the velocity of the water is $v_E$, $v_N$, $v_H$, the same as that of the air. The density of the water substance is $w$ grams/cc. We have therefore an equation of continuity of mass of water substance similar in form to the equation of continuity of total mass (Ch. 4/2), except that $w$ replaces $\rho$ and that precipitation and convection come in.

$$\frac{\partial w}{\partial t} + \operatorname{div}(w\mathbf{v}) = \left\{ \begin{array}{c} \text{increase of water per volume and per time} \\ \text{due to convection and precipitation} \end{array} \right\}. \quad \text{.........(1)}$$

This expands in spherical coordinates just as does the corresponding equation in $\rho$.

Now $w=\mu\rho$, where $\mu$ is the mass of water substance per unit mass of moist atmosphere. So

$$\operatorname{div}(w\mathbf{v})=\operatorname{div}(\mu\rho\mathbf{v})=\operatorname{div}(\mu\mathbf{m})= \quad\ldots\ldots(2)$$

$$=\mathbf{m}\nabla\mu+\mu\operatorname{div}\mathbf{m} \text{ by the well-known vector formula.}$$

But
$$\operatorname{div}\mathbf{m}=-\frac{\partial\rho}{\partial t}. \quad\ldots\ldots(3)$$

So
$$\frac{\partial w}{\partial t}+\operatorname{div}(w\mathbf{v})=\rho\frac{\partial\mu}{\partial t}+\rho\mathbf{v}\nabla\mu=\rho\left(\frac{\partial\mu}{\partial t}+\mathbf{v}\nabla\mu\right). \quad\ldots\ldots(4)$$

Now $\frac{\partial\mu}{\partial t}+\mathbf{v}\nabla\mu$ is the increase in $\mu$, following the motion of the fluid, and is customarily denoted by $\frac{D\mu}{Dt}$. Thus, on dividing through by $\rho$, we get a second form of the equation of continuity of water substance,

$$\frac{D\mu}{Dt}=\frac{\partial\mu}{\partial t}+\mathbf{v}\nabla\mu=\frac{\partial\mu}{\partial t}+v_E\frac{\partial\mu}{\partial e}+v_N\frac{\partial\mu}{\partial n}+v_H\frac{\partial\mu}{\partial h}=$$ rate of increase of water per unit mass and per unit time by convection and precipitation. ......(5)

It is in fact immediately obvious that the relative proportions of water and air by mass do not change during the motion, and therefore that $\frac{D\mu}{Dt}=0$ except for convection and precipitation. Thus the equation is confirmed. By working backwards from (5) to (1) it is found that the second member of (1) is $\rho\frac{D\mu}{Dt}$, instead of $\frac{D(\mu\rho)}{Dt}$ which might have been erroneously expected.

Equation (1), expanded in spherical coordinates, reads

$$\frac{\partial w}{\partial t}+\frac{\partial(wv_E)}{\partial e}+\frac{\partial(wv_N)}{\partial n}-\frac{wv_N\tan\phi}{a}+\frac{\partial(wv_H)}{\partial h}+\frac{2(wv_H)}{a}=\rho\frac{D\mu}{Dt}. \quad\ldots\ldots(6)$$

## Adaptation to Conventional Strata when $W$ is given

In seeking to adapt equation (6) to conventional strata characterized by $M_E$, $M_N$, $R$, we are met by the difficulty that neither $w$ nor the velocity is independent of height; so that, for example, $\int wv_E dh$, taken between the upper and lower limits of a stratum, is not simply and accurately expressible in terms of $M_E$ and $\int wdh$ between the same limits. However $w$ varies* very much more rapidly with height than do $v_E$, $v_N$ or $v_H$. Therefore $v_E$, $v_N$ are treated as constant across a stratum, when the integral is being evaluated. Taking the lowest stratum for example and integrating (6) we get

$$\frac{\partial}{\partial t}\int_G^8 wdh+\frac{\partial}{\partial e}\left\{v_E\int_G^8 wdh\right\}+\frac{\partial}{\partial n}\left\{v_N\int_G^8 wdh\right\}+[wv_H]_8+w_G\left(v_E\frac{\partial h}{\partial e}+v_N\frac{\partial h}{\partial n}-v_H\right)_G$$
$$-\frac{\tan\phi}{a}v_N\int_G^8 wdh+\frac{2}{a}v_H\int_G^8 wdh=\int_G^8\rho\frac{D\mu}{Dt}dh. \quad\ldots\ldots(7)$$

* Hann, *Meteorologie*, 3 Auf. pp. 227–234.

The quantity $\left(v_E \frac{\partial h}{\partial e} + v_N \frac{\partial h}{\partial n} - v_H\right)_G$ vanishes, because the ground is impervious to wind. Write $\int_G^8 w dh = W$ so that $W$ is the total mass of water-substance, whether liquid, solid or gaseous, per unit horizontal area of the stratum. Then (7) becomes

$$\frac{\partial W}{\partial t} + \frac{\partial}{\partial e}\left(\frac{M_E W}{R}\right) + \frac{\partial}{\partial n}\left(\frac{M_N W}{R}\right) - \frac{\tan\phi}{a}\left(\frac{M_N W}{R}\right) + \frac{2}{a} v_H W + [w v_H]_8$$

= rate of increase of water per unit horizontal area of the stratum, by convection and precipitation $= R\frac{D\mu}{Dt}$. ..........(8)

In the terms $\frac{2}{a} v_H W$ and $R\frac{D\mu}{Dt}$ we must take $v_H$ and $\mu$ as mean values for the stratum. For any stratum, except the lowest, we must subtract from the left side of (8) the value of $w v_H$ at the lower limit.

Equation (8) is convenient because it is expressed in terms of $W$, the mass of water per horizontal area of stratum. On the other hand, the vertical integrations by which (8) is derived take no account of the three main statistical facts in this connection, namely:

That for the mean of many occasions $\log \mu$ is nearly a linear* function of height. ........(9)

That between heights of 0·5 km and 8 km $m_E$ and $m_N$ are nearly independent† of height. This is sometimes known as Egnell's law. ..........(10)

That $\log p$ is also nearly a linear function of height. ..........(11)

Now from (9) and (11) it follows that on the average $\mu = Cp^B$, ..........(12)

where $C$ and $B$ are nearly independent of height. We shall probably improve the accuracy of all our operations with conventional strata if we assume the general type of relationship (12) to hold in such a way that $C$ and $B$ are independent of height in each stratum but vary from one stratum to another. The constants $C$ and $B$ are then to be determined, for each occasion, from the instantaneous values of $p$ and $\mu$. Let numerical suffixes denote heights. Taking the stratum between $h_8$ and $h_6$ as a type, it follows from (12) that

$$B = \frac{\log \mu_8 - \log \mu_6}{\log p_8 - \log p_6}, \quad \text{..........(13)}$$

$$C = \frac{\log p_8 \times \log \mu_6 - \log p_6 \times \log \mu_8}{\log p_8 - \log p_6}. \quad \text{..........(14)}$$

It is now possible to transform various integrals, in a more accurate way than before, by making use of $C$ and $B$. Thus

$$W_{86} \equiv \int_8^6 \mu\rho dh = -\frac{1}{g}\int_8^6 \mu dp = -\frac{1}{g}\int_8^6 Cp^B dp$$

$$= -\frac{1}{g}\frac{(\log p_6 - \log p_8)(\mu_6 p_6 - \mu_8 p_8)}{\log(\mu_6 p_6) - \log(\mu_8 p_8)}. \quad \text{..........(15)}$$

* Hann, *Meteorologie*, 3 Auf. pp. 227–234.

† Gold, *Geophysical Memoirs*, No. 5, p. 138, Meteor. Off. London.

By this equation $W$, the whole mass of water per horizontal area of the stratum, can be found when $\mu$, the mass-of-water-per-mass-of-atmosphere, is given at the boundaries of the stratum. Conversely, given the value of $W$ and of $\mu$ at one boundary, we can find $\mu$ at the other boundary, and so proceed to the next stratum and do likewise. This rather troublesome process has to be used to find the term $w_8 v_{H8}$ in equation (8) when $W$ is given for each stratum in place of $\mu$ at the boundaries. For a simpler method might hardly be accurate enough.

### Alternative scheme: $\mu$ given at the levels where strata meet

We have so far supposed that observation gives the masses of water in the form of vapour or cloud per horizontal area of the strata. Now let us suppose instead that the given quantities are the "specific humidities" $\mu_G, \mu_L, \mu_8, \mu_6, \mu_4, \mu_2$ at the levels where two strata meet.

It will then be necessary to express equations (5) or (6), or their derivatives, in terms of $\mu_G \ldots \mu_2$. This can be done by means of integrations with respect to height, during which much use will be made of (9) and its derivatives (12), (13), (14), (15).

Some equations integrate more neatly than others, for instance if we attempted to integrate (5) with respect to $h$, the term $\int v_H \frac{\partial \mu}{\partial h} dh$ would prove a stumbling-block. On the other hand if (5) be multiplied throughout by $\rho/\mu$ it integrates nicely with respect to $h$. However it seems best instead to transform (6), because the physical meaning of the result is simple, and in simple relation to equation (8), and to the equation expressing the diffusion of moisture by eddies.

Put then

$$\mu = B' e^{-Ah}, \quad \ldots\ldots(16)$$

where, within any stratum, $B'$ and $A$ are independent of $h$. Then, for example, in the stratum $h_G$ to $h_8$

$$-A = \frac{\log \mu_G - \log \mu_8}{h_G - h_8}. \quad \ldots\ldots(17)$$

Then

$$\int_G^8 \mu dh = \int_G^8 B' e^{-Ah} = {}_G^8\!\left[\frac{\mu}{-A}\right] = \frac{(h_G - h_8)(\mu_8 - \mu_G)}{(\log \mu_G - \log \mu_8)} = (h_G - h_8)\,\bar{\mu}, \quad \ldots\ldots(18)$$

where $\bar{\mu}$ is an appropriate abbreviation for

$$\frac{\mu_8 - \mu_G}{\log \mu_G - \log \mu_8}, \quad \ldots\ldots(19)$$

since $\bar{\mu}$ is the mean value of $\mu$ with respect to height in the stratum. Now integrate (6) term by term with respect to $h$. The term

$$\int_G^8 \frac{\partial (w v_N)}{\partial n} dh = \int_G^8 \frac{\partial (\mu m_N)}{\partial n} dh = \frac{\partial}{\partial n} \int_G^8 (\mu m_N)\, dh + \mu_G \left( m_N \frac{\partial h}{\partial n} \right)_G$$

$$= \frac{\partial}{\partial n} \{ m_N (h_G - h_8)\, \bar{\mu} \} + \mu_G \left( m_N \frac{\partial h}{\partial n} \right) = \frac{\partial}{\partial n} (M_{NG8} \,.\, \bar{\mu}) + \mu_G \left( m_N \frac{\partial h}{\partial n} \right)_G \quad \ldots(20)$$

The term in $\frac{\partial}{\partial e}$ transforms similarly. Then

$$\int_G^8 \frac{wv_N \tan\phi}{a} dh = \frac{\tan\phi}{a} m_N \int_G^8 \mu dh = \frac{\tan\phi}{a} M_N \bar{\mu}. \quad \text{.............(21)}$$

Next

$$\int_G^8 \frac{\partial (wv_H)}{\partial h} = \mu_8 m_{H8} - \mu_G m_{HG}. \quad \text{..........................(22)}$$

The term in $a^{-1}$ does not integrate nicely, since we do not know how $m_H$ varies with height. But it is a very small term. So treating $m_H$ as independent of height we get

$$\int_G^8 \frac{2}{a} (wv_H)\, dh = \frac{2}{a} M_H \bar{\mu}. \quad \text{..........................(23)}$$

The term $\int_G^8 \rho \frac{D\mu}{Dt} dh = -\frac{1}{g}\int_G^8 \frac{D\mu}{Dt} dp = -\frac{1}{g}\frac{D}{Dt}\int_G^8 \mu dp$, provided that $p$ is regarded as constant during the differentiation $\frac{D}{Dt}$. Here $\int_G^8 \mu dp$ is given by substituting in (15) above, $G$ for 8 and 8 for 6, so as to make it apply to the lowest stratum. But $\int_G^8 \rho \frac{D\mu}{Dt} dh$ may as well be left in that form until precipitation and eddy motion have been discussed; as it is from these causes that the term arises.

Lastly the term

$$\int_G^8 \frac{\partial w}{\partial t} dh = \frac{\partial}{\partial t} W_{G8} = -\frac{1}{g}\frac{\partial}{\partial t}\int_G^8 \mu dp. \quad \text{.......................(24)}$$

And here $\int_G^8 \mu dp$ is given by substituting in (15) above, $G$ for 8 and 8 for 6, so as to make the equation apply to the lowest stratum. Note that in (24) *both* $\mu$ and $p$ are affected by the differentiation $\partial/\partial t$.

Collecting terms (6) transforms into

$$\frac{\partial W_{G8}}{\partial t} = -\frac{1}{g}\frac{\partial}{\partial t}\frac{(\log p_8 - \log p_G)(\mu_8 p_8 - \mu_G p_G)}{\log(\mu_8 p_8) - \log(\mu_G p_G)}$$

$$= \int_G^8 \rho \frac{D\mu}{Dt} dh - \frac{\partial}{\partial e}(\bar{\mu} \,.\, M_{EG8}) - \frac{\partial}{\partial n}(\bar{\mu} \,.\, M_{NG8}) + \bar{\mu}\frac{\tan\phi}{a} M_{NG8}$$

$$- \mu_8 m_{H8} - \frac{2}{a} M_{HG8} \,.\, \bar{\mu}, \quad \text{....................................(25)}$$

where $\bar{\mu}$ is defined by (19) above.

Note that the term $-\mu_G m_{HG}$ has disappeared because the ground is impervious to wind, and that in the corresponding equations for the upper strata there would always be a pair of terms of the form $\mu m_H$.

It will be shown in Ch. 4/8 that the effect of eddies can be inserted in (25).

The fundamental equations on this alternative scheme are (25) and those obtainable from (25) by changing the suffixes which indicate the height. As $D\mu/Dt$, $M_E$, $M_N$, $\mu_G$, $\mu_8$ are supposed to be known they can be inserted in (25) which then yields $\partial W_{G8}/\partial t$.

In this alternative scheme $W$ is not one of the principal variables, and finding $\partial W/\partial t$ is only a stage towards finding $\partial\mu/\partial t$. For, to make any numerical scheme of prediction "self-contained," it is necessary and sufficient to find $\partial/\partial t$ of each one of the variables which have been tabulated initially. Unfortunately $\partial W_{G8}/\partial t$ does not immediately give $\partial\mu_G/\partial t$ and $\partial\mu_8/\partial t$, for it also depends on the pressure changes. It will be better then to wait until the new distribution of pressure has been found after the time step $\delta t$, and then from the new set of $W$ given by (25) to find the new values of $\mu$ at the boundaries by means of equation (15), starting from a known value $\mu_G$ given by the conditions at the ground.

The increase of accuracy, which is the aim of all these special proceedings, could of course be attained by the general, and possibly easier, method of using sufficiently thin strata.

The stratosphere does not require special treatment because there is no appreciable amount of water in it.

## CH. 4/4. THE DYNAMICAL EQUATIONS, WHEN EDDIES ARE NOT AVERAGED-OUT AND MOLECULAR VISCOSITY IS NEGLECTED

In vector* notation

$$-\nabla\psi' - \frac{1}{\rho}\nabla p = \frac{\partial \mathbf{v}}{\partial t} + (\mathbf{v}\nabla)\,\mathbf{v} + 2\,[\boldsymbol{\omega}.\mathbf{v}] + [\boldsymbol{\omega}.[\boldsymbol{\omega}.\mathbf{a}]], \quad \text{..............(1)}$$

where the operator $\partial/\partial t$ refers to time changes at a point rotating with the globe. The angular velocity $\boldsymbol{\omega}$ of the earth is here regarded as a vector, while $\mathbf{a}$ is the radius vector from the centre of the earth to the atmospheric point under consideration. The term $[\boldsymbol{\omega}.[\boldsymbol{\omega}.\mathbf{a}]]$ represents the acceleration of a point at rest relative to the rotating globe; and this acceleration is customarily included together with $\nabla\psi'$, the true acceleration of gravity; since it is possible to derive both together from a potential $\psi$, so that

$$\nabla\psi = \nabla\psi' + [\boldsymbol{\omega}.[\boldsymbol{\omega}.\mathbf{a}]]. \quad \text{..............................(1}a\text{)}$$

Further, it is shown in Vector Algebra that

$$(\mathbf{v}\nabla)\,\mathbf{v} = \nabla\frac{\mathbf{v}^2}{2} - [\mathbf{v}.\operatorname{curl}\mathbf{v}]. \quad \text{..............................(1}b\text{)}$$

So that the dynamical equation (1) may be written

$$-\nabla\psi - \frac{1}{\rho}\nabla p = \frac{\partial \mathbf{v}}{\partial t} + (\mathbf{v}\nabla)\,\mathbf{v} + 2\,[\boldsymbol{\omega}.\mathbf{v}] \quad \text{.......................(1}c\text{)}$$

$$= \frac{\partial \mathbf{v}}{\partial t} + \nabla\frac{\mathbf{v}^2}{2} - [\mathbf{v}.\operatorname{curl}\mathbf{v}] + 2\,[\boldsymbol{\omega}.\mathbf{v}]. \quad \text{...........(1}d\text{)}$$

The method of deriving the corresponding equations in spherical or spheroidal coordinates is set out in Lamb's *Hydrodynamics* in the section dealing with Laplace's theory of the tides on a rotating globe. Lamb first modifies Lagrange's dynamical equations in generalized coordinates, to suit the case when the coordinate system is rotating. He then applies the result to spheroidal coordinates. An alternative method,

* See, for example, Silberstein's *Vectorial Mechanics*, ch. VI.

probably simpler, is first to work out Lagrange's equations in *fixed* spherical coordinates $\lambda'$, $\phi$, $r$ having the same axis as the rotating coordinates $\lambda$, $\phi$, $r$; and then to transform to rotating coordinates by putting, for the eastwards velocity $d\lambda'/dt = d\lambda/dt + a\cos\phi\,.\,\omega$. Thus, in one way or the other, we get the dynamical equations for a particle moving in any manner relative to the earth. To form the corresponding equations for a fluid we must express the quantities $\frac{d^2\lambda}{dt^2}$, $\frac{d^2\phi}{dt^2}$, $\frac{d^2r}{dt^2}$ belonging to a particle moving with the fluid, as the sum of (i) the changes in the velocity of the fluid at points moving with the earth, and of (ii) changes in the velocity of the particle due to the fluid-velocity varying from point to point over the earth. The second part corresponds to the vector $(\mathbf{v}\nabla)\,\mathbf{v}$ in equation (1).

The final result is not given completely by Lamb, because, in the theory of the tides, various terms involving squares and products of velocities can be neglected, as the tidal velocities are so small. In meteorology we must retain some at least of these terms. Here in Ch. 4/4 all are retained, except that $a$ is put for $a+h$. The complete equations are given by Bigelow in his *Report on the International Cloud Observations* 1896 *to* 1897. They are of the type

$$-\frac{\partial v_E}{\partial t} = \frac{1}{\rho}\frac{\partial p}{\partial e} - 2\omega\sin\phi\,.\,v_N + \text{etc.} \quad \ldots\ldots\ldots\ldots\ldots\ldots\ldots(2)$$

But for the purposes of this book it will be necessary to have the equation giving $\frac{\partial m_E}{\partial t}$ instead of $\frac{\partial v_E}{\partial t}$, and similarly for the other components. To obtain the required forms: the dynamical equation beginning $\frac{\partial v_E}{\partial t}$ is multiplied throughout by $\rho$; the equation of continuity of mass in the form $\frac{\partial \rho}{\partial t} = \text{etc.}$ is multiplied by $v_E$; then $\rho\frac{\partial v_E}{\partial t}$ is added to $v_E\frac{\partial \rho}{\partial t}$ to give $\frac{\partial m_E}{\partial t}$. There results:

$$\underset{10^{-4}}{-\frac{\partial m_E}{\partial t}} = \underset{10^{-3}}{\frac{\partial p}{\partial e}} + \underset{10^{-4}}{\frac{\partial}{\partial e}(m_E v_E)} + \underset{10^{-3}}{\frac{\partial}{\partial n}(m_E v_N)} + \underset{10^{-3}?}{\frac{\partial}{\partial h}(m_E v_H)} - \underset{10^{-3}}{2\omega\sin\phi\,.\,m_N} + \underset{10^{-5}}{2\omega\cos\phi\,.\,m_H}$$
$$+ \underset{10^{-6}}{\frac{3m_E v_H}{a}} - \underset{10^{-4}\tan\phi}{\frac{2m_E v_N\tan\phi}{a}}, \quad \ldots\ldots\ldots(3)$$

$$\underset{10^{-4}}{-\frac{\partial m_N}{\partial t}} = -g_N\rho + \underset{10^{-3}}{\frac{\partial p}{\partial n}} + \underset{10^{-4}}{\frac{\partial}{\partial e}(m_N v_E)} + \underset{10^{-3}}{\frac{\partial}{\partial n}(m_N v_N)} + \underset{10^{-3}?}{\frac{\partial}{\partial h}(m_N v_H)} + \underset{10^{-3}}{2\omega\sin\phi\,.\,m_E}$$
$$+ \underset{10^{-6}}{\frac{3m_N v_H}{a}} + \frac{\underset{10^{-5}\tan\phi}{\tan\phi}\,\{\underset{10^{-5}\tan\phi}{m_E v_E - m_N v_N}\}}{a}, \quad \ldots\ldots\ldots(4)$$

$$\underset{10^{-6}?}{-\frac{\partial m_H}{\partial t}} = \underset{1}{g\rho} + \underset{1}{\frac{\partial p}{\partial h}} + \underset{10^{-5}?}{\frac{\partial}{\partial e}(m_H v_E)} + \underset{10^{-5}?}{\frac{\partial}{\partial n}(m_H v_N)} + \underset{10^{-5}?}{\frac{\partial}{\partial h}(m_H v_H)} - \underset{10^{-3}}{2\omega\cos\phi\,.\,m_E}$$
$$+ \frac{\underset{10^{-7}}{2m_H v_H} - \underset{10^{-5}}{m_E v_E} - \underset{10^{-5}}{m_N v_N} - \underset{10^{-6}\tan\phi}{m_H v_N\tan\phi}}{a}. \quad \ldots\ldots\ldots(5)$$

The terms in equations (3), (4), (5) have the following significance. They are all of the dimensions mass × (length)$^{-2}$ × (time)$^{-2}$, that is to say they are forces-per-unit-volume. In each equation there are three terms, such, for example, as

$$\frac{\partial}{\partial e}(m_E v_E)+\frac{\partial}{\partial n}(m_E v_N)+\frac{\partial}{\partial h}(m_E v_H) \text{ in equation (3)}$$

arising from the spatial variation of velocity and of density, and derived partly from the term $\rho(\mathbf{v}\nabla)\mathbf{v}$ in $\rho$ times equation (1), and partly from the equation of continuity of mass. Next in each equation there are either one or two terms in $\omega \cos\phi$ or $\omega \sin\phi$; these form the components of $\rho[\boldsymbol{\omega}\,.\,\mathbf{v}] \equiv [\boldsymbol{\omega}\,.\,\mathbf{m}]$, which is the vector product of the earth's angular velocity and the momentum-per-unit-volume-relative-to-the-earth. Lastly in each equation there are a number of terms in $a^{-1}$. These depend on the curvature of the earth. Those in $\tan\phi$ depend also on the curvature of the parallels of latitude, in such a way as to become formidable near the poles. The terms in $a^{-1}$ are derived partly from the centripetal accelerations and partly from the effect of the crowding together of the coordinate lines in the equation of continuity of mass.

A body rotating with the earth as if it were rigidly attached to it, experiences centrifugal forces per unit volume $\rho(\omega\cos\phi)^2(a+h)$ upwards and $-\rho\omega^2\cos\phi\sin\phi\,.\,(a+h)$ northwards. These forces are found from Lagrange's equations along with those depending on the winds, but the former are not allowed to appear explicitly in equations (5) and (4) because they are customarily included respectively in $g$ and in the component of the acceleration of gravity towards the north, which is here denoted by $g_N$. The other part of $g_N$ is a true gravity effect depending on the ellipticity of the earth. Values of $g_N$, for heights and latitudes to fit our conventional division of the atmosphere, are given in the annexed table.

$\pm g_N$

*Component of the acceleration of gravity along a line running north and south at a constant height above mean sea-level. Directed towards the equator in both hemispheres. From Helmert's constants* (1901) *quoted in Landolt Börnstein Meyerhoffer's Tabellen*

| Distance from equator | | Kilometres above mean sea-level 1·0 | 3·0 | 5·5 | 9·1 | 16·3 |
|---|---|---|---|---|---|---|
| | | Approximate mean pressures in decibars 9 | 7 | 5 | 3 | 1 |
| Degrees | Megametres | 10,000 times accelerations in cms/sec² | | | | |
| 0 | 0 | 0 | 0 | 0 | 0 | 0 |
| 9 | 1 | 3 | 8 | 14 | 23 | 41 |
| 18 | 2 | 5 | 14 | 26 | 43 | 78 |
| 27 | 3 | 7 | 20 | 36 | 60 | 107 |
| 36 | 4 | 8 | 23 | 43 | 70 | 126 |
| 45 | 5 | 8 | 24 | 45 | 74 | 133 |
| 54 | 6 | 8 | 23 | 43 | 71 | 126 |
| 63 | 7 | 7 | 20 | 36 | 60 | 108 |
| 72 | 8 | 5 | 14 | 26 | 44 | 78 |
| 81 | 9 | 3 | 8 | 14 | 23 | 41 |
| 90 | 10 | 0 | 0 | 0 | 0 | 0 |

The numbers under the terms in (5) show that the vertical equation is nearly equivalent to $0 = g\rho + \frac{\partial p}{\partial h}$, ..............................(6)

as is well known. If the largest of the small terms be included, the equation runs

$$0 = g\rho + \frac{\partial p}{\partial h} - 2\omega\cos\phi \, . \, m_E. \qquad \text{(7)}$$

The term $2\omega\cos\phi \, . \, m_E$ may produce a pressure of 0·5 millibar at sea-level under extreme conditions. Mr W. H. Dines points out that this term $2\omega\cos\phi \, . \, m_E$ may be taken to mean that air moving eastward is lighter than the same air moving westward, and suggests that this may explain the fact that westerly winds commonly increase aloft much more than do easterly winds*. If the term $2\omega\cos\phi \, . \, m_E$ really has such an important influence it ought not to be neglected when one is dealing with parts of eddies. He also adduces an observation of Nansen's as illustrating a similar effect in the horizontal—the Siberian rivers, which flow northwards, erode the land more on their eastern sides, as if the water there moved faster than on the west.

Now let us try to express the dynamics of a stratum, considered as a whole, in terms of the quantities $R$, $M_E$, $M_N$, which fitted so neatly into the equation of continuity of mass (Ch. 4/2). Integrate the longitude equation (3) with respect to height across a stratum. Taking the lowest stratum with its limits denoted, for brevity, by $G$ and 8 as example and considering terms separately

$$\int_G^8 \frac{\partial p}{\partial e} dh = \frac{\partial}{\partial e}\int_G^8 p dh + \left(p\frac{\partial h}{\partial e}\right)_G. \qquad \text{(8)}$$

Let us denote $\int_G^8 p dh$ by $P_{G8}$. Now it will probably be more convenient to use $p_8$ and $p_G$ as variables rather than $P_{G8}$. In that case we shall require to express $P_{G8}$ in terms of $p_8$ and $p_G$, and this can be done by the following formula, which is strictly true if the air is dry and if temperature and gravity are independent of height, and is in any case a good approximation†

$$P_{G8} = \frac{(h_8 - h_G)(p_G - p_8)}{\log_e p_G - \log_e p_8}. \qquad \text{(9)}$$

The important terms $-\frac{\partial m_E}{\partial t}$, $-2\omega\sin\phi \, . \, m_N$, $+2\omega\cos\phi \, . \, m_H$ yield respectively, on integration across the stratum, $-\frac{\partial M_E}{\partial t}$, $-2\omega\sin\phi \, . \, M_N$, $+2\omega\cos\phi \, . \, M_H$. The remaining terms can only be expressed as functions of $R$, $M_E$, $M_N$, $M_H$ in special cases, because they contain squares and products of velocities; but fortunately these terms are usually small.

Consider the three terms $\frac{\partial}{\partial e}(m_E v_E) + \frac{\partial}{\partial n}(m_E v_N) + \frac{\partial}{\partial h}(m_E v_H)$.

* *Manual of Meteorology*, by Sir Napier Shaw, Part IV. p. 58.

† No doubt, but a better one could easily be obtained.

On integrating with respect to $h$ they become

$$\frac{\partial}{\partial e}\int_G^8 (m_E v_E)\,dh + \frac{\partial}{\partial n}\int_G^8 (m_E v_N)\,dh + (m_E v_H)_8$$
$$+\left\{m_E\left(v_E\frac{\partial h}{\partial e}+v_N\frac{\partial h}{\partial n}-v_H\right)\right\}_{\text{at the ground,}} \qquad \text{.........(10)}$$

and the part in {} vanishes, because the ground is impervious to wind, just as did the similar expression in the equation of continuity of mass. Now in the special case in which the velocities are independent of height $v_E = \frac{M_E}{R}$, $v_N = \frac{M_N}{R}$, $v_H = \frac{M_H}{R}$, and in this case the two integrals in the last expression transform into

$$\frac{\partial}{\partial e}\left(\frac{M_E^2}{R}\right)+\frac{\partial}{\partial n}\left(\frac{M_E M_N}{R}\right).$$

I have assumed that this transformation is a fair representation in general, even if the velocities are not independent of height. The range of velocity in a stratum diminishes, as a rule, as the stratum becomes thinner. So if the above assumption should prove to be insufficiently exact, one course will be to take a larger number of thinner conventional strata. The other terms containing products have been dealt with similarly. Collecting terms, the dynamical equations are transformed into the following (11), (12), (13), which are suitable for use in computing, and which take account of the height of the ground. The differentiations $\frac{\partial}{\partial e}$ and $\frac{\partial}{\partial n}$ now follow the stratum as a whole. The terms $p\partial h/\partial e$, $p\partial h/\partial n$ of course only occur when the equations relate to the lowest stratum. In the integrations $g$ and $g_N$ have been regarded as independent of height. Mean values of $g$ and $g_N$ for the stratum should therefore be used.

$$-\frac{\partial M_E}{\partial t} = \frac{\partial P}{\partial e} + \left[p\frac{\partial h}{\partial e}\right]_G + \frac{\partial}{\partial e}\left(\frac{M_E^2}{R}\right) + \frac{\partial}{\partial n}\left(\frac{M_E M_N}{R}\right) + [m_E v_H]_8^* - 2\omega\sin\phi\,.\,M_N + 2\omega\cos\phi\,.\,M_H$$
$$+\frac{3M_E M_H - 2M_E M_N\tan\phi}{aR}. \qquad \text{.........(11)}$$

$$-\frac{\partial M_N}{\partial t} = -g_N R + \frac{\partial P}{\partial n} + \left[p\frac{\partial h}{\partial n}\right]_G + \frac{\partial}{\partial e}\left(\frac{M_N M_E}{R}\right) + \frac{\partial}{\partial n}\left(\frac{M_N^2}{R}\right) + [m_N v_H]_8^* + 2\omega\sin\phi\,.\,M_E$$
$$+\frac{3M_N M_H}{aR} + \frac{\tan\phi\,\{M_E^2 - M_N^2\}}{aR}. \qquad \text{.........(12)}$$

$$-\frac{\partial M_H}{\partial t} = gR + p_8 - p_G + \frac{\partial}{\partial e}\left(\frac{M_H M_E}{R}\right) + \frac{\partial}{\partial n}\left(\frac{M_H M_N}{R}\right) + [m_H v_H]_8^* - 2\omega\cos\phi\,.\,M_E$$
$$+\frac{2M_H^2}{aR} - \frac{M_E^2}{aR} - \frac{M_N^2}{aR} - \frac{M_H M_N\tan\phi}{aR}. \qquad \text{.........(13)}$$

* For any stratum except the lowest the corresponding quantity at the lower limit of the stratum is to be subtracted from this term.

In the example of Ch. 9 the terms $\frac{\partial P_{G8}}{\partial n}$ and $\left[p\frac{\partial h}{\partial n}\right]_G$ are opposite in sign and numerically much larger than any other terms in the northward dynamical equation. It is therefore well to use exactly the same process in approximating to $\frac{\partial P_{G8}}{\partial n}$ as in approximating to $\left[p\frac{\partial h}{\partial n}\right]_G$, lest a slight difference in their fractional errors should make a serious error in their sum. Accordingly integrate both these terms, with respect to latitude through an interval of 200 kilometres. Then

$$\int\left[p\frac{\partial h}{\partial n}\right]_G dn = \int p_G dh_G.$$

Now if we assume that the pressure at the ground is nearly an exponential function of the height of the ground, then the last integral above is of similar form to $P$ given by equation (9). So that, if quantities with one dash and two dashes refer to the beginning and end of the interval $\delta n = 200$ km, we have

$$\frac{\partial P_{G8}}{\partial n} + \left[p\frac{\partial h}{\partial n}\right]_G \text{ represented by } \frac{1}{\delta n}\left\{P_{G8}'' - P_{G8}' + \frac{(h_G'' - h_G')(p_G'' - p_G')}{\log_e p_G'' - \log_e p_G'}\right\} \quad \ldots(14)$$

And a similar transformation applies to the corresponding terms in the eastward dynamical equation.

## CH. 4/5. ADIABATIC TRANSFORMATION OF ENERGY

### CH. 4/5/0. GENERAL THEORY

This may be very simply expressed by saying that the entropy of a given mass of air is only changed by radiation, by precipitation or by eddy-diffusion.

But the theory of entropy is usually developed with reference to substances in the cylinders of idealized engines, where the weight of the working substance and its kinetic energy are both negligible. And such rigorous experimental versifications of it, as have related to gases, have mostly either been carried out in closed vessels or else have referred to wave-motion. When we have to apply such indoor results to the free atmosphere, where observation is scarcely as yet able to provide a rigorous check, there is apt to be some confusion of mind as to the precise way in which gravity and kinetic energy fit into the scheme. The following pages have been written to make this clear. The result, which is stated above, is shown to be correct.

**Continuity of Energy.** In counting up the total energy of a mass of gas we must beware lest we add on the same part twice over, under different names.

Imagining the gas as a swarm of molecules we see four kinds of energy

(i) gravitational due to earth.

(ii) energy within the molecules.

(iii) energy of forces between molecules.

(iv) kinetic.

Pressure does not appear. Now when we change to a scale on which molecules become invisibly minute, there remains simply under the same name only gravitational energy. It is denoted by $\rho\psi$ per unit volume, where $\psi$ is the gravitational potential and $\rho$ is the density. The kinetic energy is now split into two portions: (*a*) mean-motional, still called "kinetic energy" and denoted by $\frac{1}{2}\rho v^2$, (*b*) deviations from the mean. These deviations are grouped together with the energy within the molecules, and the energy of forces, other than gravity, between molecules. The group so formed is called the "intrinsic energy" and in this book is denoted by $\upsilon$ (upsilon) per mass or $\rho\upsilon$ per volume. Here $d\upsilon = \gamma_v d\theta$ where $\gamma_v$ is the specific heat at constant volume.

Schematically the classification of energy-per-mass runs thus

| MOLECULAR | | MOLAR |
|---|---|---|
| Gravitational | | $\psi$ |
| Kinetic | Mean | $\frac{1}{2}v^2$ |
| Kinetic | Deviations | $\upsilon$ |
| Within molecules | | $\upsilon$ |
| Between molecules | | $\upsilon$ |

The point to be emphasized is that as $\int$ pressure . $d$ (volume) did not appear in the molecular classification, it cannot appear in the molar classification unless we compensate for its appearance by omitting an equal amount of energy under some other name. For instance part of $\rho\upsilon$ that comes from the translational kinetic energy of the molecules is equal to $\frac{3}{2}p$ in an ideal gas. See Jeans' *Dynamical Theory of Gases*, 2nd edn. § 161. Margules* no doubt makes some such compensation when he calculates the energy associated with local irregularities of pressure. In the present scheme this energy appears under other names, and to include pressural energy in addition would be to count some of the energy twice over. It seems to be most natural to regard pressure only as concerned in the transfer across surfaces. Once across, energy appears in some other form: gravitational, intrinsic or molar kinetic.

The rate of increase of energy in unit volume is therefore in the present scheme

$$\frac{\partial}{\partial t}(\rho\psi + \tfrac{1}{2}\rho v^2 + \rho\upsilon). \quad \text{..............................(1)}$$

Let us next examine the rate at which energy crosses a very small plane. Again imagining the gas on an enlarged scale, we see only a swarm of molecules, each carrying its energy with it. To simplify the problem, let the plane move with the mean velocity of the swarm in its neighbourhood. That is to say let the fluxes of mass across the plane from its two sides be equal and opposite.

* "Mechanical Equivalent of Pressure" and "Energy of Storms" in Abbe's *Mechanics of the Earth's Atmosphere*, 3rd Collection.

Then as the gravitational energy which a molecule possesses when it is crossing the plane is independent of its speed and of the direction of its motion, so on the average no gravitational energy crosses the plane. The internal energy which a molecule possesses when crossing the plane is likely to increase with its speed relative to the plane, but is independent of the side from which it comes, if the general state of the gas is the same on the two sides, so for that reason, no internal energy crosses the plane. (Note that we have here made an assumption equivalent to neglecting part of the *conduction* of heat. Let us neglect the rest of it also, as it can be allowed for afterwards.)

Next, will the energy of inter-molecular forces be carried across the plane? Apparently not, for it also will be independent of the direction of motion relative to the plane, if conduction of heat can be neglected. But the same is not true of the molecular kinetic energy, for this has been reckoned relative to the earth and the motion of the plane produces a lack of symmetry in the flux of energy, so that the opposing parts do not balance, as will be shown.

For simplicity let the plane be set at right angles to the mean motion of the molecules and let its velocity relative to any Newtonian framework be $\bar{\dot{x}}$, the mean velocity of the particles. If we distinguish mean values by bars and deviations from the mean by dashes, then any molecule has kinetic energy, relative to the same standard, equal to

$$\tfrac{1}{2}\mathfrak{m}\{(\bar{\dot{x}}+\dot{x}')^2+(\bar{\dot{y}}+\dot{y}')^2+(\bar{\dot{z}}+\dot{z}')^2\}, \quad\text{.........................(2)}$$

where $\mathfrak{m}$ is the mass of a molecule.

The molecule traverses the plane with a velocity $\dot{x}'$. If there are $n$ molecules having this velocity in unit volume, a number $\dot{x}'n$ of them cross unit area of the plane in unit time and carry with them a kinetic energy

$$\tfrac{1}{2}\mathfrak{m}n\dot{x}'\{(\bar{\dot{x}}+\dot{x}')^2+(\bar{\dot{y}}+\dot{y}')^2+(\bar{\dot{z}}+\dot{z}')^2\}. \quad\text{.......................(3)}$$

The resultant flux of energy is the sum of this expression for all the molecules in unit volume and may be written as follows, $n$ being omitted because $\Sigma$ sums for the molecules individually,

$$\tfrac{1}{2}\mathfrak{m}\Sigma\dot{x}'\{(\bar{\dot{x}})^2+(\bar{\dot{y}})^2+(\bar{\dot{z}})^2+2\bar{\dot{x}}\dot{x}'+2\bar{\dot{y}}\dot{y}'+2\bar{\dot{z}}\dot{z}'+\dot{x}'^2+\dot{y}'^2+\dot{z}'^2\}. \quad\text{...........(4)}$$

Now the sum of the product of any mean into any deviation-from-a-mean vanishes. Also, in the molecular chaos, the correlations between $\dot{x}'$, $\dot{y}'$ and $\dot{z}'$ all vanish if the effects of molecular viscosity are negligible. Again $\Sigma\dot{x}'^3$ is negligible if we can neglect the molecular conduction of heat. Consequently the flux of energy simplifies down to $\mathfrak{m}\bar{\dot{x}}\Sigma\dot{x}'^2$. Now in works on the Kinetic Theory of Gases* it is shown that $\mathfrak{m}\Sigma\dot{x}'^2$ is equal to the pressure, at least in an ideal gas.

*Thus if a very small plane be moving at right angles to itself with a velocity v such that on the average no mass is traversing it, then the flux of energy across it is pv per area.* ...........................................................................................(5)

* Jeans' *Dynamical Theory of Gases*, 2nd ed., § 161 and § 216 *et seq.*

Thus is explained the appearance of pressure in the flux of energy and its absence in the energy contained in unit volume. The above treatment is sketchy and the reader should be referred to works in which the kinetic theory is treated thoroughly. The above theorem is a very special case of one by Maxwell set out in its generality in Jeans' book, 2nd edn. § 340.

Now if we add to the small plane, which we may imagine as square, five other planes, also moving with the fluid to form, at one instant, a unit cube, then we see that the rate of decrease of energy in the moving, distorting, swelling cube will instantaneously be

$$\frac{\partial (pv_x)}{\partial x} + \frac{\partial (pv_y)}{\partial y} + \frac{\partial (pv_z)}{\partial z} \text{ or, briefly, } \operatorname{div}(p\mathbf{v}). \quad \text{.................(6)}$$

Or if we had taken a cube containing unit mass instead of unit volume, the rate of decrease of energy in it would have been

$$\frac{1}{\rho}\operatorname{div}(p\mathbf{v}). \quad \text{.................................(7)}$$

But the energy in unit mass of fluid is $\frac{1}{2}\mathbf{v}^2 + \psi + v$, so if $D/Dt$ denote a differentiation following the motion of the fluid

$$-\frac{D}{Dt}(\tfrac{1}{2}\mathbf{v}^2 + \psi + v) = \frac{1}{\rho}\operatorname{div}(p\mathbf{v}). \quad \text{........................(8)}$$

There is another form, sometimes useful, into which (8) may be turned by means of the general theorem that if $A$ is any scalar and $\rho$ the density of a fluid moving with velocity $\mathbf{v}$ then

$$\rho\frac{DA}{Dt} = \frac{\partial(\rho A)}{\partial t} + \operatorname{div}(\rho\mathbf{v}A). \quad \text{........................(9)}$$

If we put $A = (\frac{1}{2}\mathbf{v}^2 + \psi + v)$ then equation (9) causes (8) to transform into

$$\operatorname{div}\{\mathbf{v}(\tfrac{1}{2}\rho\mathbf{v}^2 + \rho\psi + \rho v + p)\} = -\frac{\partial}{\partial t}(\tfrac{1}{2}\rho\mathbf{v}^2 + \rho\psi + \rho v). \quad \text{...........(10)}$$

*Equations* (9) *and* (10) *are two forms of the equation of continuity of energy*.* But we can change to a more convenient one.

For, purely by geometry, in the ordinary vector notation,

$$\operatorname{div}(p\mathbf{v}) = p\operatorname{div}\mathbf{v} + \mathbf{v}\nabla p. \quad \text{............................(11)}$$

And the term $\mathbf{v}\nabla p$ can be found and removed along with $(\frac{1}{2}\mathbf{v}^2 + \psi)$, by means of the dynamical equation, as follows.

When the axes of reference are fixed to the earth the dynamical equation is

$$-\nabla p = \rho\nabla\psi + \rho\frac{D\mathbf{v}}{Dt} + 2\rho[\boldsymbol{\omega}\mathbf{v}], \quad \text{........................(12)}$$

where the square bracket denotes a vector product. Now form the equation of activity-per-volume by taking the scalar product of $\mathbf{v}$ into each term of the above equation. Since $\mathbf{v}$ and $[\boldsymbol{\omega}\mathbf{v}]$ are at right angles to one another their scalar product vanishes.

* Cp. Webster, *Dynamics*, § 188; Lamb, *Hydrodynamics*, § 10, eqn. (5).

And since the gravity potential does not change at a point fixed to the earth

$$\partial\psi/\partial t = 0; \text{ so that } \mathbf{v}\nabla\psi = D\psi/Dt. \qquad (13)$$

Thus the dynamical activity per volume is expressed by either side of

$$-\mathbf{v}\nabla p = \rho \frac{D}{Dt}(\tfrac{1}{2}\mathbf{v}^2 + \psi). \qquad (14)$$

Now on substituting (14) in (11) we obtain as a form of the dynamical equation

$$\operatorname{div}(\mathbf{v}p) = p \operatorname{div} \mathbf{v} - \rho \frac{D}{Dt}(\psi + \tfrac{1}{2}\mathbf{v}^2). \qquad (15)$$

And this when combined with (8) yields

$$-\rho \frac{Dv}{Dt} = p \operatorname{div} \mathbf{v}. \qquad (16)$$

We might advantageously bring in the equation of continuity of mass in the form

$$\operatorname{div} \mathbf{v} = -\frac{1}{\rho}\frac{D\rho}{Dt}, \qquad (17)$$

which when substituted in (16) gives

$$-\frac{Dv}{Dt} = p \frac{D}{Dt}\left(\frac{1}{\rho}\right). \qquad (18)$$

That is to say, *following the motion, the increase of v, the intrinsic-energy-per-mass, is equal to the pressure multiplied by the decrease of volume per mass.* This is the ordinary adiabatic relation. We have neglected radiation, molecular conduction of heat, and molecular viscosity. The reason for deriving a familiar result by such a roundabout process, is to make it clear that neither the kinetic energy $\frac{1}{2}\rho\mathbf{v}^2$ nor the gravitational energy appears in (18). They are not neglected; they do not belong in this adiabatic equation.

Equation (18) from its mode of deduction should apply equally to clear air or to clouds moving in the most unrestricted way occurring in nature, provided that radiation and conduction and precipitation are negligible, or allowed for separately. There is a query as to the flux being $\mathbf{v}p$ when intermolecular forces act, but in meteorological applications the error, if any, must be small.

## Entropy

If unit mass, during its motion, is gaining energy by radiation at a rate $D\epsilon/Dt$ we have from (18) on bringing in the additional term and then transposing and dividing by $\theta$

$$\frac{1}{\theta}\frac{D\epsilon}{Dt} = \frac{1}{\theta}\frac{Dv}{Dt} + \frac{p}{\theta}\frac{D}{Dt}\left(\frac{1}{\rho}\right). \qquad (19)$$

Each side of this equation is the rate of increase of entropy-per-mass, $\sigma$, following the motion; that is $D\sigma/Dt$.

To see that this formula agrees with those in common use (Hertz, Neuhoff, etc.) take, as a test case, that of dry air. Then

$$Dv = \gamma_v D\theta \text{ and } p/\theta = b\rho.$$

There results

$$\frac{D\sigma}{Dt} = \frac{D}{Dt}(\gamma_v \log\theta - b\log\rho) = \frac{1}{\theta}\frac{D\epsilon}{Dt}, \text{ as usual.} \quad \text{..............(20)}$$

Thus the entropy is $\sigma = \gamma_v \log\theta - b\log\rho$ for dry air, and its constancy during the motion is entirely unaffected by changes in the molar kinetic energy, or by any gravitational effect.

**The Potential Temperature*** $\tau$ in a mass of air is defined as the absolute temperature which the air assumes when brought adiabatically to a standard pressure $p_i$.

If in the equation which follows from (20)

$$\sigma = \gamma_v \log\theta - b\log\rho + \text{const.}$$

we replace $\rho$ by $p$ by means of the equation $p = b\rho\theta$, applicable to dry air, we obtain

$$\sigma = \gamma_p \log\theta - b\log p + \text{const.}$$

Thus if the air at $\theta, p$ be brought adiabatically to a standard pressure $p_i$, its final temperature $\theta$, now called its potential temperature $\tau$, will be $\tau = \theta\left(\frac{p_i}{p}\right)^{b/\gamma_p} = \theta\left(\frac{p_i}{p}\right)^{0\cdot289}$, so that $\sigma = \gamma_p \log\tau + \text{const.}$ again independently of any kinetic energy generated en route.

In dealing with eddy-motion the potential temperature is more convenient than the entropy in this respect: *If two masses of dry air at different potential temperatures mix intimately and adiabatically by diffusion under the constant pressure of their surroundings, their mean potential temperature remains unaltered,* although their total entropy increases†. This follows immediately from the formulae. It will be sufficient to consider the mixture of two equal masses, as any other case may be reduced to that. Let the initial temperatures be $\theta', \theta''$. Then the potential temperatures are initially

$$\tau' = \theta'\left(\frac{p_i}{p}\right)^{0\cdot289}; \quad \tau'' = \theta''\left(\frac{p_i}{p}\right)^{0\cdot289}.$$

On mixing portions of dry air under constant pressure, without loss or gain of heat, there is no change in the total intrinsic energy, so the final temperature is $\frac{1}{2}(\theta' + \theta'')$ and, as the pressure is fixed, the final potential-temperature is $\frac{1}{2}(\tau' + \tau'')$; but the entropy, which is proportional to $\gamma_p \log\tau + \text{const.}$, necessarily changes.

**The Potential Density** might prove to be a more useful conception than potential temperature, for it is density that is directly connected with stability.

* *Vide* v. Helmholtz (p. 83), v. Bezold (p. 243) in Abbe's *Mechanics of the Earth's Atmosphere*, 1891; L. A. Bauer (p. 495) in Abbe's *Mechanics of the Earth's Atmosphere*, 3rd Collection, Smithsonian Institution.

† Pointed out to me by Mr W. H. Dines.

CH. 4/5/I. NUMERICAL VALUES FOR THE ENTROPY-PER-MASS, $\sigma$, OF MOIST AIR

## General

The formulae and diagrams have been worked over successively by Hann, Hertz*, von Bezold †, W. M. Davis and Neuhoff †. Revised constants have been collected by Bigelow ‡. Bigelow's constants are in kg metres and in heat units; we require them in ergs.

The constants in the characteristic equations

$$p' = b'\rho\theta, \text{ for dry air,} \qquad \text{.........(1)}$$

or

$$p'' = b''w\theta, \text{ for water vapour,} \qquad \text{.........(2)}$$

are

$$b' = 2{\cdot}870 \times 10^6, \qquad \text{.........(3)}$$

$$b'' = 4{\cdot}616 \times 10^6, \qquad \text{.........(4)}$$

when expressed in erg grm$^{-1}$ (degree C)$^{-1}$. In the same unit the specific heats $\gamma_p'$ and $\gamma_p''$ of dry air and water vapour at constant pressure are respectively

$$\gamma_p' = 9{\cdot}92 \times 10^6; \qquad \text{.........(5)}$$

$$\gamma_p'' = 20{\cdot}08 \times 10^6. \qquad \text{.........(6)}$$

Note that $b$ or $\gamma$, without dashes, refer to the atmospheric mixture.

In addition to the entropy we require, in connection with the elimination of the vertical velocity (Ch. 5), the coefficients $a$ in the expansion

$$d\sigma = a_p dp + a_\rho d\rho + a_\mu d\mu. \qquad \text{.........(7)}$$

## Entropy in the unsaturated stage

Hertz's formula for this stage is

$$\sigma = c + \{\mu(\gamma_p'' - \gamma_p') + \gamma_p'\}\log_e\theta - \{\mu(b'' - b') + b'\}\log_e p$$
$$= c + 10^6(10{\cdot}16\mu + 9{\cdot}92)\log_e\theta - 10^6(1{\cdot}746\mu + 2{\cdot}870)\log_e p, \qquad \text{.........(8)}$$

where $c$ is an arbitrary constant of integration.

The characteristic equation for moist air is

$$\log_e\theta = \log_e p - \log_e\rho - \log_e\{\mu(b'' - b') + b'\}$$
$$= \log_e p - \log_e\rho - \log_e(1{\cdot}746\mu + 2{\cdot}870) \times 10^6. \qquad \text{.........(9)}$$

Eliminating $\log\theta$ between (8) and (9) we get

$$\sigma = c' + \log_e p\,(8{\cdot}41\mu + 7{\cdot}05) \times 10^6 - \log_e\rho\,(10{\cdot}16\mu + 9{\cdot}92) \times 10^6$$
$$- (10{\cdot}16\mu + 9{\cdot}92) \times 10^6 \log_e\{(1{\cdot}746\mu + 2{\cdot}870) \times 10^6\}, \qquad \text{.........(10)}$$

where $c'$ is another arbitrary constant.

* Hertz, *Miscellaneous Papers*, No. XIX.

† Abbe's *Translations of Papers on Mechanics of the Atmosphere.*

‡ Bigelow, *Report on the International Cloud Observations*, 1896 *to* 1897, p. 488.

From (10) $\alpha_p$, $\alpha_\rho$, $\alpha_\mu$ can be found by differentiation, thus:

$$\alpha_p = \left(\frac{\partial\sigma}{\partial p}\right)_{\rho,\,\mu\text{ const.}} = +\frac{1}{p}(8{\cdot}41\mu + 7{\cdot}05)\times 10^6 = \frac{\gamma_v}{p}, \quad \text{(11)}$$

$$\alpha_\rho = \left(\frac{\partial\sigma}{\partial \rho}\right)_{\mu,\,p\text{ const.}} = -\frac{1}{\rho}(10{\cdot}16\mu + 9{\cdot}92)\times 10^6 = -\frac{\gamma_p}{\rho}, \quad \text{(12)}$$

$$\alpha_\mu = \left(\frac{\partial\sigma}{\partial \mu}\right)_{p,\,\rho\text{ const.}} = +8{\cdot}41\times 10^6 \log_e p - 10{\cdot}16\times 10^6 \log_e \rho$$
$$- 10{\cdot}16\times 10^6 \log_e\{10^6(1{\cdot}746\mu + 2{\cdot}870)\}$$
$$- 10^6(10{\cdot}16\mu + 9{\cdot}92)\frac{1}{\mu + 1{\cdot}644}. \quad \text{(13)}$$

We also require the quantity $\beta$, defined as equal to $-\frac{\alpha_\rho}{\alpha_p}$ in Ch. 5. From (11) and (12)

$$\beta = +\frac{p}{\rho}\cdot\frac{10{\cdot}16\mu + 9{\cdot}92}{8{\cdot}41\mu + 7{\cdot}05} = \frac{p}{\rho}\cdot\frac{\gamma_p}{\gamma_v}. \quad \text{(14)}$$

By means of the characteristic equation (9), $\beta$ can be expressed in terms of $\theta$ and $\mu$. The result is

$$\beta = \theta\,.\,10^6\,.\,\frac{4{\cdot}04 + 6{\cdot}593\mu + 2{\cdot}516\mu^2}{1 + 1{\cdot}193\mu}. \quad \text{(15)}$$

So that for dry air $\beta$ is simply $4{\cdot}04\times 10^6\,\theta$.

## Entropy in the Rain, Hail and Snow stages

To fit with the rest of this work we require the entropy expressed as a function of pressure, density and moisture; the temperature must not appear explicitly. However, there is a difficulty in expressing the relation in formulae, because the vapour pressure is given experimentally as a somewhat complicated function of the temperature. On this account it is simpler to proceed by way of graphs or numerical tables, of which those by Neuhoff* appear to be the best, although they need to be converted to millibar units. If $p'$ is the partial pressure of the dry air, Neuhoff gives the adiabatic relation in the form

$$\log p' - \frac{F_2(\theta)}{p'} - F_1(\mu)\log\theta + \text{const.} = 0,$$

where $F_1$ is a function only of the amount of water present, and varies slowly, while $F_2$ depends only on the temperature, with which it rapidly increases. Following Bezold, Neuhoff treats also the "pseudo-adiabatic" case in which the water is precipitated as soon as it is condensed.

* Abbe's *Translations*, 3rd Collection, pp. 430—494.

### CH. 4/5/2. THE CONVEYANCE OF HEAT ON A LARGE SCALE

Hot winds from the desert and cool breezes from the sea are well known. F. M. Exner* has published a prognostic method based on the source of air supply. V. Bjerknes and his assistants, J. Bjerknes, Bergeron and Solberg, attach great importance to the conveyance of heat. They go so far as to find a "polar front" where cold air from the polar regions meets warm air from the Tropics in European latitudes, and thereby causes cyclones. The present writer regards the "polar front" as a sketch, in black and white, of a reality, which these authors deserve much credit for having discovered, but which requires, for its proper representation, many delicate gradations of half tones, as well as the occasional sharp outlines. In other words we may not take the entropy-per-mass $\sigma$ as unchanging following the motion, because $\sigma$ must be altered by radiation, precipitation, and mixing. If these effects are known, then we know $D\sigma/Dt$, the rate of change following the motion. And then the rate of change $\partial\sigma/\partial t$ at a point fixed relatively to the earth is given, in the usual way, by

$$\frac{\partial\sigma}{\partial t} = -v_E\frac{\partial\sigma}{\partial e} - v_N\frac{\partial\sigma}{\partial n} - v_H\frac{\partial\sigma}{\partial h} + \frac{D\sigma}{Dt}. \quad \ldots\ldots\ldots\ldots\ldots\ldots\ldots(1)$$

This equation is used in finding the vertical velocity. Having served that purpose it is found in the lower strata to be no longer necessary; so that we need not integrate it to make it apply to a stratum as a whole. The uppermost stratum receives a special treatment in Ch. 6.

### CH. 4/6. UNIFORM CLOUDS AND PRECIPITATION

When the dependent variables are tabulated numerical quantities, it is as easy as account-keeping to deduct a numerical quantity of water from the amount present in any stratum, and to transfer it to the ground.

Knowing the total amount of water-substance present, and also the temperature, we can find the amount of liquid or solid by means of a table, for it has been found that supersaturation does not occur†. As a general indication of what may be expected, the following table shows the limiting amount of water, which can just exist as gas, in the several conventional strata, for an average English temperature distribution‡.

| Pressure in decibars at limits of strata | 10–8 | 8–6 | 6–4 | 4–2 | 2–0 |
|---|---|---|---|---|---|
| Gaseous water in grams per square centimetre of horizontal area of stratum | 1·34 | 0·67 | 0·25 | 0·04 | ? |
| | water vapour | vapour in equilibrium with ice | | | |

* Exner, *Dynamische Meteorologie* (Teubner, 1917), § 70.
† Sir Napier Shaw, *Principia Atmospherica*, p. 83.
‡ Mean temperature British Isles, 1908 to 1911. *Geophysical Journal*, 1913, p. 91.

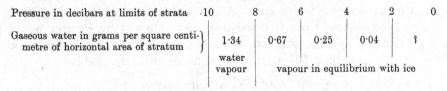

Of the condensed water some will be precipitated and some will float as cloud. Several observers* have measured the amount of liquid in clouds on mountains and their results indicate a maximum mass of liquid of about $4 \times 10^{-3}$ per mass of atmosphere.

For the high cloud alto-stratus the maximum value of the volume of the particles per horizontal area has been estimated† by photometric measurements to be of the order of three times the diameter of the particle, regarded as spherical. The mean of value for the diameter found by Pernter, from measurements of coronae, was about $1 \times 10^{-3}$ cm (Hann, *Meteorologie*, 3 Auf. p. 257). The mean thickness of such a cloud comes to about 500 metres (Hann, *l.c.* p. 284), and its mean height above ground to say 4 km (Hann, *l.c.* p. 280). From these data it follows that the ratio of mass of water to mass of air is of the order of $7 \times 10^{-5}$. This is remarkably less than the ratio $4 \times 10^{-3}$ found at lower levels.

However, only a few alto-stratus clouds show coronae. C. K. M. Douglas, who has studied these clouds while flying through them, writes: "The snow particles of alto-stratus or false cirrus are usually of the order of 1 m.m. in length and of elongated form. Occasionally much smaller particles are met with, perhaps of the order of 0·1 m.m. These are usually thinly scattered and cause halos readily. They are occasionally met with quite low down. A layer of dense grey alto-stratus usually consists of larger particles. Occasionally alto-stratus consists of water drops, especially in summer, or with warm damp upper currents in the autumn or early winter. The water drops may of course be super-cooled, but the water drop clouds more often have the appearance of alto-cumulus. The water drops are usually too small to be felt on the face." According to Douglas' observations we might take the volume of the particle to be that of a sphere of about 0·03 cm in diameter, and it would then follow, from the photometric observations, that the mass of water or of ice per mass of air is of the order of $2 \times 10^{-3}$, which is about half of the maximum value observed on mountains. Thus the limiting amount of condensed water which can float in any one conventional stratum will be about 0·8 gram per square centimetre of horizontal area.

What has been said above applies to the formation of extended sheets of cloud. Detached clouds are discussed in Ch. 5 below.

In order to save labour I have supposed there to be a sharp distinction between rain which falls and clouds which float. Actually there is a gradual transition. All sizes of particles occur. Even the smallest ones observed by Pernter would fall, according to Stokes' formula, at a rate of 70 metres per day. Generally speaking a small sphere of ice or water falls relatively to the air at a rate of about‡

$$\mathfrak{V}_H = -0{\cdot}3 \times 10^6 \text{ (diameter in cms)}^2 \text{ cm sec}^{-1}. \quad \ldots\ldots\ldots\ldots\ldots\ldots(1)$$

* Hann, *Meteorologie*, 3 Auf. p. 306.

† Recent measurements similar to L. F. Richardson's *Roy. Soc. Proc.* A, 96 (1919), p. 22.

‡ Wrong sign of power of ten in *Roy. Soc. Proc.* A, 96, p. 15.

In discussing the conveyance of water in Ch. 4/3 above, we supposed that the water had the same velocity as the air. Let us now correct that assumption. The total mass of solid, liquid, and gaseous water, per mass of atmosphere has been called $\mu$. Let the mass of the solid and liquid parts jointly per mass of atmosphere be $\nu$.

$$\left.\begin{array}{l}\text{Then } (\mu-\nu) \text{ has the velocity } v_H \text{ of the air} \\ \text{And } \nu \text{ has the velocity } v_H+\mathfrak{V}_H\end{array}\right\} \text{relative to the earth.}$$

The horizontal velocities may be considered the same in both cases.

So that the indestructibility of water-substance leads to

$$0=\frac{\partial\mu}{\partial t}+v_E\frac{\partial\mu}{\partial e}+v_N\frac{\partial\mu}{\partial n}+v_H\frac{\partial(\mu-\nu)}{\partial h}+(v_H+\mathfrak{V}_H)\frac{\partial\nu}{\partial h}$$

$$=\left(\frac{\partial\mu}{\partial t}+v_E\frac{\partial}{\partial e}+v_N\frac{\partial}{\partial n}+v_H\frac{\partial}{\partial h}\right)\mu+\mathfrak{V}_H\frac{\partial\nu}{\partial h}. \quad \text{.......................(2)}$$

If there are particles of different sizes and velocity the last term would have to be summed for all sizes so that

$$-\frac{D\mu}{Dt}=\Sigma\,\mathfrak{V}_H\frac{\partial\nu}{\partial h}, \quad \text{..................................(3)}$$

where $D$, as usual, denotes a change following the motion *of the air*.

Similarly in the equation of continuity of mass we have taken $m_H$ as equal to $\rho v_H$. But if $v_H$ is the velocity of the air, and if $\rho$ is the total density, then the mass per volume of all the gases jointly is $\rho(1-\nu)$, so that, strictly

$$m_H=\rho(1-\nu)\,v_H+\rho\nu\,(v_H+\mathfrak{V}_H)$$

$$=\rho v_H+\rho\nu\mathfrak{V}_H. \quad \text{.........................................(4)}$$

Or, if the sizes are various,

$$m_H=\rho v_H+\rho\,\Sigma\,\nu\mathfrak{V}_H, \quad \text{...............................(5)}$$

which should be substituted throughout.

But we cannot use this equation in a scheme of numerical prediction until we can predict the number of particles of each size. The size of particles depends on the amount of water to be condensed and on the number of nuclei available*. The nuclei are partly dust and partly ions. So that before (5) could be used, it would be necessary to bring into the scheme, as an eighth dependent variable, the amount of dust. For the present that will not be attempted.

Rain in falling through layers of varying temperatures must effect a **transference of heat**. If the raindrops were always at the same temperature as the air

* Sir J. J. Thomson, *Conduction of Electricity through gases*, Chs. VI and VII.

which momentarily surrounds them, then the downward flux of heat would be equal to

$$\theta \times (\text{specific heat of water}) \times (\text{rate of precipitation as mass per time per area}) = f \text{ say.} \quad \ldots(6)$$

The rate of accumulation of heat at any level would then be $\partial f/\partial h$ per unit volume. Mathematically the problem would then be very similar to the transference of heat in the soil by the percolation of water, a problem which is discussed in Ch. 4/9. But it is doubtful whether the assumption that the rain is at the same temperature as the air, is a sufficiently good approximation. Descending streaks of rain sometimes appear to evaporate. A column for the gain of heat by precipitation has been provided on one of the computing forms.

## CH. 4/7. RADIATION

### CH. 4/7/0. GENERAL

For solar radiation I have relied on the observations of Abbot, Fowle and Aldrich and on their reduction by L. V. King; for atmospheric radiation on the observations of Anders Ångström, Fowle and W. H. Dines with reference also to the theory and collection of data given by E. Gold.

As the temperature of the sun is a large multiple of that of the earth, it is possible to make a division at a wave length of about 5 microns, such that nearly all the solar energy is in the shorter wave lengths and nearly all the atmospheric and terrestrial in the longer ones. That explains the meaning of short and long in the present connection. But our computing processes are scarcely affected by the fact that there is a slight overlap.

### CH. 4/7/1. LONG-WAVE RADIATION

The radiation from a full radiator at all temperatures and wave lengths is satisfactorily expressed by the excellent formula of Planck. The simple connection between emission and absorption has been made clear by Kirchhoff. The experimental researches of Paschen, of Rubens and Aschkinass, and more recently of Fowle have shown that the absorption of the atmospheric gases is concentrated in certain ranges of wave length. But owing to the opacity of rock salt prisms for wave lengths greater than 15 to 20 microns, the absorption in the important region beyond this limit remains unexplored, save for isolated measurements by means of "reststrahlen." Some of the leading facts for our present purpose are set out in the diagram and Tables. The upper curve shows the spectrum of the radiation emitted by a full radiator at 270° A. The ordinate is expressed in gram calories per day emitted in the totality of directions to one side of a square centimetre and reckoned per micron of wave length. This curve was computed from the constants given in Winkelmann's *Physics*, 3rd edition, III. p. 402. The lower curves give some rough idea of the percentage absorptions from a vertical ray in traversing a conventional stratum, 200 millibars

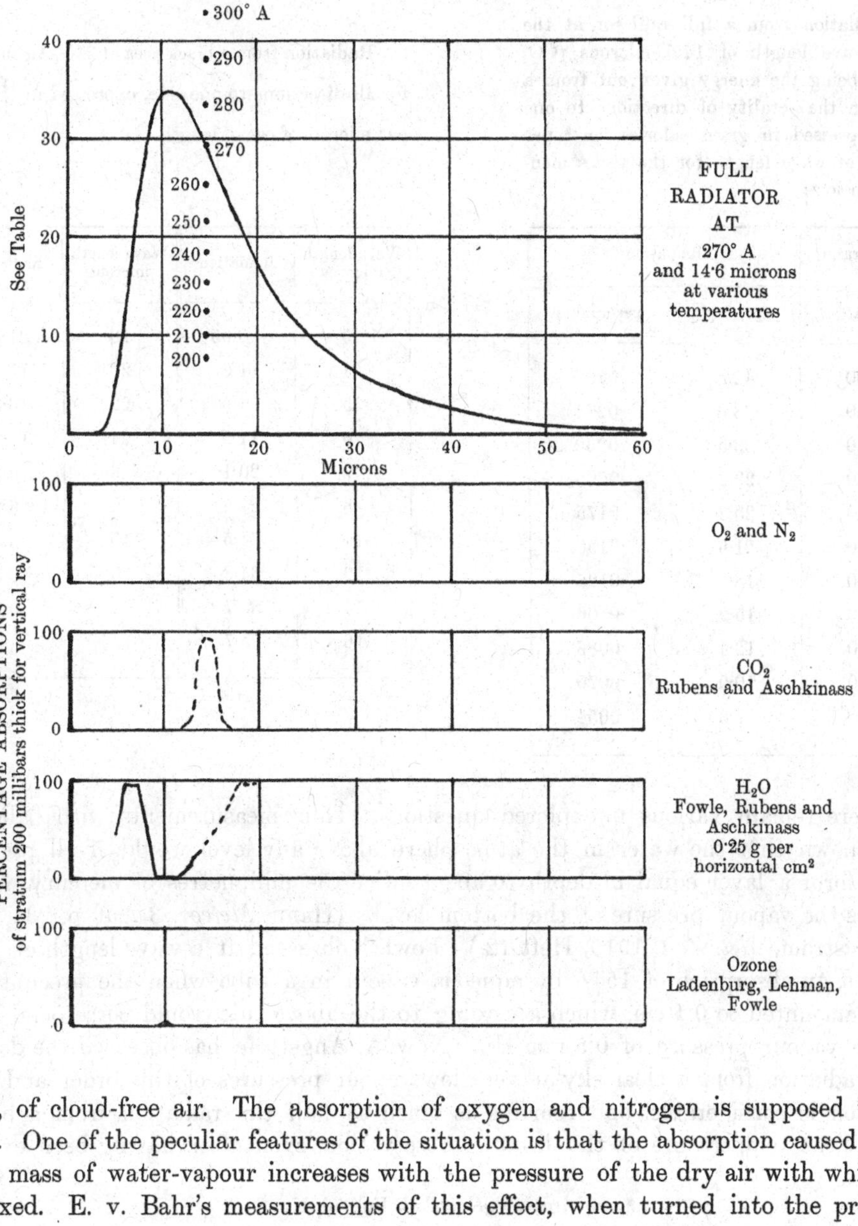


thick, of cloud-free air. The absorption of oxygen and nitrogen is supposed to be small. One of the peculiar features of the situation is that the absorption caused by a given mass of water-vapour increases with the pressure of the dry air with which it is mixed. E. v. Bahr's measurements of this effect, when turned into the present notation, appear to mean that the absorptivity-per-density, due to water vapour alone, is proportional to $\mu\rho^{\frac{1}{2}}$.

Between wave lengths of 9 and 11·5 microns there is a transparent region where none of the atmospheric gases absorb, except ozone, and that only slightly.

Radiation from a full radiator at the fixed wave length of 14·6 microns ($CO_2$ band), being the energy given out from a plane in the totality of directions to one side expressed in gram calories $cm^{-2}$ per micron of wave length for the times mentioned below:

| Temperature | Radiation | |
|---|---|---|
| °A | per day | per minute |
| 300 | 42·8 | ·0297 |
| 290 | 38·0 | ·0264 |
| 280 | 33·5 | ·0232 |
| 270 | 29·2 | ·0203 |
| 260 | 25·2 | ·0175 |
| 250 | 21·6 | ·0150 |
| 240 | 18·2 | ·0126 |
| 230 | 15·2 | ·0106 |
| 220 | 12·4 | ·0086 |
| 210 | 10·0 | ·0070 |
| 200 | 7·8 | ·0054 |

Radiation from a black area at 270° A., summed for all directions to one side, expressed in $\frac{\text{g. cal.}}{\text{day cm}^2}$ per micron of wave length.

| Wave length microns | Radiation | Wave length microns | Radiation |
|---|---|---|---|
| 3 | 0·05 | 18 | 21·3 |
| 4 | 1·0 | 20 | 17·1 |
| 5 | 4·9 | 22 | 13·8 |
| 6 | 11·9 | 24 | 11·4 |
| 7 | 20·1 | 26 | 9·2 |
| 8 | 27·2 | 30 | 6·2 |
| 10 | 34·5 | 40 | 2·6 |
| 12 | 34·3 | 50 | 0·8 |
| 14 | 30·7 | 60 | 0·4 |
| 16 | 25·7 | | |

There remain various unexplored questions. Thus measurements with balloons have shown that the water in the atmosphere above any level would, if all precipitated, form a layer equal in depth to about twice the millimetres of mercury which express the vapour pressure at the bottom level. (Hann, *Meteor.* 3 Auf. p. 231, also A. Ångström, *Met. Zeit.* 1916, Heft 12.) Fowle* observed at a wave length of 13·5 microns, an absorption of 15 °/₀ by aqueous vapour in a tube when the precipitable water amounted to 0·1 cm, which according to the above rule would correspond to a surface vapour pressure of 0·5 mm Hg. Now A. Ångström has observed the downward radiation from a clear sky at very low vapour pressures of this order and the radiation is considerable, and remains as much as half the radiation from a black body at the temperature of the level of observation, even after having been extra-

* *Smithsonian Misc. Coll.* Vol. 68, No. 8.

polated, by means of other observations, to absolute dryness. On referring to the diagram, it looks as though ozone and carbon dioxide were not the only absorbers in clear dry air. The suggestion is that dry clear air absorbs strongly between 20 and 40 microns. But the laboratory observations do not harmonize together as well as they might, and there is need for more experimental work, especially relating to wave lengths exceeding 15 microns.

In the meantime meteorologists must carry on business on premises which are, so to speak, in the hands of the builders.

If the distribution of absorption in the spectrum were well known, we could then group together all ranges of wave length which had similar absorptions, so as to form three or more groups. The energy of a full radiator in each group would be obtainable from Planck's formula. The energy in each group would be propagated independently of that in the others.

However, in the meantime, for comparison with observation, let us work out the distribution of radiation in height and in direction, making two simplifying assumptions:

(i) The absorptivity of dry air independent of the wave length.

(ii) No scattering in cloud-free air.

The experience thus gained in treating height and direction will be useful when, at a later stage, wave lengths can be grouped.

The first task is to work out the absorptivity of dry clear air (for the existing estimates vary greatly), and simultaneously to develop a method for computing the flux of radiation at any time and place. Both objects are attained in the following discussion of some observations by A. Ångström. We shall treat first the observational material, next the general theory of the connection between the absorptivity and the radiation, and lastly, the approximate computation of the absorptivity.

**Observations.** At the summit of Mount Whitney, California, on the dry clear night of 11 Aug. 1913 Ångström* found that the radiation falling from above, in all directions, on to a fully exposed horizontal surface was at the rate of 0·218 cal $cm^{-2}$ $min^{-1}$. The temperature at the place of observation was 270·5° A and the vapour pressure there only 0·6 mm of mercury. An extrapolation to absolute dryness, by means of fig. 5 in the same publication*, reduces the radiation only to 0·210 cal $cm^{-2}$ $min^{-1}$. The pressure at the top of the mountain at the time of observation is not stated, so I have taken a normal pressure of 586 millibars. The temperatures of the air radiating down on to the mountain are known approximately, from balloon ascents made at Avalon, California. During the ascent on 28 July 1913, the balloon recorded a temperature at the level of the summit, equal to the temperature at the summit on the occasion of the radiation observation quoted above, and this ascent has therefore been selected to give the temperatures. Ångström expresses his observed radiations as functions of the temperature and humidity at the point of observation,

* *Smithsonian Miscellaneous Collections,* Vol. 68, No. 8 (1917).

ignoring the fact that the levels of his stations varied by 4 kilometres. I have carefully avoided the mixing of data belonging to different levels.

To come now to the **general theory**: the **absorptivity per density,** $\Lambda$, may be defined by the statement that a fraction $\Lambda\rho dl$ of the incident radiation is absorbed in passing through a length $dl$ of air. ..............................................................(1)

For dry clear air, for the long waves radiated from the air itself, it is assumed as a temporary expedient that $\Lambda$ does not depend on the wave length. ...............(2)

It may be recalled* that a beam of diffuse radiation requires for its limitation two areas, say $A_1$ and $A_2$, not in the same plane. If $A_1$ and $A_2$ are plane and equal to $dA_1$, $dA_2$ and are small in comparison with $r^2$, the square of their distance apart, then the amount of radiation which will pass through both of them when they are surrounded by a black enclosure at $\theta°$ is

$$\supset \frac{\theta^4}{\pi} \frac{dA_1 \,.\, dA_2 \,.\, \cos\zeta_1 \,.\, \cos\zeta_2}{r^2}, \quad \text{..........................(3)}$$

where $\zeta_1$, $\zeta_2$ are the angles between the normal to $dA_1$, $dA_2$ respectively and the line joining $dA_1$ to $dA_2$; and where $\supset = 7{\cdot}68 \times 10^{-11}$ cal cm$^{-2}$ min$^{-1}$ (Kurlbaum's value).

The symbol $\supset$ is chosen because of its resemblance to the enclosure with a small aperture which is used in determining the constant.

Discussions about diffuse radiation would be made much clearer if we had a technical term for this important geometrical figure formed by two closed curves not in one plane. I propose to call it a "parcel"—which we may think of as tied with a loop of string at each end.

The quantity

$$\frac{dA_1 \,.\, dA_2 \,.\, \cos\zeta_1 \,.\, \cos\zeta_2}{r^2}$$

is of the dimensions of an area. For brevity it will be referred to as **"the area of the elementary parcel."** The area of a finite parcel has to be found by integrating its elements. Then $\supset\theta^4\pi^{-1}$ is the full radiation per parcel of unit area. Note that $\supset\theta^4/\pi$ is also in a special case the full radiation per cm$^2$ per unit solid angle, and that in finding the full radiation in all directions from one side of a black surface, although the solid angle is $2\pi$, we must on account of obliquity multiply only by $\pi$.

According to Kirchhoff's law, a portion of air, if placed so as to be traversed by a length $dl$ of the straight line joining $dA_1$ to $dA_2$, will radiate through the parcel defined by $dA_1$, $dA_2$ in either direction, a fraction $\Lambda\rho dl$ of the radiation through the same parcel in a black enclosure at the same temperature as the air. ....................(4)

Now $dl = dh \,.\, \sec\zeta$, where $\zeta$, without a subscript, is the zenith distance of the ray. Also $\rho dh = -dp/g$.

So
$$\Lambda\rho dl = -\frac{\Lambda \sec\zeta \,.\, dp}{g}. \quad \text{..........................(5)}$$

Note that $\sec\zeta \,.\, dp$ is negative, whether the radiation goes upwards or downwards.

* See, for example, equation (17) in M. Planck's *Vorlesungen über die Theorie der Wärmestrahlung*, published by J. A. Barth, Leipsig.

Now if $E$ is the activity in a parcel of radiation in the direction $\zeta$, and limited by the two areas $dA_1$, $dA_2$ and falling on a horizontal layer of air $dp$ thick, then the radiation emerging from the layer is

$$E\left(1+\frac{\Lambda \sec \zeta}{g}\zeta \,.\, dp\right)-\frac{\Lambda \sec \zeta \,.\, dp}{g}\,.\,\frac{Ↄ\theta^4}{\pi}\frac{dA_1 \,.\, dA_2 \,.\, \cos \zeta_1 \,.\, \cos \zeta_2}{r^2}\,.$$

So that the differential equation for the radiation passing in either direction through the parcel is

$$\frac{dE}{dp}=\frac{\Lambda \sec \zeta}{g}\left\{E-\frac{Ↄ\theta^4}{\pi}\frac{dA_1 \,.\, dA_2 \,.\, \cos \zeta_1 \,.\, \cos \zeta_2}{r^2}\right\}. \quad \text{......(6)}$$

Now to apply these general theorems to Ångström's observations, let $dA_1$ be one square centimetre of the horizontal receiving surface of his instrument. Then $\zeta_1$ is the zenith distance of the narrow beam received from the direction $dA_2$. Let $dA_2$ be an infinitesimal zone $d\zeta$ of the sphere-at-infinity concentric with the instrument. Then $dA_2 \cos \zeta_2/r^2$ is the solid angle of the zone as seen from the instrument, and is equal to $2\pi \sin \zeta \,.\, d\zeta$. So (6) becomes

$$\frac{dE}{dp}=\frac{\Lambda \sec \zeta}{g}\{E-Ↄ\theta^4 2 \sin \zeta \,.\, \cos \zeta \,.\, d\zeta\}. \quad \text{......(7)}$$

Now let $I$ be the "brightness" of the sky for this invisible radiation, that is to say let $I$ be the radiation per parcel of unit area. Then

$$E=I \cos \zeta \,.\, 2\pi \sin \zeta \,.\, d\zeta. \quad \text{......(8)}$$

So that (7) becomes

$$\frac{dI}{dp}=\frac{\Lambda \sec \zeta}{g}\left\{I-\frac{Ↄ}{\pi}\theta^4\right\}. \quad \text{......(9)}$$

But $Ↄ\theta^4/\pi$ is the full radiation for a parcel of unit area.

Expressed in words (9) means that: *in traversing an infinitely thin horizontal lamina, of mass $dR$ per unit area, the radiation gains $\Lambda \sec \zeta \,.\, dR$ of its deficit from the full radiation which would pass through a parcel of the same area in an enclosure at the temperature of the air.*

It would not be convenient to integrate (9) analytically even for clear air, because we should have to take account of the variations of $\Lambda$ and $\theta$ with height. And when it is remembered that clouds have a great influence on $I$, and that clouds can hardly be treated analytically, the necessity for a treatment by strata and finite differences becomes evident. However, within the narrow range of single stratum there is much advantage in integrating (9) analytically. The integrals for the successive strata can then be fitted together at the boundaries. Within a stratum it is assumed that $\Lambda$ may be considered independent of height, and that $\theta^4$ may be considered to increase linearly with $p$ at a known rate. Thus $d\theta^4/dp$ is about 65 (degrees)$^4$ cm$^2$ dyne$^{-1}$ in the troposphere and is zero in the stratosphere.

Under these restrictions, (9) may be solved by the general analytical method for linear equations. When this has been done, and when the arbitrary constant of integration has been eliminated between the values of $I$ and $\theta$ at the boundaries of the stratum, which, for illustration, has been taken as the one lying between $h_8$ and $h_6$, the result is

$$I_8 = e^{\frac{\text{A} \sec\zeta (p_8 - p_6)}{g}} \left\{ I_6 - \frac{\text{Ɔ}}{\pi}\left(\theta_6^4 + \frac{g}{\text{A}\sec\zeta}\frac{d\theta^4}{dp}\right)\right\} + \frac{\text{Ɔ}}{\pi}\left(\theta_8^4 + \frac{g}{\text{A}\sec\zeta}\frac{d\theta^4}{dp}\right). \quad \ldots\ldots(10)$$

This equation is suitable for reducing observations made simultaneously in mountainous country at different levels. Ångström's excellent Californian observations do not refer, unfortunately, to an isolated zenith distance, and so cannot be used thus.

In (10), as in (5), $\sec\zeta$ is positive for ascending radiation, negative for descending. ......(11)

The variation of $\theta$ with height occurs in (10) in two ways: by the term in $\frac{d\theta^4}{dp}$ and by the separate appearance of $\theta_8$ and $\theta_6$. In the following calculation of A from Ångström's observations, the variation of $\theta$ with height has been treated by using two different sizes of coordinate differences. There was consequently little to be gained by retaining the temperature variation in (10). It was neglected so that (10) became simply

$$I_8 = (1-\eta)\, I_6 + \eta \frac{\text{Ɔ}\theta^4}{\pi}, \quad \ldots\ldots(12)$$

where

$$\eta = 1 - e^{-\text{A}\sec\zeta \,.\, R_{86}} \quad \ldots\ldots(13)$$

In (13) $\sec\zeta \,.\, R$ is taken as always positive, so that (12) refers to descending radiation. For ascending radiation $I_8$ and $I_6$ change places. The quantity $\eta$ may be called the "absorptance" of the stratum for rays in the direction $\zeta$.

The energy absorbed by the stratum per unit time is, for the descending radiation $I_6 - I_8$, for the ascending radiation $I_8 - I_6$. ......(14)

The two terms on the right of (12) may be called: $(1-\eta) I_6$ the "transmitted" radiation; and $\eta \frac{\text{Ɔ}\theta^4}{\pi}$ the "emitted" radiation. The emitted radiation $\eta \frac{\text{Ɔ}\theta^4}{\pi}$ is less than $\text{A}\sec\zeta \,.\, R \frac{\text{Ɔ}\theta}{\pi}$ for thick layers, on account of the absorption by the layer of its own radiation. For very thin layers $\eta = \text{A}\sec\zeta \,.\, R$.

We want the radiation from the whole sky falling on a horizontal unit area; it is

$$\int_{\zeta=0}^{\zeta=\pi/2} I\cos\zeta \,.\, 2\pi\sin\zeta \,.\, d\zeta = \Sigma E, \text{ say.} \quad \ldots\ldots(15)$$

For the uppermost stratum this integral would be obtainable analytically from (10), but to repeat such integrations for the successive lower strata would not be easy. Instead an approximation has been obtained by a summation.

For purposes of **computation** let the infinitesimal zone $d\zeta$ be replaced by a finite zone, first of $\zeta = 30°$; that is to say, let the hemisphere above the instrument be divided into three parts by the cones $\zeta = 30°, 60°$. The solid angle of each part is then calculated as $2\pi \int \sin\zeta \, . \, d\zeta$ between the limits and is multiplied by a mean value of $\cos\zeta$. For example, in the case of the conical shell lying between $\zeta = 30°$ and $\zeta = 60°$ the expression $2\pi \cos\zeta \, . \, \sin\zeta \, . \, d\zeta$ in (7) is approximately represented by

$$2\pi \cos 45° \, (\cos 30° - \cos 60°).$$

Similarly in calculating $\eta$ from (5), for this conical shell, $\zeta$ is put equal to its mean value 45°. In calculating $\eta$ a trial value of $\Lambda$, namely $\Lambda = 1{\cdot}601 \times 10^{-3}$ C.G.S. units is first assumed. The mean temperatures of the layers of air are found from the balloon observation and from them the full radiation $\mathfrak{I}\theta^4/\pi$ is deduced for each layer. Then the descending radiation in each cone or conical shell can be traced downwards by making successive applications of (12), from the top, where it is assumed to be zero, to the instrument. The computation is set out in the accompanying table. It gives 0·2267 cal cm$^{-2}$ min$^{-1}$ at the instrument, for the stated trial value of the absorptivity-per-density $\Lambda$.

*Computation of Radiation from the Atmosphere*

**Assuming $\Lambda = 1{\cdot}601 \times 10^{-3}$ C.G.S. units and temperatures from balloon at Avalon, California on 28 July 1913**

| Pressure millibars | Mean temperature of stratum. °A | Full Radiation per parcel of unit area. cal cm$^{-2}$ min$^{-1}$ | | $\zeta=0°$ to 30° | 30° to 60° | 60° to 90° | Sum for all zones |
|---|---|---|---|---|---|---|---|
| | | | mean $\zeta$ | 15° | 45° | 75° | |
| | | | $1-\eta$ | ·718 | ·636 | ·291 | |
| | | | $\eta$ | ·282 | ·364 | ·709 | |
| | | | approx. area of parcel per cm² at instr. | ·813 | 1·626 | ·813 | |
| | | | **VERTICAL CONTRIBUTION OF ZONES** cal cm$^{-2}$ min$^{-1}$ | $\zeta=0°$ to 30° | 30° to 60° | 60° to 90° | |
| | 214° | ·0510 | | | | | |
| | | | transmitted | 0 | 0 | 0 | |
| $\frac{1}{3} \times 586$ | | | emitted | ·0117 | ·0302 | ·0294 | |
| | | | total | ·0117 | ·0302 | ·0294 | ·0712 |
| | 231° | ·0701 | | | | | |
| | | | transmitted | ·0084 | ·0192 | ·0086 | |
| $\frac{2}{3} \times 586$ | | | emitted | ·0161 | ·0415 | ·0405 | |
| | | | total | ·0245 | ·0607 | ·0491 | ·1343 |
| | 260° | ·1118 | | | | | |
| | | | transmitted | ·0176 | ·0386 | ·0143 | |
| 586 | | | emitted | ·0256 | ·0661 | ·0645 | |
| | | | total | ·0432 | ·1047 | ·0788 | ·2267 |

Next to estimate the errors due to finite differences, the calculation was repeated in exactly the same manner, but taking 6 layers of equal mass instead of 3, and dividing the hemisphere into 6 parts by the cones $\zeta = 15°, 30°, 45°, 60°, 75°$.

This gave 0·2210 cal $cm^{-2}$ $min^{-1}$ at the instrument. Now the errors are usually proportional to the square of the coordinate differences, when these are small enough. If this rule holds in the present case, the radiation at the instrument would be 0·2192 cal $cm^{-2}$ $min^{-1}$, if calculated by infinitesimal steps for the same absorptivity-per-density $\text{A} = 1{\cdot}601 \times 10^{-3}$ C.G.S units. The observed value of the radiation was 0·210 cal $cm^{-2}$ $min^{-1}$, which is slightly less than the calculated. To correct A to the observed value of the radiation, the variation of the radiation was recalculated with a slightly different value of A, namely $\text{A} = 1{\cdot}501 \times 10^{-3}$ C.G.S. units.

In this calculation the larger of the two afore-mentioned sizes of differences was used, to save trouble, and it was assumed that the small correction would be practically the same, whether determined by comparing two coarse difference tables or two fine difference tables. For $\text{A} = 1{\cdot}501 \times 10^{-3}$ C.G.S. units coarse differences gave a radiation at the instrument of 0·2195 cal $cm^{-2}$ $min^{-1}$. The final result is that, with infinitesimal differences, the observed radiation would be given when

$$\text{A} = 1{\cdot}48 \times 10^{-3} \text{ C.G.S. units,}$$

for dry clear upper air, for its own radiation.

We are now in a position to find the radiation emitted by a horizontal lamina of clear air $dp$ thick.

This may be shown to be

$$\mathfrak{I}\theta^4 \int_{\zeta=0}^{\zeta=\pi/2} 2\cos\zeta \sin\zeta \left(1 - e^{-\frac{\text{A} \sec\zeta dp}{g}}\right) d\zeta = 2\mathfrak{I}\theta^4 \frac{\text{A}}{g} dp = 2\mathfrak{I}\theta^4 \text{A} \rho \, dh, \quad \ldots\ldots(16)$$

in which both $dp$ and $dh$ are to be taken as positive. Putting in the value of A and taking $g = 980$ cm/sec² it follows that: *a horizontal lamina of air* $\delta p$ *millibars thick emits from each side radiant energy at a rate* 0·00302$\delta p$ *times that from an ideally black plane to one side, for the same area.*

For comparison with the experimental constants quoted by Gold on page 53 of *Proc. Roy. Soc.* Vol. 82, we require $\frac{\partial}{\partial \theta}$ of the radiation emitted by a layer of air 1 cm thick. Gold does not state the temperature or the pressure of the layer, so I have assumed them to be normal, that is 273° A and 1013 mb. It then follows from (16) that the increase in the radiation from the layer per degree is $0{\cdot}143 \times 10^{-5}$ cal $cm^{-2}$ per *hour*, when A is given the value $1{\cdot}48 \times 10^{-3}$ C.G.S. units. Thus the figures quoted by Gold correspond to much greater absorptivities than the one deduced above. In another paper* Gold has criticized the deduction of the absorptivity from observations of the cooling of surface air. In what follows I have preferred to rely on Ångström's observations as reduced above.

It would be interesting to repeat the computation of the absorptivity on the assumption that dry air is perfectly transparent except between wave lengths of 13 and 16 microns, which is the position of the band due to carbon dioxide.

* *Q. J. R. Met. Soc.* Oct. 1913.

**Moist Clear Air.** No satisfactory treatment will be possible until we are able to group wave lengths into ranges according to the absorptivity. In the meantime we may note that the effect of the moisture is usually less than that of the dry air. This has been shown by A. Ångström. He observed the total radiation falling from a clear sky upon a horizontal plane. Then having corrected all the observations empirically to a standard air temperature of 293° A at the instrument, he plotted them against vapour pressure at the instrument. A range of vapour pressure of 0 to 10 mm of Hg caused the radiation to vary from 0·27 to 0·41 g cal $cm^{-2}$ $min^{-1}$. That was when observations at all levels were combined in a single diagram. In view of the result which has been obtained above, that a stratum of dry air 200 millibars thick allows as much as 0·7 of vertical radiation to pass through, it does not appear to be permissible to combine the results from levels differing by several kilometres. And for the same reason the humidity and temperature at the instrument are only of interest in so far as they are a guide to the whole distribution above.

If we treat the levels separately it is found that the effect of humidity is as follows:

| Station | Height (metres) | Increase of radiation per 1 mm of vapour pressure at the instrument. Radiation being in g cal $cm^{-2}$ $min^{-1}$. Temperature constant at 293° C. |
|---|---|---|
| Mt Whitney ... | 4420 } | ·016 |
| Mt San Gorgonio ... | 3500 } | |
| Mt San Antonio ... | 3000 } | ·017 |
| Lone Pine Canyon | 2500 } | |
| Bassour ... ... | 1160 | ·007 |
| Lone Pine ... ... | 1140 | ·009 |
| Indio ... ... | 0 | ·005 |

Thus the down-coming radiation is most increased by water vapour in the higher levels. That is as we should expect, knowing that the radiation of water vapour is concentrated in bands. For as soon as the full temperature radiation is attained in the band, increases will only follow the increase of temperature.

It would be desirable to know the absorptivity $\mathrm{A}$ as a function of $\mu$, the mass of water-vapour per mass of atmosphere. The relationship could be found, if of the form

$$\mathrm{A} = (\text{value for dry air})\{1 + (\text{constant})\,\mu\},$$

by pursuing step-by-step computations like those made above for dry air. It would be necessary to have both $\mu$ and the temperature given as functions of height. A very rough calculation of this kind, based on the mean of the above data, indicates that the constant is of the order of 40, so that

$$\mathrm{A} = 1{\cdot}48 \times 10^{-3}\,(1 + 40\mu)\ \mathrm{cm^2\,grm^{-1}}. \qquad \text{.....................(17)}$$

The definition of $\Lambda$ is contained in equation (1) above. The absorptances of the conventional strata might be taken to be increased in the same ratio $(1+40\mu)$, where $\mu$ is put equal to $W/R$, the total mass of water divided by the total mass of atmosphere in the stratum. But no treatment will be satisfactory until the wave lengths can be grouped.

**Approximate simplified process.** It is proposed to diminish the amount of computing by treating the hemisphere as a whole, instead of in separate zones. For the resultant flux of radiation is almost invariably vertical. Near the edge of a clouded area, the resultant could have a horizontal component, but this exceptional case has been neglected. The whole radiation falling on one side of a horizontal unit area has already been denoted by $\Sigma E$ which is defined by equation (15). We require to find, for $\Sigma E$, an equation corresponding to (12), for the brightness $I$. Now if we multiply (12) throughout by $\cos\zeta \,.\, 2\pi \sin\zeta \,.\, d\zeta$ and integrate over the hemisphere, the left-hand side transforms to $\Sigma E_8$. The term in $-\eta I_6$ does not however transform into one in $\Sigma E_6$. But if $\eta$ were independent of $\zeta$, and equal to $\bar{\eta}$, then equation (12) would transform into

$$\Sigma E_8 = (1-\bar{\eta})\,\Sigma E_6 + \bar{\eta}\,\mathfrak{O}\theta^4, \quad \text{..........................(18)}$$

which is accordingly taken as an empirical equation used to define a mean absorptance $\bar{\eta}$ for descending radiation distributed in all directions in the actual manner. In the last term of (18) $\mathfrak{O}\theta^4$ is the radiation emitted by unit area of a perfectly black plane, to one side. Now $\Sigma E$ is given in the last column of the table on p. 53, which is based on Ångström's observations. From this column, by use of (18), the values of $\bar{\eta}$ have been deduced for the three strata shown in the table. They may be stated by giving the mean zenith distance $\bar{\zeta}$ which must be inserted in equation (13) in order to make $\eta$ equal to $\bar{\eta}$. The advantage of giving the angle is that it is probably not affected by the small error in $\Lambda$ in the table. The result is, for strata approximately the same as our three upper conventional ones of 2 decibars thickness,

| | $p_0$ | | $p_2$ | | $p_4$ | | $p_6$ |
|---|---|---|---|---|---|---|---|
| $\bar{\zeta}$ = | | 57°·1 (1 radian nearly) | | 54°·5 | | 54°·7 | |

It is seen that the descending radiation is more horizontal above, more vertical below. The mean angle $\bar{\zeta}$ may also be different for ascending radiation. More observational values are needed.

This simplified process will be illustrated in connection with the data for 20 May 1910, which will be discussed in Ch. 9. There $\bar{\zeta}$ is taken as 55° throughout, $\Lambda$ is taken as $1{\cdot}48 \times 10^{-3}(1+40\mu)$ cm$^2$ grm$^{-1}$, $\Lambda$ is calculated by putting $\bar{\zeta}$ in place of $\zeta$ in (13), and $\Sigma E$ is traced from stratum to stratum by means of (18).

Mr W. H. Dines* has used this simplified process extensively to compute the radiation at all heights according to various trial values of the absorptance.

* *Q. J. R. Met. Soc.* April 1920.

*Clouds.* A. Ångström in the introductory summary to his paper makes a statement which implies that low and dense clouds behave almost as black bodies to the long-wave nocturnal radiation. The recent observations of W. H. Dines confirm this

*Surface of Sea.* The radiation from water may be calculated by Kirchhoff's law from its reflectivity, which in turn may be deduced, by Fresnel's formulae, from its refractive index. The reflectivity varies from unity for grazing incidence to 0·02 for normal incidence. But for rays distributed as they would be in all directions A. Ångström has computed, in this way, that a flat water surface gives out 94 °/。 of the radiation from a black surface.

## Some Papers on Long-Wave Radiation


E. GOLD. "The Isothermal Layer...and Atmospheric Radiation." *Roy. Soc. Proc.* A, Vol. 82, pp. 43 to 70.

E. GOLD. *Q. J. R. Met. Soc.* Oct. 1913.

A. ÅNGSTRÖM. "A Study of the Radiation of the Atmosphere." *Smithsonian Miscellaneous Collections,* Vol. 65, No. 3 (1915).

A. ÅNGSTRÖM. "Über die Gegenstrahlung der Atmosphäre." *Meteor. Zeit.* Heft 12 (1916) and Heft 1 (1917).

F. E. FOWLE. "Water-vapour Transparency to Low-temperature Radiation." *Smithsonian Miscellaneous Collections,* Vol. 68, No. 8 (1917).

A. ÅNGSTRÖM. "On the Radiation and Temperature of Snow and the convection of the air at its surface." *Arkiv för Mat. Astr. och Fysik.* Stockholm, 1918.

A. ÅNGSTRÖM. "Determination of the constants of Pyrgeometers." *Arkiv för Matematik.* Stockholm, 1918.

C. G. ABBOT. "Terrestrial Temperature and Atmospheric Absorption." *Proc. National Academy of Sciences, U.S.A.*, Vol. 4, pp. 104–106 (1918).

A. BOUTARIC. "Contribution à l'étude du pouvoir absorbant de l'atmosphère terrestre." *Thèses à la faculté des sciences de Paris.* Gauthier-Villars et Cie, 1918.

W. H. DINES. "Atmospheric and Terrestrial Radiation." *Q. J. R. Met. Soc.* April 1920.


CH. 4/7/2. SOLAR RADIATION

## General

What we require to find is the energy absorbed by each stratum of the atmosphere and by the soil, vegetation or sea. Scattering must be taken into account.

The fraction of the direct solar beam transmitted by a horizontal layer, $\delta p$ units of pressure in thickness, is

$$\operatorname{Exp}\{-\delta p \,.\, \sec\zeta\,(f(l) + Bl^{-4})\}, \quad \text{.........................(1)}$$

where $B$ is a quantity depending only on water and dust and $l$ is the wave length.

The part in $f(l)$ represents the loss by absorption. The loss varies irregularly with wave length, from a large value in certain bands due to oxygen or water vapour, to a much smaller value in the continuous spectrum between the bands. The other part in $Bl^{-4}$ represents the loss due to scattering. Scattering does not warm the air. The scattered radiation spreads in all directions, but not equally so. Its intensity in a direction making an angle $A$ with the incident beam has been found* by Rayleigh, by Kelvin and by Schuster, to be proportional to $1+(\cos A)^2$. ....................(2)

The important point about this formula, for present purposes, is its symmetry: the radiations scattered in any two opposite directions are equal. In its course through the atmosphere, the scattered radiation is partly absorbed and partly repeatedly scattered. Some of it eventually goes off to space and the rest comes to earth as skyshine.

It will be convenient to consider firstly the radiation reaching the upper atmosphere from the sun and secondly its distribution in space.

## Sunshine above the Atmosphere

Abbot, Fowle and Aldrich† by extrapolating, for each wave length separately, from observations made at large and small zenith distances of the sun, found the distribution of energy in the solar spectrum outside the atmosphere; and then by integrating with respect to wave length they found the total energy. Their mean value for this total is 1·93 calories $\text{cm}^{-2}\ \text{min}^{-1}$ which is equivalent to

$$1{\cdot}16\times10^{11}\ \text{ergs cm}^{-2}\ \text{per mean solar day.} \quad\ldots\ldots\ldots\ldots\ldots\ldots(3)$$

By simultaneous observations at different stations Abbot‡ has found evidence of small variations in the total radiation, variations having a period of the order of a few days, weeks or months. Until some way of predicting these variations can be found, we must just neglect them. The small effect of the seasonal change of the sun's distance from the earth is easily taken into account by reference to astronomical tables. We require the radiation falling on a unit horizontal area. At any instant this is proportional to $\cos\zeta$ where $\zeta$ is the sun's zenith distance. The mean value of $\cos\zeta$ during a time-step can be found by expressing $\zeta$ in terms of the sun's hour angle and declination and then integrating. The result is that the mean value of $\cos\zeta$, between two times $t_2'$ and $t_1'$, reckoned in mean solar days from local apparent noon, is

$$\sin(\text{north decl.})\times\sin\phi+\frac{1}{2\pi}\cos(\text{north decl.})\times\cos\phi\,\frac{\sin 2\pi t_2'-\sin 2\pi t_1'}{t_2'-t_1'}\,.\quad\ldots\ldots(4)$$

From this formula the following numerical values have been calculated. They refer to the example of Ch. 9 which is based on observations made at about $8^h$ local apparent time ($=7^h$ G.M.T.) on 20 May 1910.

* *Vide* L. V. King, *Phil. Trans.* A, 1913, p. 376.

† *Smithsonian Astrophysical Annals*, Vol. III.

‡ *Nature*, July 29, 1920.

| Kilometres north of equator | Mean value of cosine of sun's zenith distance 1910 May 20 d. | |
|---|---|---|
| | $5^h$ to $11^h$ local apparent time | $11^h$ to $17^h$ local apparent time |
| 6200 | 0·522 | 0·693 |
| 6000 | 0·523 | 0·706 |
| 5800 | 0·528 | 0·718 |
| 5600 | 0·531 | 0·729 |
| 5400 | 0·535 | 0·740 |
| 5200 | 0·537 | 0·750 |
| 5000 | 0·539 | 0·759 |
| 4800 | 0·541 | 0·766 |

## Transmission of Solar Radiation through the Atmosphere

To be exact we should have to follow the Smithsonian observers in treating each wave length separately, and L. V. King in considering scattering as occurring repeatedly. All this might be practicable if the atmosphere could be treated always as a single stratum, but such a treatment would break down as soon as clouds intervened. Cloudiness or transparency are of course the most important atmospheric properties in connection with solar radiation, and everything else must be subordinated to the distinction between them. Thus it becomes necessary to keep account, separately, of the radiation passing through the several conventional strata; and then, to economize arithmetic, approximations must be sought elsewhere. The following approximations, while greatly reducing the work, would appear to permit an accuracy of a few per cent in the final result.

(i) Dust is left to be estimated statistically. That is to say it is not included, as water is, among the dependent variables, whose history we attempt to trace throughout.

(ii) We neglect the radiation passing from any horizontal coordinate chequer, of 200 kilometres of latitude times 2° 49′ of longitude, to the neighbouring chequers. This point may be illustrated by the fact that the curvature of the earth prevents a cloud, above the centre of one of these chequers, from casting a shadow on the centre of the next chequer, unless the cloud is more than 3 kilometres above sea-level. Even for the highest clouds it would only be possible if the sun were so low as to have a negligible heating effect.

(iii) As the zenith distance $\zeta$ of the sun changes largely during our time-step of six hours, a mean value of sec $\zeta$ must be inserted in all transmission formulae such as (1). A mean value of cos $\zeta$, for the same interval, has in any case to be calculated from (4), and its reciprocal may be used for mean sec $\zeta$. A much better mean value of the factor $e^{-B\sec\zeta}$ would be $\int \cos\zeta\, e^{-B\sec\zeta}\,dt \div \int \cos\zeta\, dt$, in which the "weight" $\cos\zeta$ is introduced because the radiation on the top of the atmosphere is proportional to it. But the essential difficulty is that $B$ varies with time, owing to its dependence on the amount of cumulus cloud, and that six hours is a long interval when dealing with the latter. Three-hour steps would be much better.

(iv) Instead of treating the scattered radiation as distributed in all directions each with its appropriate absorption, the absorption is taken to be that corresponding to a single mean direction. For radiation scattered once only, this direction is taken as 45°, pending more exact estimates.

(v) Repetition of scattering may be allowed for, in a rough way, by slightly increasing the above angle. For consider that, if scattered radiation were not absorbed by the atmosphere, we should hardly be concerned with the repeated scattering; for although the repetitions may alter the relative brightness of various parts of the sky, they can not affect the fact that half the scattered radiation would make its exit to space, and the other half would reach the ground, in accordance with the symmetry of expression (2). The fraction reflected by soil or foliage can hardly depend at all upon the distribution of brightness over the sky, but the fraction reflected by a perfectly calm sea would do so in a known manner. As scattered radiation is slightly absorbed by the atmosphere, and as repeated scattering increases the average aggregate path of the radiation in the air, we may allow for the repetitions by increasing the above-mentioned mean zenith distance from 45° to say at a guess 55°.

(vi) Lastly, instead of treating each wave length separately, the whole range of wave lengths has been divided into interlacing portions, according to the strength of the absorption. It seems best to group together wave length ranges which have similar absorptions, rather than those which have similar scatterings, because the variations of absorption are much greater than the variations of scattering, and because we are only concerned with scattering in so far as it prevents or produces absorption. For simplicity the range of wave lengths has been divided into only two groups—that of the atmospheric absorption bands, and the remainder. According to Abbot, Fowle and Aldrich the energy lost in the bands is not more than 15 °/ₒ of the total amount. The total is therefore divided into 15 °/ₒ and 85 °/ₒ, and these amounts are entered at the head of the columns marked "bands" and "remainder" in the computing form. See Ch. 9.

The above are the simplifying approximations.

We must next consider the observed values of the constants for scattering and absorption, for these two groups of wave lengths. The absorption in the bands has been determined by Abbot, Fowle and Aldrich* by measuring areas on graphs in which the observed intensity is recorded as a function of the wave length. The

* *Smithsonian Astrophysical Annals*, Vol. III.

continuous spectrum was taken as the base line from which the areas were measured. The absorption in the continuous spectrum has been found by L. V. King* by treating the logarithm, of the fraction of the radiation which is transmitted, as the sum of two terms: one term varies inversely as the fourth power of the wave length and is therefore due to scattering; the other term is presumably due to absorption, and is found to be independent of wave length. In the bands these two absorptions must apparently have been superposed, but whether the coefficients should be added or whether two separate exponential terms with different coefficients will be required, is hardly clear from the experimental evidence. So, for simplicity, the former course has been taken. Again the total observed scattering must be divided between the two groups of wave lengths. Now the bands are mostly in the red or ultra-red, where the scattering, for the same incident energy, is only about one-half what it is for the remainder of the spectrum. So instead of taking 15 °/$_\circ$ of the total scattering as occurring in the bands, only $7\frac{1}{2}$ °/$_\circ$ has been taken.

From the above-mentioned sources, some of the most interesting data have been collected into the following table.

In the table on p. 62 all the activities are expressed as fractions of the radiation incident on the top of the atmosphere. Before being used to deduce constants for the air between the levels of Mt. Whitney and Mt. Wilson they require to be expressed as fractions of the energy incident upon this layer for the particular group of wave lengths concerned.

Now for any wave length, expression (1) may be put in the form:

$$\frac{\text{radiation directly transmitted}}{\text{radiation incident}} = e^{-\int \rho (\Lambda + \not{s}) \sec \zeta dh}, \quad \text{..............(5)}$$

where $\Lambda$ is the absorptivity-per-density and where $\not{s}$, which measures the scattering, may be called the scatterivity-per-density. The integral is to be taken across the layer. If $\Lambda$ and $\not{s}$ are independent of height, (5) becomes:

$$\frac{\text{radiation directly transmitted}}{\text{radiation incident}} = e^{-(\Lambda \times \not{s}) \sec \zeta \,.\, R}, \quad \text{..............(6)}$$

where $R$ is the mass per unit area of the layer.

To determine the constants $\Lambda$ and $\not{s}$ from the observations quoted above, for layers which are not thin enough to permit the approximation $e^{-(A+B)} = 1 - A - B$, we find first the sum $(\Lambda + \not{s})$ from (6). Next this sum, $(\Lambda + \not{s})$, is divided into two parts in the ratio of the total absorption to the total scattering in the layer. This ratio equals $\Lambda/\not{s}$ because if the activity in the direct beam at any level is $E$, then the amounts absorbed and scattered in the layer $i$ to $i'$ are respectively $\Lambda \int_i^{i'} \rho E \sec \zeta \, dh$ and $\not{s} \int_i^{i'} \rho E \sec \zeta dh$.

* *Phil. Trans.* A, 1913, p. 425 etc., also *Nature*, July 1914.

*Absorption and Scattering of Solar Radiation by Clear Air*

| | Date | Height, metres above M.S.L. $h$ | Pressure, millibars $p$ | Precipitable water (Fowle) grm cm$^{-2}$ | After reduction to zenith sun. Fractions of energy incident on top of atmosphere. Expressed as fractions of total energy: transmitted in direct beam | absorbed in continuous spectrum | absorbed in bands | scattered = not transmitted minus absorbed | Expressed as Fractions of energy in band region (15 %): absorbed = absorption in bands + 15 % of continuous | scattered = $7\frac{1}{2}$ % of total scattered | Expressed as Fractions of energy in remainder (85 %): absorbed = 85 % of absorbed in continuous | scattered = $92\frac{1}{2}$ % of total scattered |
|---|---|---|---|---|---|---|---|---|---|---|---|---|
| Column number | I | II | III | IV | V | VI | VII | VIII | IX | X | XI | XII |
| Mt. Whitney | mean 1909, 1910 | 4420 | 595 | 0·082 | 0·884 | 0·014 ± 0·003 | 0·017 (17 Aug. 1910) | 0·085 | 0·127 | 0·042 | 0·014 | 0·092 |
| Mt. Wilson | mean 1910 | 1780 | circa 831 | 0·765 | 0·808 | 0·023 ± 0·001 | 0·047 (Oct. 1906) | 0·122 | 0·337 | 0·061 | 0·023 | 0·133 |
| difference between Mt. Whitney and Mt. Wilson | | 2640 | 236 | 0·683 | 0·076 | 0·009 | 0·030 | 0·037 | 0·210 | 0·010 | 0·009 | 0·041 |
| Energy descending at level of Mt. Whitney = unity minus absorption and scattering above Mt. Whitney | | | | | | | | | 0·831 | | 0·894 | |

The results are set out in the table below, and refer to clear air

| | Bands | | Remainder | | Water substance per mass $=\mu$ |
|---|---|---|---|---|---|
| | *Absorptivity* per density $=\Lambda$ | *Scatterivity* per density $=\mathcal{F}$ | *Absorptivity* per density $=\Lambda$ | *Scatterivity* per density $=\mathcal{F}$ | *See note below* |
| Air above Mt. Whitney | $2{\cdot}28\times10^{-4}$ | $0{\cdot}76\times10^{-4}$ | $0{\cdot}24\times10^{-4}$ | $1{\cdot}60\times10^{-4}$ | $0{\cdot}135\times10^{-3}$ |
| Air between levels of Mt. Whitney and Mt. Wilson | $12{\cdot}8\times10^{-4}$ | $0{\cdot}6\times10^{-4}$ | $0{\cdot}43\times10^{-4}$ | $1{\cdot}98\times10^{-4}$ | $2{\cdot}83\times10^{-3}$ |

*Note.* The measurements of the absorption in the bands were made on isolated occasions when the water content may have differed from the mean value given in the last column.

*Clouds.* Abbot states that thick clouds reflect about 65 °/$_\circ$ of the solar radiation falling upon them. Photometric measurements* indicate that the fraction of red light transmitted by uniform stratus is about 25 °/$_\circ$ for an ordinary cloud and 8 °/$_\circ$ for a very dark one. Thus the effect of clouds appears to be mainly a scattering. How much they absorb is apparently not yet known exactly.

**Routine Process.** The foregoing principles and data have been applied to the observations in Ch. 9. The procedure was as follows in any group of ranges of wave lengths. Take the "bands" group for illustration.

First $\Lambda$ and $\mathcal{F}$ were obtained for each stratum, by using the values given in the table above, and assuming that $\Lambda$ and $\mathcal{F}$ were linear functions of the water-per-mass, $\mu$. It was concluded that the sky was cloudless, because the water-content was everywhere low, and the period considered, $5^h$ to $11^h$ by local time, too early in the day for much cumulus cloud. Next the fraction of energy not transmitted

$$1-\text{Exp}\,\{R\,(\Lambda+\mathcal{F})\sec\zeta\}$$

was formed for each stratum, and thence the fraction absorbed

$$\frac{\Lambda}{\Lambda+\mathcal{F}}[1-\text{Exp}\,\{R\,(\Lambda+\mathcal{F})\sec\zeta\}],$$

and the fraction scattered

$$\frac{\mathcal{F}}{\Lambda+\mathcal{F}}[1-\text{Exp}\,\{R\,(\Lambda+\mathcal{F})\sec\zeta\}].$$

These are all fractions of the energy incident on the particular stratum. They were entered on the computing form which is printed in Ch. 9. Next 15 °/$_\circ$ of the incident radiation was entered opposite $h_0$ in the column headed "Direct beam, flux." The direct beam was then traced down stratum by stratum, deducting the amounts absorbed or scattered, until the vegetation was reached. The downward flux of diffuse

* *Roy. Soc. Proc.* A, Vol. 96, pp. 23 and 31 and similar measurements now in progress.

radiation was next traced, starting from zero at the top $h_0$. In each stratum one half of the energy scattered from the direct beam was added to the downward flux of diffuse radiation. From this sum was deducted the amount of diffuse radiation absorbed by the stratum, the amount absorbed being taken as a known fraction, corresponding to $\zeta = 55°$, of the diffuse radiation entering the stratum from above. Thus diffuse radiation was brought down to the vegetation film. The portions of the descending radiation, direct or diffuse, which were considered as reflected by the vegetation, were added and transferred to the foot of the column of ascending diffuse radiation, at the level $h_L$. The ascending radiation was then traced upwards, subtracting in each stratum the fraction absorbed, and adding half the energy scattered from the direct beam in that stratum, until $h_0$ was reached, where the radiation left the earth for interplanetary space. In the last column of the computing form, the energy absorbed by the several strata, and by the earth, was added up. Finally the total energy received from the sun was checked against the total of all absorbed plus the loss to space.

## Publications on Solar Radiation

ABBOT, FOWLE and ALDRICH. *Smithsonian Astrophysical Annals*, Vol. III. and *Smithsonian Miscell. Collections*, Vol. 65, No. 4.

F. LINDHOLM. "Extinction des Radiations solaires dans l'atmosphère terrestre." *Nova Acta Upsaliensis*, Ser. IV. Vol. 3, No. 6 (1913).

L. V. KING. *Phil. Trans.* A, Vol. 212 and *Nature*, July 30, 1914.

R. S. WHIPPLE. "Instruments for the Measurement of Solar Radiation." *Trans. Optical Society*, London, 1915.

W. H. DINES. "Heat Balance of the Atmosphere." *Q. J. R. Met. Soc.* Apr. 1917.

SIR NAPIER SHAW. "Memorandum on Atmospheric Visibility," Feb. 1918. Hydrographic Dept., Admiralty, London.

F. E. FOWLE. "The Atmospheric Scattering of Light." *Smithsonian Misc. Collections*, Vol. 69, No. 3 (May 1918).

C. G. ABBOT. "Terrestrial Temperature and Atmospheric Absorption." *Proc. Nat. Acad. Sci., U.S.A.*, April 1918.

A. BOUTARIC. "Contribution à l'étude du pouvoir absorbant de l'atmosphère terrestre." *Thèses à la faculté des sciences de Paris.* Gauthier-Villars et Cie, 1918.

F. LINDHOLM. "Sur l'insolation dans la Suède Septentrionale." *Kungl. Svenska Vet. Hand.* Bd. 60, No. 2 (Wm Wesley & Son, 28 Essex St., Strand).

A. ÅNGSTRÖM. "Über die Schätzung der Bewölkung." *Met. Zeit.* Heft 9/10, 1919.

C. G. ABBOT. *Nature*, July 29, 1920.

## CH. 4/8. THE EFFECTS OF EDDY MOTION

*Note:* Numerical references attached to persons' names in Ch. 4/8 refer to the bibliography on p. 92.

### CH. 4/8/0. GENERAL

The kinetic theory of gases has revealed to us that the properties of viscosity diffusion, and conduction of heat, which are attributable to a gas enclosed in a small vessel, are in reality due to molecular motions which we cannot follow in detail. In the same way in the atmosphere, many varieties of motion which we cannot or do not wish to record in detail, can be ignored; provided that their general statistical effect is taken into account by adding to the equations, describing the general motion, appropriate additional terms. This was very clearly brought out by Osborne Reynolds (3) in connection with viscosity. He took the mean value of the dynamical equations and found that they were of exactly the same form in the mean velocities as in the actual velocities; except that the quadratic terms introduced, on taking the mean, the body force derived from a system of "eddy stresses." This was true however large was the interval of time or the volume over which the mean was taken, with the limitation that the eddies within this time and space must be sufficiently numerous*.

Our theory and constants must be appropriate to the size of the element of the fluid which we treat as a "differential" in that we ignore the details of any motions taking place wholly within it †. The upper limit to the size of an eddy is, like the length of a piece of string, a matter of human convenience. When an airman says that the wind is "bumpy" he is thinking in terms of a differential element probably comparable in size with one wing of a flying machine. For the present purpose any motion which disappears on taking the average over our coordinate intervals ($\delta t = 6$ hours, $\delta\lambda = 3°$, $\delta\phi = 200$ km, $\delta h = \frac{1}{5}$ mass upwards) has of necessity to be ignored. If it occurs in large numbers—as cumulus eddies do for example—then its general effect can be satisfactorily represented by additional terms; although unfortunately this does not help us for example to say whether it will hail or not on Mr X's field.

It has been customary to separate circulatory motions in the atmosphere into two main groups according as they derive their energy from the local heating of the lower layers of air ("convection") or from the kinetic energy of the wind ("dynamical instability"). It is easy to classify a thunder-cloud as belonging to the convectional type, or a gust in a gale as being an example of the other extreme. But recent researches have shown the existence of numerous intermediate forms, which derive their energy from both sources jointly. Thus Åkerblom (8) found that the eddy viscosity at the Eiffel Tower was greater in summer, when the supply of heat for convection was greater. Hesselberg (13) obtained a similar result for the wind 500 metres above Lindenberg. Again Taylor (21) found a very large seasonal variation, in the same sense, for the diffusion of heat as deduced from the Eiffel Tower temperatures.

* See also Ch. 4/9/5 below.

† Bryan, *Thermodynamics* (Teubner), Art. 46; Jeans, *Dynamical Theory of Gases*, Art. 11.

Exceptionally low diffusivities have been measured at night by L. F. Richardson (32) in the cold air near the earth. Airmen are very familiar with the increased bumpiness of the wind caused by sun shining on the ground below them. All these facts show that the production of eddies in the wind is greatly facilitated when the thermal equilibrium becomes less stable, although we may not suppose that actual thermal instability is reached in the majority of cases, because such an event is unusual among the collected observations made either by registering balloons or from aeroplanes.

A quantitative theory of the criterion of turbulence has been given by L. F. Richardson (32).

On the other hand we find that convectional motions are hindered by the formation of small eddies resembling those due to dynamical instability. Thus C. K. M. Douglas writing of observations from aeroplanes remarks: "The upward currents of large cumuli give rise to much turbulence within, below, and around the clouds, and the structure of the clouds is often very complex." One gets a similar impression when making a drawing of a rising cumulus from a fixed point; the details change before the sketch can be completed. We realize thus that: big whirls have little whirls that feed on their velocity, and little whirls have lesser whirls and so on to viscosity—in the molecular sense.

Thus, because it is not possible to separate eddies into clearly defined classes according to the source of their energy; and as there is no object, for present purposes, in making a distinction based on size between cumulus eddies and eddies a few metres in diameter (since both are small compared with our coordinate chequer), therefore a single coefficient is used to represent the effect produced by eddies of all sizes and descriptions. We have then to study the variations of this coefficient. But first we must consider the differential equation. In doing so the aim has been to lay down theoretically only so much as can be determined with strictness, leaving all uncertainties to be decided by observation.

In hydrodynamics or aerodynamics it is customary to speak of the motions of "definite portions" of the fluid, portions which may be marked by a dot of milk in water or of smoke in air. The capital $D$ in $D/Dt$ is commonly used to denote a time differentiation following such a definite element. It is customary to ignore the fact that molecules are constantly passing in and out of the element called "definite." When we have to deal with eddies, the interchanges are more conspicuous, for boundaries marked by smoke would rapidly fade and disperse. Yet some way must be found of specifying an element which follows the *mean* motion. The fundamental idea seems to be the following. When there are no eddies we are accustomed to compute the flow of entropy or water across a plane from the flow of mass across the plane. As the effect of eddies is to be treated as additional, it should not include any flow due to the mean motion of mass across a plane. Accordingly we should adopt some such definition as the following:

Draw a sphere in the fluid. Let the radius be as large as is necessary to include a considerable number of eddies, but no larger. Let the sphere move so that the whole momentum of the fluid inside it is equal to the mass of the same fluid multiplied by the vector velocity of the centre of the sphere. The centre may then be said to be "a point moving with the mean-motion." ..........(1)

As bars denote mean values the capital letter with a bar over it in $\bar{D}/\bar{D}t$ may suitably denote a differential following the mean motion.

Here we may usefully bear in mind the analogy with the conduction of heat in a solid. The total water in a portion of air, or the mean potential temperature of it, is not altered by gently mixing it at constant pressure, provided precipitation does not occur. If, then, $Z$ stands for either of these quantities or for any other quantity, which

Has its total, for a secluded portion of air, unchanged by the internal rearrangement of that portion, ..........(2)

Or by delay, ..........(3)

then the rate of increase of the total $Z$ in the portion must be due to $Z$ flowing in over the sides.

As the variations of moisture, entropy and velocity in the atmosphere are usually much more rapid vertically than horizontally, the ensuing treatment is confined to the vertical diffusion.

Consider a large horizontal surface moving so that there is no net flow of mass across it. Let us define the "upward eddy flux" of the quantity $Z$ across any such surface as the ratio of the amount of $Z$ rising across the surface in unit time to the area of the surface, which is supposed to intersect many eddies. ..........(4)

Then rate of increase of $Z$ per area in the layer $dh$ bounded by two surfaces which rise and fall with the mean motion must be

$$-dh\frac{\partial}{\partial h}(\text{upward flux}). \quad \text{..........(5)}$$

If now $\chi$ be defined as the amount of $Z$ per unit mass of atmosphere, then

The amount of $Z$ per area in $dh$ is $\chi\rho\, dh$, ..........(6)

so that

$$\frac{\bar{D}(\chi\rho\, dh)}{\bar{D}t} = -dh\frac{\partial}{\partial h}(\text{upward flux}). \quad \text{..........(7)}$$

Now we can define a coefficient $c$ such that

$$\text{upward flux of } Z = -c\frac{\partial \chi}{\partial h}, \quad \text{..........(8)}$$

where $c$ might be called the "eddy-conductivity"* when the quantity transferred is moisture or energy; or it might be called the "eddy-viscosity" when we have to do with exchanges of momentum. The coefficient $c$ corresponds closely but not exactly to the "Austausch" which W. Schmidt of Vienna denotes by the symbol $A$. When $c$ is defined by (8) it will be for observation to show whether $c$ is a constant, as would

* Since I adopted this name (Bibliog. 32, p. 2) I see that G. I. Taylor (12, p. 4) had previously used it to denote another quantity. To avoid muddles it is best to refer always to the differential equation for diffusion.

be nice, or whether $c$ depends on various factors, and in particular on $\partial\chi/\partial h = 0$. But, if $c$ is to be useful, it must not be infinite when $\partial\chi/\partial h = 0$. Therefore we had better confine ourselves to conditions in which

The upward flux vanishes when $\partial\chi/\partial h = 0$. .................(9)

Now this implies that:

$\chi$ must be unchanged by the simple transportation of air to a different level. ...(10)

Otherwise expressed $D\chi/Dt = 0$ where $D$ denotes a differentiation following the eddying motion. ..........................................................(10$a$)

Now $\chi$ is the total $Z$ per unit mass of a definite portion of air. So, as the mass is unchanged by transport, (10) implies that:

The total $Z$ in a definite portion of air must be unchanged by its removal to a different level. ..........................................................(11)

Fortunately, (11) is satisfied by the same two quantities which previously satisfied (2), so that (10) is satisfied when $\chi$ is either mass-of-water-per-mass-of-atmosphere or else potential-temperature. ........................(11$a$)

If, in place of (8), we had made the flux proportional to $c'\partial(\rho\chi)/\partial h$ where $\rho\chi$ is the total amount of $Z$ per *volume* instead of per *mass* of atmosphere then, to avoid an infinite value of $c$, the flux would have to vanish when $\partial(\rho\chi)/\partial h$ vanished. That would not be possible for the two given meanings of $\chi$, because rising air changes its volume. So it appears that we had better keep to the definition (8).

Inserting (8) in (7) we have $\dfrac{\bar{D}(\rho\chi\, dh)}{\bar{D}t} = dh\dfrac{\partial}{\partial h}\left(c\dfrac{\partial\chi}{\partial h}\right)$. ........................(11$b$)

Now $$\frac{\bar{D}(\rho\, dh)}{\bar{D}t} = 0, \qquad \text{.................................(12)}$$

for the layer is bounded by surfaces moving with the mean motion.

Then (11$b$) reduces, without approximation, to

$$\frac{\bar{D}\chi}{\bar{D}t} = \frac{1}{\rho}\frac{\partial}{\partial h}\left(c\frac{\partial\chi}{\partial h}\right). \qquad \text{..............................(13)}$$

It is sometimes more convenient to use pressure as a measure of height by means of the transformation

$$dp = -g\rho\, dh.$$

At the same time, in order to get a single constant, let us put

$$g^2\rho c = \xi, \qquad \text{.......................................(14)}$$

then (13) becomes

$$\frac{\bar{D}\chi}{\bar{D}t} = \frac{\partial}{\partial p}\left(\xi\frac{\partial\chi}{\partial p}\right), \qquad \text{..............................(15)}$$

where $\chi$ may have either of the meanings (11$a$).

The dimensions of $\xi$ are (mass)$^2$ (length)$^{-2}$ (time)$^{-5}$.

Pressure is peculiarly convenient if we can neglect any lateral convergence of wind, for then an isobaric surface moves up and down with the mean motion, so that when $t$ and $p$ are the independent variables, $\bar{D}/\bar{D}t$ is the same as $\partial/\partial t$ and (15) becomes

$$\frac{\partial \chi}{\partial t} = \frac{\partial}{\partial p}\left(\xi \frac{\partial \chi}{\partial p}\right). \qquad (16)$$

The flux is given by (5), (8), and (14), that is to say, if $\chi$ is the amount of $Z$ per unit mass of atmosphere, then the amount of $Z$ rising across a large horizontal surface per area per time, when no mass traverses the surface on the average, is

$$\frac{\xi}{g}\frac{\partial \chi}{\partial p} \text{ or equivalently, } -c\frac{\partial \chi}{\partial h}. \qquad (17)$$

The quantity $\chi$ may have either of the meanings (11$a$).

The conductivity $c$ or the turbulivity $\xi$ are left to be determined by observation, but we know beforehand that they are finite when $\partial\chi/\partial h = 0$. No attempt is here made to show that they are the same for various meanings of $\chi$.

It should be pointed out* that entropy per mass $\sigma$ cannot stand for $\chi$ in the above equations. Because, when the variegated structure produced by eddies is smoothed out by molecular diffusion at constant pressure, then the entropy increases and so does not satisfy (2) or (3); although the mean potential temperature of a secluded portion of air remains unchanged. The equation for diffusion of entropy per mass $\sigma$ has to be derived from that for potential temperature $\tau$.

For dry air

$$d\sigma = \gamma_p d \log \tau, \qquad (18)$$

so that (13) leads to

$$\frac{\bar{D}}{\bar{D}t} e^{\sigma/\gamma_p} = \frac{1}{\rho}\frac{\partial}{\partial h}\left(c\frac{\partial e^{\sigma/\gamma_p}}{\partial h}\right). \qquad (19)$$

Or alternatively

$$\frac{\bar{D}\sigma}{\bar{D}t} = \frac{1}{\tau}\frac{\partial}{\partial p}\left(\xi\tau\frac{\partial\sigma}{\partial p}\right). \qquad (20)$$

### The case of velocity

Can velocity stand for $\chi$ and momentum for $Z$ in the preceding equations? The answer will be yes, provided that (2), (3), and (9) are satisfied. Suppose that $Z$ and $\chi$ are both in the same fixed horizontal direction. Now (3) is satisfied because the external forces on a lamina, due to local pressure differences, are normal to its broad surfaces, and therefore cannot alter its momentum parallel to any line in its own plane. So that it can gain horizontal momentum only by air moving into it. Condition (2) is satisfied because forces between the parts cannot alter the momentum of the whole. In examining (9) we must distinguish between the mean velocity $\bar{v}$, which here plays the part of $\chi$, and $v'$, the deviation from the mean. The question is whether, if $\partial\bar{v}/\partial h$ vanish at any level, an upward flux of momentum can cross that level?

* In view of a mistake of mine in *Roy. Soc. Proc.* A, Vol. 96 (1919), p. 9 from which paper the above theory is taken, with improvements.

In other words, can there be an eddy-shearing-stress where there is no rate-of-mean-shear? The converse, of course, occurs where there are no eddies. Or to put the question over again in another way: imagine two adjacent horizontal layers each say 100 metres thick and having the same mean velocity; would it be possible for the faster moving portions to sort themselves out and to flock together into one layer, leaving the slower moving portions in the other? If molecules did that sort of thing, the occurrence would be one of the exceedingly rare exceptions to the second law of thermodynamics. We may reasonably expect that there is a corresponding law, on an enlarged scale so to speak, applicable to the statistical mechanics of eddy motion.

The only other possibility seems to be that portions of air should wander in from quite distant regions, giving a part of the vertical flux of momentum proportional to $\partial^3\bar{v}/\partial h^3$ and a term in (15) proportional to $\partial^4\bar{v}/\partial h^4$.

When $v_E$ plays the part of $\chi$ then $c$ plays the part of viscosity, as may be seen by imagining (13) as a dynamical equation from which pressure gradient and geostrophic wind have been omitted because they just balanced each other.

**To sum up:** we may expect equations (13) and (15) to be an exact expression of the effect of mixing when $\chi$ is either:

potential temperature (but not entropy per mass),

mass of water in all its phases jointly per mass of atmosphere,

horizontal velocity in a fixed azimuth,

or any other quantity satisfying (2), (3) and (9).

But there is no evidence from this theory as to whether the coefficients $c$ and $\xi$ are the same for these different meanings of $\chi$.

**The Eddy-Flux of heat** behaves in a manner remarkably different from the flux of heat in a solid. If $\chi$ in the above general theory be given its permissible special meaning of the potential temperature $\tau$, then $Z$ is (potential temperature)×(mass) and the amount of $Z$ rising per second across a unit horizontal area which moves with the mean motion is $-c\,.\,\partial\tau/\partial h$. The flux of heat may be said to be the flux of

(mass)×(ordinary temperature)×(specific heat at constant pressure),

and this will be $\gamma_p\,(p/p')^{0\cdot 29}$ times the flux of $Z$.

That is to say the flux of heat will be, in dry air,

$$-\gamma_p\left(\frac{p}{p'}\right)^{0\cdot 29} c\frac{\partial\tau}{\partial h}. \qquad \text{.............................(21)}$$

Let us consider the simplest case. Suppose that the temperature does not change at points moving with the mean motion, and that the pressure does not change either. Then $D\tau/Dt=0$ and from (13) it follows that, as far as eddies are concerned

$$\frac{\partial}{\partial h}\left(c\frac{\partial\tau}{\partial h}\right)=0, \qquad \text{so that } c\frac{\partial\tau}{\partial h} \text{ is independent of height.}$$

Therefore from (21) we see that in this steady state the flux of heat depends on the height. The explanation of this distribution—so different from that in a solid—is that heat is being transformed into the kinetic energy of eddies or *vice versa* as it goes. The matter has been treated by the author (33, p. 354). If $\tau$ increases upwards, then heat flows down; but then also the atmosphere is stable, and as it is stirred kinetic energy is drawn from the mean wind and becomes, partly immediately heat, and partly first eddying kinetic energy. The latter part ultimately, by molecular dissipation, is converted to heat. If, on the contrary, $\tau$ increases downwards, then the flow of heat is upwards; but also the equilibrium is unstable even in the absence of wind, and heat is being converted into eddying kinetic energy. Ultimately this is also dissipated as heat. But in the meantime, provided $D\tau/Dt = 0$, then whichever sign $\partial\tau/\partial h$ may have, the flux of heat, in conformity with equation (21), is greater below.

### CH. 4/8/1. COLLECTION OF OBSERVATIONS OF VISCOSITY AND CONDUCTIVITIES

Although these are put down as values of $c$ in equation (13) yet very many of the observers have gone on the unreliable hypothesis that $c$ was independent of height so that $\frac{\partial}{\rho\partial h}\left(c\frac{\partial\chi}{\partial h}\right)$ was taken by them as $\frac{c}{\rho}\frac{\partial^2\chi}{\partial h^2}$.

The most striking fact is the large value of the viscosity which eddies give to the atmosphere. Thus several observers have found viscosities of the order of 100 C.G.S. units at a height of 100 to 1000 metres above land. Glycerine in the same units has a viscosity of 46 at 0° C., Lyle's Golden Syrup 1400 at 12° C. However, when the forces resisted by viscosity are produced by the inertia of the fluid, we should compare fluids by their ratio of viscosity to density. For the atmosphere this ratio attains values of the order of $10^5$ C.G.S. units, a figure which lies between $10^3$ for golden syrup at 12° C. and $5\times10^6$ for shoemakers' wax at 8° C. (The laboratory data are taken from Kaye and Laby's tables.)

In the adjoining columns are set out various circumstances which are likely to have influenced the state of eddying. They will be examined later. I have endeavoured to pick them out to fit the same ranges of height, place and time as the observations of turbulence.

The coefficients $c_v$, $c_\mu$, $c_\tau$ mean the values of $c$ in equation (13) which apply when the diffusing quantity is respectively $v$, $\mu$ or $\tau$. Thus $c_v$ is the eddy-viscosity and is not to be mistaken for the specific heat at constant volume, for which, following G. H. Bryan, I use the symbol $\gamma_v$.

## *Observed values of the Eddy-viscosity $c_v$ for* **velocity**

The $x$ axis is directed with the wind at the mean level

| $c_v$ | $\left(\frac{\partial v_x}{\partial h}\right)^2$ | $\left(\frac{\partial v_y}{\partial h}\right)^2$ | $\frac{g}{\gamma p}\frac{\partial \sigma}{\partial h}$ | Height above ground | Wind near surface: Velocity | Wind near surface: At height | Character of surface | Remarks |
|---|---|---|---|---|---|---|---|---|
| $\frac{\text{grm}}{\text{sec cm}}$ | sec$^{-2}$ $10^{-4}\times$ | sec$^{-2}$ $10^{-4}\times$ | sec$^{-2}$ $10^{-4}\times$ | Metres | cm sec$^{-1}$ | cms | | |
| '. ÅKERBLOM (8). From variation of wind with height, allowing for possible variation of barometric gradient. | | | | | | | | |
| 83 | 6·8 | 0·4? | 2·4 | 21 to 305 | 239 | 2100 | Paris, winter | Eiffel Tower winds |
| 113 | 4·1 | 0·8? | 1·6 | | 205 | 2100 | Paris, summer | |
| G. I. TAYLOR (11) and (12). From variation of wind with height, not allowing for variation of barometric gradient. | | | | | | | | |
| 70 | | | | 0 to 1700 | 950 | 3000 | Salisbury Plain | G. M. Dobson's Balloons |
| 56 | | | | | 590 | | | |
| 32 | | | | | 330 | | | |
| 0·9 o 8·0 | mean 2·0 | mean 1·8 | mean 6·0 | 0 to 200 | | | Sea, Newfoundland | "Scotia" kites |
| 9 to 4·8 | 1·8 | 3·4 | 0·9 | 0 to 200 | 440 | 1000? | Sea | 1913, Aug. 2d 11h |
| HESSELBERG and SVERDRUP (14). From variation of wind and barometric gradient with height. | | | | | | | | |
| 0·9 | | | | 0 to 9 | | | Stevenson and Hellmann obsns. | |
| 40 | 2·6 | 1·7 | 2·6 for 0 to 500 at 8 a.m. | 9 to 209 | 564 | Anem. 900 | Lindenberg | |
| 50 | 0·5 | 0·8 | | 109 to 309 | | | | |
| 50 | 0·1 | 0·2 | | 209 to 409 | | | | |
| 60 | 0·0 | 0·0 | | 309 to 509 | | | | |
| not allowing for variation of barometric gradient | | | | | | | | |
| 50 | | | | 9 to 3009 | | | | |
| W. SCHMIDT (18). Assuming that very near the ground the vertical flux of momentum is independent of height and therefore $c_v\partial v/\partial h$ constant. | | | | | | | | |
| 30 | assumed as basis at height of 200] | | | | | | | |
| 2·6 | | | | 9 | | | | Hellmann, Anemometers |
| 1·3 | | | | 2 | | | | |
| 1·1 | | | | 1 | | | | |
| H. U. SVERDRUP (19). Variation of velocity with height in the North Atlantic Trade Wind, allowing for variation of pressure gradient. | | | | | | | | |
| 40 | 0·08 | 0·00 | 1·7 (zero near sea) | 0 to 1200 | 640 | 1000? | Sea | Trade wind |
| 60 | 0·01 | 0·32 | 2·6 | 1200 to 1800 | | | | Intermediate layer |
| 90 | 0·07 | 0·02 | 2·6 | 1800 to 2000 | | | | Anti-trade wind |

## *Observed values of the Eddy-viscosity $c_v$ for* **velocity** *(continued)*

The $x$ axis is directed with the wind at the mean level

| $c_v$ | $\left(\frac{\partial v_x}{\partial h}\right)^2$ | $\left(\frac{\partial v_y}{\partial h}\right)^2$ | $\frac{g}{\gamma_p}\frac{\partial \sigma}{\partial h}$ | Height above ground | Wind near surface | | Character of surface | Remarks |
|---|---|---|---|---|---|---|---|---|
| | | | | | Velocity | At height | | |
| $\frac{\text{grm}}{\text{sec cm}}$ | sec$^{-2}$ $10^{-4}\times$ | sec$^{-2}$ $10^{-4}\times$ | sec$^{-2}$ $10^{-4}\times$ | Metres | cm sec$^{-1}$ | cms | | |
| L. F. RICHARDSON (32). By integrating the loss of momentum with respect to height from a level chosen so that $\partial v/\partial h$ and consequently the stress there both vanish. Neglecting variation of pressure gradient. | | | | | | | | |
| 15 | parallel to wind | | | 400 | | | Lindenberg General mean | |
| 110 | horizontally across wind | | | | | | | |
| L. F. RICHARDSON (32). From the speed of ascent of cumuli, the height through which they rise and the fraction of the sky covered. | | | | | | | | |
| 1030 | Zero | Zero | Nearly zero | 0 to 2000 | Zero | | Land | Midday |

## *Observed values of the Eddy-conductivity c for* **dust, smoke or other floating solids**

| $c$ | $\left(\frac{\partial v_x}{\partial h}\right)^2$ | $\left(\frac{\partial v_y}{\partial h}\right)^2$ | $\frac{g}{\gamma_p}\frac{\partial \sigma}{\partial h}$ | Height above ground | Wind near surface | | Character of surface | Remarks |
|---|---|---|---|---|---|---|---|---|
| | | | | | Velocity | At height | | |
| $\frac{\text{grm}}{\text{sec cm}}$ | sec$^{-2}$ $10^{-4}\times$ | sec$^{-2}$ $10^{-4}\times$ | sec$^{-2}$ $10^{-4}\times$ | Metres | cm sec$^{-1}$ | cms | | |
| L. F. RICHARDSON (32). From irregularities in the flight of a free balloon observed by Capt. Cave with two theodolites. | | | | | | | | |
| 6 | 0·2 | 0·1 | $10^{h}$ in February | 1000 | 700 | | Hills 150 m. high | |
| L. F. RICHARDSON (32). Dispersal of smoke. | | | | | | | | |
| 120 | | | Aug. $20^{h}$ | 250 | | | Argonne Forest, wooded hills | |
| 6 | | | July $7\frac{1}{2}^{h}$ overcast | 9 | | | Moor | |
| 0·16 | 13 | ? | 13? | 3·4 | 120 | 340 | Flat fields | Bibliog. (32) fig. 3 |
| 25 | 225 | ? | − 2 ? | 5 | 370<br>310 | 500<br>100 | Steam plough obstructions subtend 1/15 radian | 1920 Sept. [illegible] $10^{h}$ $45^{m}$, 2/10 cum. |
| 7 | 2 | ? | − 1 ? | 2 | 200 | 200 | | 1920 Sept. [illegible] $13^{h}$. Firework 6/10 ci-str. 2/10 cum. |

## *Observed values of the Eddy-conductivity $c_\mu$ for* **water**

| $\mu$ | $\left(\frac{\partial v_x}{\partial h}\right)^2$ | $\left(\frac{\partial v_y}{\partial h}\right)^2$ | $\frac{g}{\gamma_p}\frac{\partial \sigma}{\partial h}$ | Height above ground | Wind near surface | | Character of surface | Remarks |
|---|---|---|---|---|---|---|---|---|
| | | | | | Velocity | At height | | |
| rm / cm | sec$^{-2}$ $10^{-4}\times$ | sec$^{-2}$ $10^{-4}\times$ | sec$^{-2}$ $10^{-4}\times$ | Metres | cm sec$^{-1}$ | cms | | |
| . SCHMIDT (25). Evaporation in the court of the Bureau Central Meteorologique, and difference of water per mass between Parc St Maur and the summit of the Eiffel Tower. | | | | | | | | |
| 1 | | | | 2 to 302 | | | | Dec. to Feb. 9[h] to 12 |
| 1 | | | | | | | | " " 21[h] to 24 |
| 4 | | | | | | | | July 3[h] to 6 |
| 0 | | | | | | | | " 21[h] to 24 |
| L. F. RICHARDSON (28). By comparing precipitation with the up-grade of water per mass of atmosphere. Means for whole globe, including as eddies any up and down motion however extensive. | | | | | | | | |
| 06 to 38 | 0·15* | 0·0 | 1·5 | 8500 | | | | |
| 20 | | | | 500 | | | | |
| r less | | | | 0·5 | | | | |

## *Observed values for* **carbon dioxide**

W. SCHMIDT (18, p. 36). The observed distribution of carbon dioxide is a balance between eddy diffusion, gravity, and molecular diffusion

| c | | | | | | | | |
|---|---|---|---|---|---|---|---|---|
| 001? | 0·15* | 0·0 | 2·8 | 10000 | | | | Wigand's Observation, *Met. Zeit.*, 33, 433 (1916). |

*Note.* Schmidt himself rejects this value as incredibly small, but I am inclined to believe it.

* This figure is a rough estimate of the mean *square* of $\partial v/\partial h$ derived from G. M. B. Dobson's diagram in *. R. Met. Soc.*, Jan. 1920, p. 55.

*Observed values of the Eddy-conductivity $c_\tau$ for* **potential temperature**

| $c_\tau$ | $\left(\frac{\partial v_x}{\partial h}\right)^2$ | $\left(\frac{\partial v_y}{\partial h}\right)^2$ | $\frac{g}{\gamma_p}\frac{\partial \sigma}{\partial h}$ | Height above ground | Wind near surface | | Character of surface | Remarks |
|---|---|---|---|---|---|---|---|---|
| | | | | | Velocity | At height | | |
| $\frac{\text{grm}}{\text{sec cm}}$ | $\text{sec}^{-2}$ $10^{-4}\times$ | $\text{sec}^{-2}$ $10^{-4}\times$ | $\text{sec}^{-2}$ $10^{-4}\times$ | Metres | cm $\text{sec}^{-1}$ | cms | | |
| G. I. TAYLOR (11). From the life-history of the air currents, and the sudden bends in the curve showing the distribution of temperature with height. | | | | | | | | |
| | | | | | | | | 1913 |
| 1·6 | 11·3 | 0·3 | 8·7 | 0 to 170 | 440 | 1000? | Sea Newfoundland banks | July 29 |
| 3·0 | 1·8 | 3·4 | 0·9 | 0 to 200 | 440 | 1000? | | Aug. $2^d$ $11^h$ |
| 3·1 | zero | zero? | 8·7 | 0 to 370 | 220 | 1000? | | Aug. $4^d$ $19^h$ |
| 4·1 | | | 1·3 | 370 to 770 | 220 | | | |
| G. I. TAYLOR (21). From Eiffel Tower temperatures, neglecting radiation. | | | | | | | | |
| 180 | | | | 18 to 123 | | | Paris | Year |
| 20 | | | | 197 to 302 | 247 | 2100 | Paris | Feb. |
| 364 | | | | | 208 | | | July |
| 54 | 8·7 | 0·8 | 1·8 | 18 to 302 | 238 | | Paris | Jan. |
| 221 | 3·6 | 0·2 | 0·7 | | 206 | | | June |
| W. SCHMIDT (25). From Eiffel Tower temperatures, allowing for radiation. | | | | | | | | |
| 31·2 | | | 1·77 | 197 to 302 | 228 | 2100 | Paris | Spring |
| 46·5 | | | 1·61 | | 205 | | | Summer |
| 10·6 | | | 1·98 | | 290 | | | Autumn |
| 9·9 | | | 2·47 | | 238 | | | Winter |
| 9·2 | | | | 2 to 123 | | | | Year |
| W. SCHMIDT (25). From Allahabad temperatures in the same way. | | | | | | | | |
| 0·6 | | | | 1·2 to 14 | | | | cold |
| 0·7 | | | | | | | | hot |
| 2·8 | | | | | | | | rainy |
| 4·6 | | | | 31·7 to 50·6 | | | | cold |
| 3·3 | | | | | | | | hot |
| 7·4 | | | | | | | | rainy |

*A. Ångström* (27) *by observing, during the polar night, the net radiation leaving a surface of snow*, which he asumes to be a full radiator, and by observing also the temperature gradient in the first 0·6 metre of air above the surface, and by allowing for the conduction through the snow, found an eddy flux of heat equal to

$$-0{\cdot}03\,\partial\theta/\partial h \text{ grm cal cm}^{-2}\text{ sec}^{-1}.$$

As the up-grade of temperature was so large—at the rate of about 4400° C per kilometre—we may neglect pressure variations, and by taking the standard pressure as that at the surface, we may put $\partial\tau/\partial h = \partial\theta/\partial h$. It follows that

$$-\gamma_p\,.\,c\,.\,\partial\theta/\partial h = -0{\cdot}03\,\partial\theta/\partial h$$

if energy is expressed in calories.

Whence as $\gamma_p = 0{\cdot}24$ g-cal grm$^{-1}$ degree$^{-1}$, it appears that

$$c = 0{\cdot}12.$$

The mean wind velocity was 2·8 metres per second at a height of 15 metres above ground.

$$\frac{g}{\gamma_p}\frac{\partial\sigma}{\partial h} \text{ comes to } 1650\times10^{-4}\text{ sec}^{-2}.$$

**Critique of the Observational data.** From geometrical considerations* we must admit the possibility that $c$ or $\mu$ at a fixed point may be different in three directions at right angles. The preceding data however refer only to the vertical direction.

Comparing the values of $c$ given in the first column of the tables, we notice that it increases with height up to a level of about a kilometre, and thereafter decreases.

At the same height the coefficient appears to be of the same order, whatever it may be that diffuses. This was observed and explained by G. I. Taylor (12), and later W. Schmidt (18) gave the general notion behind any mathematical argument in the form that an exchange of mass goes on, and that the measure $c$ will be the same for any properties which this mass simply carries with it. This argument fails if the atmosphere has a diversified structure in the diffusing substance, provided the diversity tends to advance or to retard the motions which produce diffusion. For instance the value $c = 1030$ grm sec$^{-1}$ cm$^{-1}$ for the viscosity produced by rising cumuli does not apply to the diffusion of heat, because cumuli are patches of air which are rising because they are warm. Again horizontal velocities at right angles to one another are to be considered as two different diffusing "substances" and observational or theoretical proof is required to show whether the eddy-viscosity is the same for both. A single computation quoted on page 73 shows $c_v$ at a height of 400 metres to be seven times greater across the wind than along it†. This problem deserves further attention. The columns containing $(\partial v_X/\partial h)^2$, $(\partial v_Y/\partial h)^2$ and $g/\gamma_p\,.\,\partial\sigma/\partial h$ have been added because of their relation to the criterion of turbulence. Here $\sigma$ is the entropy per mass, so $g/\gamma_p\,.\,\partial\sigma/\partial h$ is negative for a thermally unstable atmosphere, zero in an

* Richardson, L. F. (32), p. 10.

† Errata in the paper referred to, *Phil. Trans.* A, Vol. 221, p. 5, Table II. Delete the data at a height of 1·0 and 0·8 kilometres. The unit for the rates of mean shearing should be $10^{-3}$ not $10^{+3}$.

adiabatic atmosphere, and increasingly positive as stability (in the absence of wind) increases. When there is wind, the stability is affected by the rates of mean shear. Let the kinetic energy of eddies be $\Theta$ per mass of atmosphere, then from considerations of energy it has been deduced (L. F. Richardson (33), p. 354)* that

$$\iiiint\left[\frac{\partial(\rho\Theta)}{\partial t}+\frac{2\kappa\pi^2}{l^2}\Theta\right]dx\,dy\,dh$$
$$=\iiiint\left[c_v\left(\frac{\partial\bar{v}_X}{\partial h}\right)^2+c_v\left(\frac{\partial\bar{v}_Y}{\partial h}\right)^2-c_\tau\frac{g}{\gamma_p}\frac{\partial\sigma}{\partial h}\right]dx\,dy\,dh, \quad \text{.........(22)}$$

where the integrals are taken over an atmospheric block so large that we may neglect the eddying energy traversing its boundaries. In the first member $l$ is the linear dimension of a "typical" eddy, $\kappa$ is the molecular viscosity. The bar over the velocity denotes a mean. If $c_v=c_\tau$, then we should expect the kinetic energy of eddies $\Theta$ to decrease on the average provided that

$$\iiiint\left[\left(\frac{\partial\bar{v}_X}{\partial h}\right)^2+\left(\frac{\partial\bar{v}_Y}{\partial h}\right)^2<\frac{g}{\gamma_p}\frac{\partial\sigma}{\partial h}\right]dx\,dy\,dh \quad \text{....................(23)}$$

throughout the block. Local changes within the block might be caused by the diffusion of eddies. G. I. Taylor had arrived at a similar criterion, but with a different numerical coefficient, from a theory of the oscillations of superposed laminae. To explain the different numerical coefficient we can easily imagine that the atmosphere might be stable for a special type of disturbance at a time when a more general disturbance would tend automatically to increase itself. No limitation of type is made in the "energy" theory of stability which leads to equations (22) and (23) above. On the other hand there may conceivably be limitations in nature, depending on the character of the earth's surface.

In comparing the theory with observational data we must remember that $c$ is quite different from $\Theta$ although they are alike in so far as they both vanish in the absence of eddies.

Now looking at pp. 72 to 75 we see that, at a given place and height, for example the Eiffel Tower, $c_\tau$ or $c_v$ increases as $g/\gamma_p \,.\, \partial\sigma/\partial h$ decreases, that is to say as the atmosphere becomes less stable in the absence of wind. The same applies to the Newfoundland banks.

Again for Salisbury Plain G. I. Taylor has shown that $c_v$ is roughly proportional to the gradient velocity. Now, at fixed heights near the ground $\partial v/\partial h$ increases with $v$ and, from the point of view of the criterion, it is more convenient to express this result of Taylor's by saying that $c_v$ increases with $\partial v/\partial h$.

On the Newfoundland banks where, on the average of the observations, $g/\gamma_p \,.\, \partial\sigma/\partial h$ was nearly double $(\partial v_X/\partial h)^2+(\partial v_Y/\partial h)^2$ so that the motion was very stable, the eddy-viscosity and $c$ for potential temperature were both found by Taylor to be remarkably low for the height in question (0 to 200 metres). It is proper to deal only with the

* Note, in the paper referred to, a wrong sign before the second term of (5.2) and (5.3) and in italicised statement on p. 362 and that in equation (6.4) $\partial\sigma/\partial p$ should be squared.

average here, because $c_v$ was deduced from a distribution which required many hours to establish itself.

It might be thought that the low eddy-viscosity was due to the smoothness of the sea surface. But by contrast over the sub-tropical sea, Sverdrup (19) has found eddy-viscosities which are abnormally large, even for the greater height 0 to 2000 metres to which they refer. The peculiarity of the latter situation—the North Atlantic Trade Wind—is that $g/\gamma_p \,.\, \partial\sigma/\partial h$ is eight to twenty times greater than $(\partial v_X/\partial h)^2 + (\partial v_Y/\partial h)^2$ for the general range of height, so that we should expect the atmosphere to flow without eddies. Just at the lower part of the range from 0 to 500 m $\partial\sigma/\partial h$ is zero so that eddies would certainly form there. May we suppose that the large eddy-viscosity between 1200 and 1800 metres is due to eddies, which acquire their energy between 0 and 500 m and which diffuse themselves upwards?

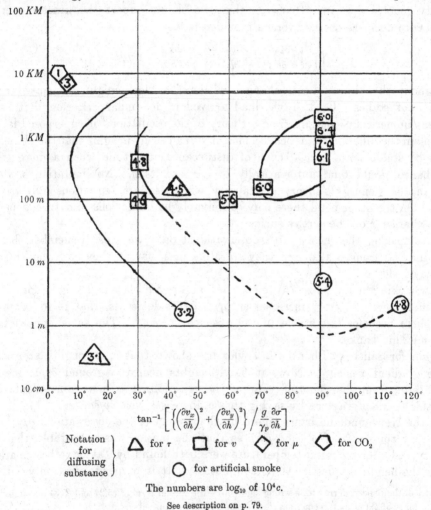

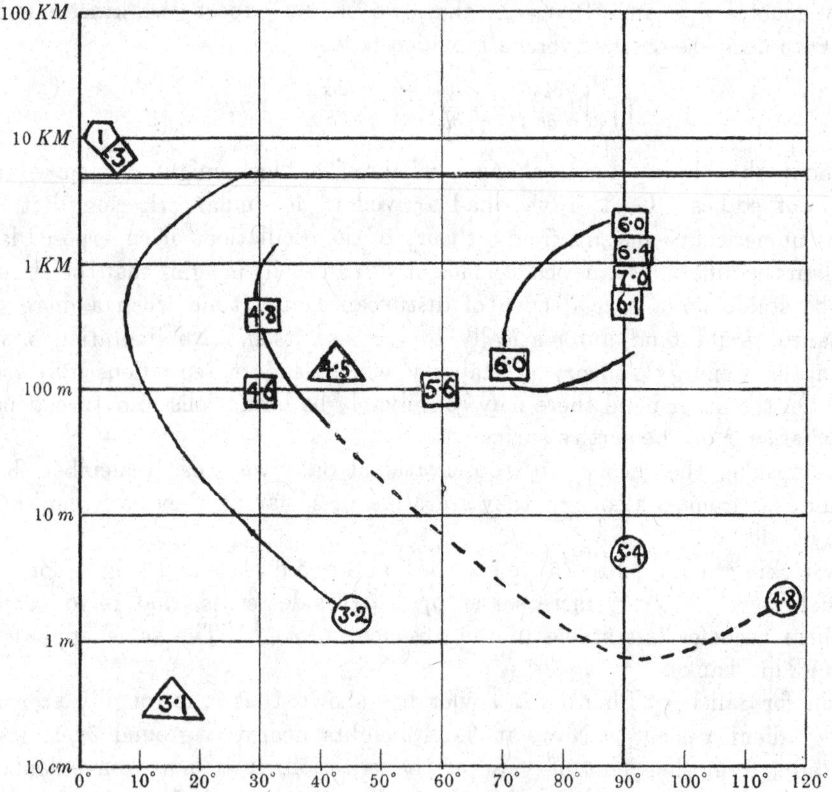


$$\tan^{-1}\left[\left\{\left(\frac{\partial v_x}{\partial h}\right)^2 + \left(\frac{\partial v_y}{\partial h}\right)^2\right\} \Big/ \frac{g}{\gamma_p}\frac{\partial\sigma}{\partial h}\right].$$

Notation for diffusing substance: △ for $\tau$; □ for $v$; ◇ for $\mu$; ⬠ for $CO_2$; ○ for artificial smoke

The numbers are $\log_{10}$ of $10^4 c$.

See description on p. 79.

Rising eddies would die away on reaching the stable upper layers, but the supply of new eddies arriving there would be maintained. The observations indicate that the upward diffusion of eddy motion has an important effect, and in consequence that the stability or instability of the motion next the ground is often the deciding factor in determining the turbulence at much greater heights.

The road to a fuller knowledge of the variations of viscosity appears to lie through a study of the diffusion of eddies*.

The observations in the tables have, as far as possible, been shown in the diagram. Special scales have been necessary. The numbers in the body of the diagram are $4+\log_{10} c$, so that the molecular viscosity would be represented by 0·2. The ordinate is the logarithm of the height. The peculiar function chosen for the abscissa has the following advantages: (i) it does not commit us in the derivation of the criterion to supposing that the eddy-viscosity $c_v$ is equal to the eddy-conductivity for potential temperature; (ii) it increases steadily through 90° when the up-grade of entropy per mass changes sign; (iii) it is zero in circumstances such that we should expect no eddies. The stability used in the diagram is that of the lowest layer which happens to be given in the tables. This interim procedure is due to be improved when there is a better understanding of how eddies diffuse. In particular the observations at 8·5 and 10 km have been plotted against the stability at their own height. Many of the observations refer to a considerable vertical range. In this case they have been plotted against the mean height, not against the mean of its logarithm.

## CH. 4/8/2. APPLICATION TO THE UPPER CONVENTIONAL STRATA

From 1 km upwards the change of $\chi$ with height is usually sufficiently gradual to permit of significant analysis by coordinate differences as large as 2 decibars. Within the kilometre next the surface the change of both $\xi$ and $\chi$ with height is so rapid that our coordinate differences have no longer any resemblance to infinitesimals, and a different process has to be adopted. This will be discussed in Ch. 4/8/3 and the following sections. For the present let us confine our attention to heights exceeding one kilometre.

**Velocity.** In order to fit into the dynamical equations beginning

$$-\frac{\partial M_E}{\partial t} = \quad ; \qquad -\frac{\partial M_N}{\partial t} = \qquad \text{(Ch. 4/4 \# 11 and 12)}$$

we require the rate at which horizontal momentum is crossing unit area of the boundaries of the strata by means of eddies. In other words we require the eddy-shearing-stress at the interface.

Momentum is also carried across the interface by the *mean* vertical velocity, if any. This flux is represented by the terms $[m_E v_H]_8$, $[m_N v_H]_8$ in the aforesaid dynamical equations. But that general transport receives separate attention, so that we need not deal with it here.

* L. F. Richardson (33), p. 372.

Now in the theory of Ch. 4/8/0 we may take $Z$ successively as either of the horizontal components of momentum, and then $\chi$ is $v_E$ or $v_N$, and the eddy-shearing-stress on a horizontal plane has components (Ch. 4/8/0 # 17)

$$\frac{\xi}{g}\frac{\partial v_E}{\partial p}, \quad \frac{\xi}{g}\frac{\partial v_N}{\partial p}. \quad \text{.................................(1)}$$

Now $M/R$, the ratio of the momentum of a stratum to its mass, gives some kind of mean velocity, which we may use to find $\partial v_E/\partial h$, $\partial v_N/\partial h$ by taking differences with height. For instance, where suffixes refer to height, we should have a stress at the $h_6$ level having an eastward component

$$\frac{\xi_6}{g}\frac{\left(\frac{M_{86}}{R_{86}}-\frac{M_{64}}{R_{64}}\right)}{p_7-p_5}, \quad \text{.................................(2)}$$

where $p_7-p_5$ may be taken as $\frac{1}{2}(p_8-p_4)$ or as $\frac{g}{2}(R_{86}+R_{64})$. Now $R$ is approximately the same, and equal to 2 decibars, for all conventional strata except the lowest, so that the shearing stress at the $h_6$ level is nearly

$$\frac{\xi_6}{g^2R^2}(M_{86}-M_{64}). \quad \text{.................................(3)}$$

And the differences between the shearing stresses at the different levels are the additional terms required in the dynamical equations quoted above.

**Water, when $W$ is given.** Putting $W$, the mass of water per horizontal area, in place of $M$ in (2) we have for the water crossing unit area of a horizontal plane at height $h_6$

$$\frac{\xi_6}{g^2}\frac{\left(\frac{W_{86}}{R_{86}}-\frac{W_{64}}{R_{64}}\right)}{\frac{1}{2}(R_{86}-R_{64})}. \quad \text{.................................(4)}$$

This is the effect of eddies alone. The change due to any general motion of air up or down is dealt with under the "conveyance of water," Ch. 4/3.

**Water, alternative scheme, $\mu$ given at boundaries of strata.** (Compare part of Ch. 4/3.) In the equation

$$\frac{\overline{D}\mu}{\overline{D}t}=\frac{\partial}{\partial p}\left(\xi\frac{\partial\mu}{\partial p}\right), \quad \text{.................................(5)}$$

it is now most simple to put

$$\frac{\partial\mu}{\partial p} \text{ approximately equal to } \frac{\mu_8-\mu_6}{p_8-p_6} \quad \text{......................(6)}$$

for the stratum $h_8$ to $h_6$, which may serve as a type.

Similarly for the stratum above. So that (5) transforms into

$$\frac{\overline{D}\mu_8}{\overline{D}t}=\frac{1}{p_7-p_5}\left\{\xi_7\frac{\mu_8-\mu_6}{p_8-p_6}-\xi_5\frac{\mu_6-\mu_4}{p_6-p_4}\right\}, \quad \text{......................(7)}$$

giving the time change of $\mu$ correctly centered at the boundary where we need it. Note that $\xi$ is now required at the levels, with odd suffixes, corresponding to the mean pressure of the strata. As $p_7$ and $p_5$ are not tabulated variables we must put

$$p_7 - p_5 = \tfrac{1}{2}(p_8 - p_4). \quad \text{..............................(8)}$$

Or indeed, in view of the uncertainty of $\xi$, it may be near enough to make both the $\delta p$ simply equal to their average value of 200 millibars.

It hardly seems worth while making use of the fact that, on the average of many occasions, $\mu = Cp^B$ where $C$ and $B$ vary slowly with height, as was done in Ch. 4/3, until we know more about the variations of $\xi$ with height, as they also are involved.

In the equation for the conveyance of water (Ch. 4/3 # 25) there is a term in $D\mu/Dt$, which is shown in that section to transform thus

$$\int_G^8 \rho \frac{D\mu}{Dt} dh = -\frac{1}{g} \frac{(\log p_8 - \log p_G)\left(p_8 \frac{D\mu_8}{Dt} - p_G \frac{D\mu_G}{Dt}\right)}{\frac{D\log\mu_8}{Dt} - \frac{D\log\mu_G}{Dt}}. \quad \text{...........(9)}$$

In so far as this term arises, not from precipitation, but from the eddy-flux, the values of $D\mu_8/Dt$ for insertion in it are given by (7) above.

**Potential Temperature.** The mean potential temperature of a stratum follows from the pressures at its boundaries and from its water content per horizontal area. Having found this mean it is to be inserted in place of $W/R$ in the numerator of (4). The result is the rate at which temperature × mass is crossing unit horizontal area. Again the effect of any *general* vertical motion has been separately formulated in Ch. 4/5/2.

As to the *method of finding the mean potential temperature of the stratum* when dry or at least clear. In the troposphere, on the average, $\tau$ only varies by 5 or 10 °/₀ of itself in the thickness of a stratum, so that it does not much matter whether we take an arithmetic, geometric, or harmonic mean. *For dry air* exactly, and very closely for clear air, $\tau = \theta\,(p'/p)^{0\cdot289}$, ..............................................(10)

where $p'$ is the standard pressure used in defining $\tau$, so that the vertical static equation may be written*

$$\frac{\partial p}{\partial h} = -g\rho = -\frac{gp}{b\theta} = -\frac{gp}{b\tau}\left(\frac{p'}{p}\right)^{0\cdot289}, \quad \text{.....................(11)}$$

whence

$$\tau = -0\cdot289\,\frac{g}{b}\,(p')^{0\cdot289} \cdot \frac{dh}{d\,(p^{0\cdot289})}. \quad \text{.......................(12)}$$

* *Vide* Exner, *Dynamische Meteorologie*, § 70.

It follows that a mean potential temperature is given by the following fairly simple formula:

$$\bar{\tau}_{64} = -0{\cdot}289\,\frac{g}{b}\,(p')^{0{\cdot}289}\cdot\frac{h_6 - h_4}{p_6^{\,0{\cdot}289} - p_4^{\,0{\cdot}289}}\cdot \qquad\ldots\ldots\ldots\ldots\ldots\ldots\ldots(13)$$

Tables of 0·289th powers would be a useful aid.

For cloudy air we come in for more complications, as in Ch. 4/1 and Ch. 4/5/1.

### Ch. 4/8/3. EDDY SHEARING STRESS ON THE EARTH'S SURFACE

When $\chi$ is velocity we require the eddy shearing stress on the ground to be given by statistics as a function of such variables as are available, namely of: the position on the map, the date, the mean temperature $\bar{\theta}_{G8}$ of the lowest stratum, the temperature of the sea, of the bare earth or of the air in the vegetation film, and the mean momentum per volume of the lowest stratum. For brevity the last-named quantity will be denoted by $\bar{m}$. It has components

$$\bar{m}_E = M_{EG8}/(h_8 - h_G); \quad \bar{m}_N = M_{NG8}/(h_8 - h_G). \qquad\ldots\ldots\ldots\ldots\ldots\ldots(1)$$

Åkerblom appears to have been the first to get a fairly reliable measure of the surface stress. He deduced it from Angot's statistics of wind observations at the top of the Eiffel Tower (305 metres) and at the Bureau Central (21 m) by assuming the eddy-viscosity to be independent of height. We now know that this assumption is hardly permissible.

Åkerblom, F. (8).

| Shearing stress on surface dyne $cm^{-2}$ | Velocity cm $sec^{-1}$ at Bureau Central 21 m | Velocity at 305 m Eiffel Tower cm $sec^{-1}$ | |
|---|---|---|---|
| 0·40 | 234 | 982 | Winter |
| 0·29 | 209 | 777 | Summer |

Hesselberg and Sverdrup (14), treating the same data by a different method, confirm the above result.

G. I. Taylor (16) illuminated the subject by comparing, by aid of the theory of dimensions, the friction on the earth with that obtained on small metal plates at much higher velocities in laboratories. By analysing G. M. Dobson's Pilot balloon observations made over Salisbury Plain, Taylor showed that the surface stress was more or less proportional to the square of the velocity "near the surface." Like Åkerblom he assumed the eddy-viscosity to be independent of height.

G. I. Taylor (16).

| Stress dynes $cm^{-2}$ | Gradient wind cm $sec^{-1}$ | $\overline{m}$ grm $sec^{-1}$ $cm^{-2}$ |
|---|---|---|
| 0·31 | 460 | — |
| 1·37 | 910 | — |
| 2·44 | 1560 | 1·70 |

As velocity varies so rapidly with height near the surface it may well be that the upper wind represented by $\overline{m}$ provides a standard more suitable than the wind near the surface. In the present scheme at any rate we have to use $\overline{m}$, for it is the only measure of wind available.

W. V. Ekman (6) in 1905 pointed out, with reference to the sea when a steady state has been attained, that the total momentum, produced by a tangential stress on the surface, is directed at right angles to the stress, and is equal in magnitude to the stress divided by $2\omega \sin \phi$. Thus the momentum produced by the stress is quite independent of the value of the viscosity or of its variation with height. This is true provided that the quadratic terms in the dynamical equations produce only a negligible disturbance, and that will be so if we take, as our standard of what the momentum would be in the absence of stress, not the wind deduced from the isobaric map, but the actual upper wind. Even that approximation will fail near the equator. A high level should be chosen at which the stress vanishes because $\partial \mathbf{v}/\partial h$ vanishes there, and deviations of momentum from the aforesaid standard should be computed from the ground to the level of no-stress. One must either select observations in which the upper momentum becomes independent of height—as I have done—or else, preferably, compute the variation with height of the standard momentum from observations of horizontal temperature-gradients. Here read the table on p. 84.

Another way of measuring the eddy-shearing-stress is provided by O. Reynolds' theorem to the effect that the stress dragging the lower air in the direction of $x$ increasing is

$$-\rho\, \overline{v_X' v_H'}, \quad \text{.....................................(2)}$$

where $v_X'$ and $v_H'$ are simultaneous deviations, at the same point, of velocity from the mean; and where the bar denotes a mean, over a long interval of time, of the product of these deviations. By observing the motion of thistledown and using Reynolds' formula, there was obtained on one occasion (L. F. Richardson (32), pp. 10 to 15 and p. 21)

| Height | Mean velocity | Stress dragging lower air to right of wind | Probable error |
|---|---|---|---|
| 200 cm | 145 cm $sec^{-1}$ | + 0·48 dyne $cm^{-2}$ | 0·20 |

| Place | Latitude | $\overline{m}$ $\frac{\text{grm}}{\text{cm}^2\,\text{sec}^2}$ | Resultant stress $\div \overline{m}^2$ $\frac{\text{cm}^3}{\text{grm}}$ | Stress on ground is veered* from wind near surface | Stress dragging ground in the direction of $\overline{m}$. dynes cm$^{-2}$ | Stress dragging ground perpendicularly to left of $\overline{m}$ | References and Notes | $\bar{\xi}_{GS}$ for use in Ch. 4/8/5 grm$^2$ cm$^{-2}$ sec$^{-5}$ |
|---|---|---|---|---|---|---|---|---|
| Eskdalemuir ... | 55·3° | 0·72 | 6·7 | −8° | +2·9 | +1·9 | Mean of 39 observations in 1913 and 1914. Light winds omitted | $2·9 \times 10^5$ |
| Lindenberg ... | 52·2° | 1·24 | 1·0 | 26° | +1·5 | +0·05 | General mean. E. Gold, Met. Office, *Geophys. Mem.*, v. p. 143 | $1·1 \times 10^5$ |
| Upavon ... | 51·3° | 1·70 | 1·0 | assumed to be zero to fit Taylor's | +2·9 | +0·8 | Strong winds. Dobson, *Q. J. R. Met. Soc.*, April, 1914 | $1·5 \times 10^5$ |
| Batavia ... | −6·2° | 0·26 | 1·3 | −21° | +0·08 | −0·03 | May and June ... (Van Bemmelen; Batavia, *Verhandelingen*, vol. I, 1911, pp. 13, 14) | $0·3 \times 10^5$ |
| | | 0·47 | 1·1 | 30° | +0·04 | −0·24 | July to September ... (Van Bemmelen; Batavia, *Verhandelingen*, vol. I, 1911, pp. 13, 14) | $0·11 \times 10^5$ |
| | | 0·47 | 0·6 | ±180° | −0·12 | −0·05 | December to February (Van Bemmelen; Batavia, *Verhandelingen*, vol. I, 1911, pp. 13, 14) | meaningless |
| The above were obtained by the method described in the preceding page. Below follow two results taken from G. I. Taylor's paper, 1915. | | | | | | | | |
| Upavon ... | 51·3° | as above | 0·83 | assumed to be zero | — | — | The same observations as above | |
| Newfoundland Banks | 52° | — | about 0·06 | — | — | — | | about $0·06\overline{m} \times 10^5$ |
| L. H. G. Dines by a similar method has obtained the following results. Winds of all velocities were included except those of $5^m/s$ or less at the anemometer. | | | | | | | | |
| Cahirciveen ... | 51·9° | 1·38 | 1·3 | −56° | +1·37 | +2·03 | 33 ascents. Wind from S. 20° W. to S. 80° W. from Atlantic, partly crossing Valentia Island | |
| Cahirciveen ... | 51·9° | 1·08 | 5·1 | +19° | 5·97 | −0·3 | 15 ascents. Wind from S. 20° E. to S. 20° W. over hills | |

* Veering here means turning in the sense of north to east, and this statement applies to both sides of the equator.

On clear nights the air over land is often calm near the ground while the wind blows uninterruptedly at a height of some hundred metres. How does this calm air remain unmoved under the joint action of friction from above and of the all-pervading barometric gradient? The answer must apparently be that the upper surface of the calm is not level, but is tilted so as to balance these forces. Seeing that the calm air always has a distinctly greater "potential density" than the air above it, a tilt would produce a horizontal pressure gradient. Observations of the height of the calm at neighbouring stations would be interesting, and would give a measure of the surface stress, which may be very small*.

After sunrise the wind descends. Its junction with the calm is sometimes very abrupt, as is shown by shooting spheres up to various heights. The arrival of the wind at the level of the anemometer is often quite sudden. One gets the impression that the upper wind planes away the night calm a shaving at a time. Suppose that the lower boundary of the wind descends at rate of $B$, that the velocity of the wind is $v$ and that the air-density is $\rho$. Then a mass of air equal to $B\rho$ grams per horizontal $cm^2$ acquires a velocity $v$ in one second, so that the surface stress is $B\rho v$. .........(3)

The mean stress will have this value whatever be the order in which the parts of the calm air are accelerated, provided the interval of time and their final velocity are given.

Observation at Benson.

| Date | Early data | | Anemometer data | | Mean $v$ say cm sec$^{-1}$ | $B$ | Stress $\frac{\text{dyne}}{\text{cm}^2}$ |
|---|---|---|---|---|---|---|---|
| | Calm reaches to | Velocity cm sec$^{-1}$ | Wind first reached anem. | Upper velocity reached anem. | | | |
| 1919 Oct. 22 | 70 m. at $7^h\ 12^m$ | 500 at 120 m | $10^h\ 55^m$ | $12^h\ 30^m$ | 500 | $\frac{120\text{ m}}{5^h\ 18^m} = 0{\cdot}63$ cm sec$^{-1}$ | 0·39 |
| 1920 Feb. 23 | 170 m at $6^h\ 50^m$ | 300 at 250 m | $13^h\ 50^m$ | $15^h\ 10^m$ | 300 | $\frac{250\text{ m}}{8^h\ 30^m} = 0{\cdot}83$ cm sec$^{-1}$ | 0·31 |

To these stresses must be added any due to the (as yet unmeasured*) slope of the upper surface of the calm.

A *cornfield* provides a uniform elastic surface by observing which we can measure the shearing stress. For example at Benson, 1920 Aug. 16th to 22nd, observations were made on a field of ripe wheat. The wind acts mainly on the ear, but partly also on the stalk, as may be shown by lowering a glass jar over the ear. The centre of the forces due to the wind appears to be about 5 cm below the ear. The springiness of a number of growing stalks was measured on a calm evening and it was found that 1 cm deflection of the centre of force corresponded on the average to a horizontal

* Since this was set in type I learn that W. Georgii has made extensive observations (*Ann. Hydr. Berlin*, 1920, pp. 207—222 and 241—262).

force of about 130 dynes applied at that level. The number of ears per square metre was about 175. Therefore a deflection of the centre of force of about 1·5 cm on the average corresponded to a stress of

$$1{\cdot}5 \times 130 \times 175 \times 10^{-4} = 3{\cdot}4 \text{ dynes cm}^{-2}.$$

| Date Aug. 1920 | Stress $\frac{\text{dyne}}{\text{cm}^2}$ | Wind 30 cm above ears cm sec$^{-1}$ | Anem at 26 m cm sec$^{-1}$ | At 600 m by balloon cm sec$^{-1}$ | $\bar{m}$ $\frac{\text{grm}}{\text{cm}^2\,\text{sec}}$ | Other conditions | $\frac{\text{Stress}}{(\bar{m})^2}$ cm$^3$/grm |
|---|---|---|---|---|---|---|---|
| 17$^d$ 10$^h$ | 3·4 | 130 | 350 | 550 | 0·6 | Lapse rate nearly normal. Overcast | 9 |
| 22$^d$ 16$^h$ | 16 | 400 | 550 | — | 1·2? | 2/10 cumulus | 11? |

Thus a cornfield appears to be rougher than the average of the country-side.

Thus estimates of the stress, by five or six different methods, give values clustering round 1 dyne cm$^{-2}$.

Suppose we have found the surface shearing stress on the air, for the particular locality, with due reference to the strength of the wind and to the vertical gradient of entropy-per-unit-mass. The stress will be specified by its components directed with $M$, and perpendicularly to the left of $M$. Denote these by $A$ and $B$ respectively. The components of stress on the air to the east and the north are then

$$\frac{1}{M}(AM_E - BM_N), \quad \text{..........} (4)$$

$$\frac{1}{M}(AM_N + BM_E), \quad \text{..........} (5)$$

where, as usual, $M = (M_E^2 + M_N^2)^{\frac{1}{2}}$.

### CH. 4/8/4. THE EDDY-FLUX OF HEAT AT THE SURFACE

We have already noted on pp. 70, 71 that, when temperature remains steady, the flux of heat increases downwards.

The curve showing the relation of temperature to height reminds one of a fishing rod, held aloft with its thin end pointing downwards, and shaken. The upper part is fairly steady both in slope and in position. The lower part switches to and fro, and its slope alternates between large positive and negative extremes. Measurements of $\partial\theta/\partial h$ made at Benson by carrying an Assmann aspirated thermometer up a steel mast to a height of 16 metres show that on a sunny summer morning $\partial\theta/\partial h$ may be at the rate of $-45^\circ$ C per km; while just before sunrise it may be at the rate of $+130^\circ$ C per km. Such up-grades as these are unheard of, if the differences are taken over a whole kilometre. We see thus that the relations of potential temperature to height are as detailed as the relations of wind to height would be, if the earth's surface had the habit of slipping to and fro parallel to itself, with a daily period, and with a velocity attaining 30 metres per second.

The stratum from the ground to 2 km was rather thick for dealing with the variations of wind, but the case of potential temperature is worse. Yet as thinner strata would mean more arithmetical toil, let us for the sake of economy try first to make the thick stratum do. The origin of most of our difficulties lies in the fact that the up-grade of potential temperature in the first 50 metres is frequently of opposite sign to the mean up-grade over the first two kilometres, so that the latter alone is of no use in estimating the flow of heat at the surface.

Although the variation between day and night of heat produced at the surface from the balance of radiation has a large effect, yet we cannot ignore the consequences of the warmth or coolness of the upper air relative to the daily mean at the ground.

We might attempt to separate the mean eddy-flux from the daily oscillation. There are some good observations of a damped periodic wave of temperature propagated with an amplitude which diminishes, and a time of maximum which lags, as it goes upwards. For instance there are Angot's statistics for the Eiffel Tower, and J. Regers' for Lindenberg*. The amplitude of the daily wave decreases to $1/e$ of its surface value at a height of about 400 metres above Lindenberg. Such statistics have been compared with theory by G. I. Taylor (21) and by W. Schmidt (25), (34), (35) with a view to finding a measure of turbulence. Schmidt separates the effect of radiation absorbed by the air from the effect of eddy-conduction. Schmidt also treats of the ratio in which heat, generated from radiation at the interface, divides itself between the two media. These theories no doubt have some relation to fact and give us some new insight into it. But they begin by assuming $\xi$ to be independent of time and of height—a treacherous assumption, which we must try to avoid, seeing that $\xi$ is observed to vary in a range of 100 to 1. However $\xi$ may vary, it is so far possible to separate out an oscillatory part, as in consequence of Ch. 4/8/0 # 16, which reads

$$\frac{\partial \chi}{\partial t} = \frac{\partial}{\partial p}\left(\xi \frac{\partial \chi}{\partial p}\right), \quad \text{.................................(1)}$$

being linear in $\chi$, it follows that the sum of any two integrals of this equation is itself an integral. But there are other difficulties which will now be explained.

W. Schmidt (18, p. 26) concludes that the flux is on the average downwards because, on the average, potential temperature increases upwards. His argument is convincing when it refers to a height to which the daily variation hardly penetrates. Referring to layers near the ground, W. H. Dines †, on the contrary, concludes that the eddy-flux of heat is on the average upwards, because heated air rises freely on sunny days, while on cold nights the air next the ground, by becoming stagnant, prevents downflow.

Mr Dines' description suggests that the deviation of potential temperature from its daily mean will not in general satisfy the diffusion equation satisfied by the actual potential temperature. For, take the time-mean of equation (1). At any level let a time-mean be denoted by a bar, a deviation from a mean by a dash, so that

$$\xi = \bar{\xi} + \xi'; \quad \chi = \bar{\chi} + \chi'. \quad \text{.................................(2)}$$

* Lindenberg annual volume for 1912.

† "Heat Balance of the Atmosphere," *Q. J. R. Met. Soc.*, April 1917, p. 155.

Be it noted that $\xi$ and $\chi$ are both intrinsically mean values. We are now performing, over a longer interval, a second averaging.

Then (2) implies that $$\frac{\partial\chi}{\partial p}=\frac{\partial\bar{\chi}}{\partial p}+\frac{\partial\chi'}{\partial p}. \qquad \text{.................................(3)}$$

So that the diffusion equation (1), when averaged with respect to time, becomes

$$\frac{\partial\bar{\chi}}{\partial t}=\frac{\partial}{\partial p}\overline{\left\{(\bar{\xi}+\xi')\left(\frac{\partial\bar{\chi}}{\partial p}+\frac{\partial\chi'}{\partial p}\right)\right\}}, \qquad \text{.........................(4)}$$

and as the product of a mean into a deviation becomes zero when its mean is taken

$$\frac{\partial\bar{\chi}}{\partial t}=\frac{\partial}{\partial p}\left\{\bar{\xi}\frac{\partial\bar{\chi}}{\partial p}+\overline{\xi'\frac{\partial\chi'}{\partial p}}\right\}. \qquad \text{............................(5)}$$

Now there is no expectation that $\overline{\xi'\dfrac{\partial\chi'}{\partial p}}$ will vanish, at any rate if $\chi$ is potential temperature, for the wind tends to become more turbulent as the lapse-rate approaches or exceeds the adiabatic. In fact on a sunny midday, when $\partial\tau/\partial p$ is positive, the turbulivity $\xi$ at a height of about a kilometre was estimated by the present writer* to be

$$1{\cdot}1\times10^6 \text{ grm}^2\text{ cm}^{-2}\text{ sec}^{-5}$$

for velocity—and it would be greater still for potential temperature—which is about ten times greater than the average turbulivity found by various authors at this height. Again smoke observations made at a height of 2 metres above ground show that $\xi$ is much greater on clear days than on clear nights. Also, as $\xi$ tends to increase with increase of velocity aloft, so $\overline{\xi'\dfrac{\partial\chi'}{\partial p}}$ will not vanish if $\chi$ is the horizontal velocity. The case when $\chi=\mu$ is not so clear.

Thus we conclude: *the daily mean of potential temperature does not satisfy the diffusion equation satisfied by the instantaneous value.*

Subtract (5) from (1), then it follows that

$$\frac{\partial\chi'}{\partial t}=\frac{\partial}{\partial p}\left\{\xi'\frac{\partial\bar{\chi}}{\partial p}+\bar{\xi}\frac{\partial\chi'}{\partial p}+\xi'\frac{\partial\chi'}{\partial p}-\overline{\xi'\frac{\partial\chi'}{\partial p}}\right\}. \qquad \text{.....................(6)}$$

Thus the deviation from the daily mean follows the course described by this complicated equation. The effects of radiation, conveyance and precipitation are to be regarded as additional.

On these grounds it appears useless to attempt to separate the daily oscillation from the mean distribution. Another reason against such a course is to be found in the non-periodic irregularities of the thermograph record which are produced by cloud.

Turning away from harmonic analysis we come back to an extensive empiricism. From data uncomplicated by the effects of eddies, we know the rate at which heat is being produced or destroyed at the surface. This heat diffuses both upwards and downwards. Let the ratio of the upward flux to the downward flux be denoted by

* *Phil. Trans.* A, Vol. 221, p. 26.

$\approx$ and let us make an empirical study of the value and variations of $\approx$. The **partition coefficient,** as we may call $\approx$, is not enough to fix the flux of heat, for sometimes there may be no net radiant energy to be divided in parts, and yet, when for instance colder air overlies a warmer sea, there will be an upward flux of heat. We need a second empirical coefficient, say $\mathfrak{w}$, defined so that the upward flux of heat when there is no net radiation is

$$\frac{\mathfrak{w}}{g}(\bar{\tau}_{G8}-\tau_G)/\tfrac{1}{2}(p_8-p_G), \qquad \text{.............................(7)}$$

where $\tau_G$ is the potential temperature of air in thermal equilibrium with the surface of the sea or land.

It will be interesting to compare the numerical value of $\mathfrak{w}$, when it has been observed, with that of the corresponding quantity for the diffusion of water, which is discussed in the next section under the symbol $\bar{\xi}_{G8}$, and of which some estimates are given in Table on p. 84. See Ångström's observation quoted in Ch. 4/8/1.

The relation of $\approx$ to $\mathfrak{w}$ and their combination or alteration to give the flux of heat, is a question which is deferred to Ch. 8/2/15.

W. Schmidt (35), from whom I get the idea of the partition-coefficient, has written of it mostly from the point of view of integrals of $\partial\chi/\partial t = B\partial^2\chi/\partial h^2$. Going instead directly to Homen's observations*, of the heat penetrating the ground, and of radiation, it is seen that in latitude 60° on 14th and 15th August $\approx$ had the following values:

| Surface........... | Granite rock | Sandy heath | Grass moor |
|---|---|---|---|
| By day $6^h$ to $18^h$ ... | 0·84 | 2·34 | 2·80 |
| By night $18^h$ to $6^h$... | −0·35 | 0·60 | 1·86 |

The two coefficients $\approx$ and $\mathfrak{w}$ might be classified as functions of such variables as are available in the scheme of numerical prediction, namely of $\bar{v}_{G8}$, of $\tau_G-\bar{\tau}_{G8}$ and of the rate of production of heat at the interface. It is probable that $\approx$ increases as each of these variables increases. Homen's observations show an increase with either $\tau_G-\bar{\tau}_{G8}$, or with the rate of production of heat, or with both of these variables. The classification should also take account of position on the map, for $\approx$ is much less over water than over land.

W. Schmidt (35) estimates that $\approx = 0{\cdot}0037$, on the average, over the sea. But that estimate is deduced from a theory of a simple harmonic oscillation.

Perhaps the surest way to obtain $\approx$ and $\mathfrak{w}$ would be to analyse air-temperatures observed from kites, aeroplanes and balloons. The gradual "depolarization" of the first two kilometres of "polar air" as it flows southwards would, when compared with the radiation, give us the information we need.

* Hann, *Meteorologie*, 3rd edn. p. 49.

## CH. 4/8/5. EDDY-FLUX OF WATER FROM THE SURFACE

The eddy-flux of water upward from the surface must, on the yearly average of the whole globe, be equal to the rainfall. But here we need to take a detailed view of the matter. The records of evaporation from tanks are not supposed to be applicable to lakes or seas, and do not help us much. The botanists and agriculturalists have measured the evaporation from plants and that will be referred to in Ch. 4/10.

The large daily oscillation of *relative* humidity at head level above ground is nearly compensated by the variation of temperature, so that the diffusing quantity $\mu$ varies only some 15 °/₀ from its mean. What concerns us more is that $\partial\mu/\partial h$, like $\partial v/\partial h$ but unlike $\partial\tau/\partial h$, does not ordinarily change sign near the ground during the course of a summer day. Thus Angot's statistics for the difference between the vapour pressure at the summit of the Eiffel Tower, and at Parc St Maur some 250 metres below it, show that $\partial\mu/\partial h$ is normally negative, and throughout an average July day varies only 23 °/₀ on either side of its mean.

Again such variations of moisture as occur in the atmosphere have much less effect on the density than have the variations of temperature. Thus the distribution of $\mu$, unlike that both of $\tau$ and of $v$, is not so much a contributory cause of turbulence. Thus the exchange of moisture is likely to depend on the exchange of mass as if the instantaneous deviations of $v_H$ and of $\partial\mu/\partial h$ were not correlated*. In other words observations on the dispersal of smoke should yield a value of the turbulivity more applicable to the diffusion of water than to that of heat.

Possessing observed values of $\xi$ near the ground we might combine them with $\partial\mu/\partial p$ at the same level, to get the flux $\xi/g \,.\, \partial\mu/\partial p$; but unfortunately our coordinate difference is too big to serve as $\partial p$ in $\partial\mu/\partial p$ near the ground. So the best we can do is to express $\partial\mu/\partial p$ as an empirical function of the excess of the mean value $\bar{\mu}_{GS}$ for the first stratum over the value near the earth's surface say $\mu_G$. Or when there is vegetation $\mu_L$ may advantageously replace $\mu_G$.

Let us now examine **the compensation which must be made for the greatness of the coordinate difference.**

We have, from Ch. 4/8/0 # 15, when $\overline{D}/\overline{D}t$ is expanded

$$\frac{\partial\mu}{\partial t} = \frac{\partial}{\partial p}\left(\xi\frac{\partial\mu}{\partial p}\right) + \text{terms due to precipitation and transport,} \quad \text{.........(1)}$$

$$\frac{\partial v_X}{\partial t} = \frac{\partial}{\partial p}\left(\xi\frac{\partial v_X}{\partial p}\right) + 2\omega\sin\phi\,.\,v_Y - \frac{1}{\rho}\frac{\partial p}{\partial x} \quad \text{.........(2)}$$

Now the rapid vertical changes in $\partial\mu/\partial p$ or $\partial v_X/\partial p$ occur mostly in the first 50 metres or less above the surface where they are due to the decrease of turbulence as the surface is approached. In this thin layer the mixing is probably more important than either precipitation or transport. And if we choose the $x$ axis to be parallel to the isobar, then $\partial p/\partial x$ is zero and $v_Y$ is small. Also it is probable that the flow by

* L. F. Richardson, *Phil. Trans.* A, 221 (1920), pp. 9, 10.

way of eddies through this thin layer is more important than the accumulation in it. In other words we may neglect $\partial\mu/\partial t$. If the same be done with $\partial v_X/\partial t$ then the above equations (1) and (2) both reduce to

$$0 = \frac{\partial}{\partial p}\left(\xi \frac{\partial \chi}{\partial p}\right), \quad \text{.................................(3)}$$

which implies a flux of $\chi$-times-mass which is independent of height. See Ch. 4/8/0 # 17. By integrating (3) and putting in the limit $G$

$$\frac{\partial \chi}{\partial p} = \xi_G \left(\frac{\partial \chi}{\partial p}\right)_G \frac{1}{\xi}. \quad \text{.................................(4)}$$

Integrating again and putting in the same limit

$$\chi - \chi_G = \left(\xi \frac{\partial \chi}{\partial p}\right)_G \int_G^p \frac{dp}{\xi}. \quad \text{.................................(5)}$$

Integrating once more from $p_8$ to $p_G$ and dividing by $p_G - p_8$ so as to obtain the mean denoted by

$$\bar{\chi} = \frac{1}{p_G - p_8} \int_8^G \chi \, dp, \quad \text{.................................(6)}$$

we get

$$\chi - \chi_G = \frac{1}{p_G - p_8} \left(\xi \frac{\partial \chi}{\partial p}\right)_G \int_8^G \int_G^p \frac{dp^2}{\xi}. \quad \text{.......................(7)}$$

Solving for the flux, which we have assumed to be independent of height

$$\frac{\partial \chi}{\partial p} \cdot \frac{\xi}{g} = \frac{\bar{\chi} - \chi_G}{\frac{1}{2}(p_8 - p_G)} \cdot \frac{1}{g} \cdot \frac{(p_G - p_8)^2}{\int_8^G \int_G^p \frac{dp^2}{\xi}} \quad \text{..........................(8)}$$

which shows that *when we replace* $\partial\chi/\partial p$ *by its mean* value $(\bar{\chi} - \chi_G)/\frac{1}{2}(p_8 - p_G)$ then, to get the flux, we must also replace $\xi$ by the peculiar mean value

$$\bar{\xi}_{G8} = (p_G - p_8)^2 \Big/ \int_8^G \int_G^p \frac{dp^2}{\xi}. \quad \text{..........................(9)}$$

The idea that $\xi$ was the same for all meanings of $\chi$ was first suggested by Taylor (12), and has been widely used by W. Schmidt (18), (25). Although it is almost certainly not exact, yet we may try it here in order to obtain some preliminary estimate of the upward flow of water, for comparison with direct measurements. Now if $\xi$ is the same for the diffusion of velocity as of water per mass, and if the assumptions on which (8) is based are good, then for these two diffusing quantities $\bar{\xi}_{G8}$ will also be the same.

W. Schmidt (18) using values of the Austausch, $\xi/g^2\rho$, deduced from wind at a height of 1/2 km, in combination with up-grades of water there, has estimated the mean flux of water at this level. Similarly we see that from the surface stress and wind distribution we can estimate $\bar{\xi}_{G8}$, and we can then use $\bar{\xi}_{G8}$ in combination with the changes of $\mu$ given by our large coordinate differences, to find the upward flux of water at the ground. For, replacing $\chi$ by $v_X$ in (8) and (9), the flux becomes the surface stress and

$$\bar{\xi}_{G8} = \frac{(\text{surface stress parallel to the isobar})\,(p_G - p_8)\,g}{(v_{XG} - \bar{v}_{XG8}) \times 2}. \qquad \text{.........(10)}$$

Some values of $\bar{\xi}_{G8}$ deduced in this way are set out in the last column of the Table on p. 84, ready for use in computing the flux of water. Here $V_{XG}$ has been taken as zero.

For reasons already given, the analogy cannot be extended to the flux of heat. It would be desirable to obtain $\bar{\xi}_{G8}$ directly from observations of humidity.

### Ch. 4/8/6. CONCLUSION ON EDDY MOTION NEAR THE EARTH'S SURFACE

For the sake of economy in arithmetic, an effort has been made to use a thick lower stratum extending to a height of 2 km above sea-level. Whether this plan will be successful or not depends on whether four empirical quantities, namely the stress $\widehat{xh}_G$ for velocity, ≋ and ꝏ for heat, and $\bar{\xi}_{G8}$ for water, can be expressed with sufficient accuracy as functions of the data available in the numerical process. There is some hope of it, but in order to settle the question many more observations, and reductions of existing observations, are required.

If this plan should fail, then an alternative is to be found in taking thinner strata near the surface. If for example we had divisions at heights of about 50 m, 200 m, 800 m, all our processes would be much more exact. Although $\widehat{xh}_G$, ≋, ꝏ and $\bar{\xi}_{G8}$ would still be required for the new lowest layer, since the first metre differs remarkably from the first 50 metres, yet we should then have the velocity and the up-grade of temperature near the surface as aids to the estimation of that very variable quantity, the turbulence. These thinner layers would need to rise and fall with the height of the land above sea, and consequently the dynamical equations in them would need to be furnished with extra terms depending on the slope of the ground. As the land sometimes rises above 2000 metres, these surface layers would project above the $h_8$ horizontal. In avoiding that difficulty we might end by making all the conventional strata rise and fall with the height of the land.

### Ch. 4/8/7. LIST OF PUBLICATIONS ON ATMOSPHERIC EDDIES


1. Kelvin and Tait on work obtainable from an unequally heated solid. Quoted in Preston's *Heat*, 2nd edn., § 341.
2. Guldberg and Mohn. "Studies on the Movements of the Atmosphere." 1876, revised 1883. *Abbe's Translations*, 3rd Collection.
3. Reynolds, O. "On the Dynamical Theory of Incompressible Viscous Fluids and the Determination of the Criterion." *Phil. Trans.* A, 186 (1894), *Papers*, Vol. II. quoted in Lamb's *Hydrodynamics*, IVth edn., § 369.
4. Numerous other references to theoretical papers by Kelvin, Rayleigh, and others in Lamb's *Hydrodynamics*.

5. MARGULES, M. "Energy of Storms," 1901. *Abbe's Translations*, 3rd Collection.

6. EKMAN, W. V. "On the Influence of the Earth's Rotation on Ocean Currents." *Arkiv för Mat. Astr. och Fysik*, Stockholm, 1905.

7. BERNARD, HENRI. "Les Tourbillons Cellulaires." *Ann. Chim. Phys. Paris*, Sér. 7. 23, *J. Phys. Paris*, Sér. 3. 10.

8. ÅKERBLOM, F. "Sur les Courants les plus bas de l'atmosphère." *Nova Acta Reg. Soc. Upsaliensis*, 1908.

9. SANDSTRÖM, *Kungl. Svensk Vet. Akad. Handl.* Bd. 45, No. 10 (1910), or *Bull. Mt. Weath. Obs.*, Vol. 3, part 5 (1911).

10. DINES, W. H. "The Vertical Distribution of Temperature in the Atmosphere and the work required to alter it." *Q. J. R. Met. Soc.*, July, 1913.

11. TAYLOR, G. I. In "Report of Work carried out on S.S. *Scotia*, 1913." Published by H.M. Stationery Office.

12. TAYLOR, G. I. "Eddy Motion in the Atmosphere." *Phil. Trans.* A, 215, pp. 1—26 (1914).

13. HESSELBERG, TH. "Die Reibung in der Atmosphäre." *Met. Zeit.*, May, 1914.

14. HESSELBERG, TH. and SVERDRUP, H. U. "Die Reibung in der Atmosphäre." *Geo. Inst. Leipzig*, Ser. 2, Heft 10 (1915).

15. Numerous other references in Exner's *Dynamische Meteorologie*, §§ 38 to 41.

16. TAYLOR, G. I. "Skin Friction of the Wind on the Earth's Surface." *Roy. Soc. Lond. Proc.* A, Jan. 1916.

17. DOUGLAS, C. K. M. "Weather Observations from an Aeroplane." *J. Scott. Met. Soc.* 1916.

18. SCHMIDT, W. "Der Massenaustausch bei der ungeordneten Strömung in freier Luft und seine Folgen." *Akad. Wiss. Wien*, 1917.

19. SVERDRUP, H. U. "Der nordatlantische Passat." *Geo. Inst. Leipzig*, Ser. 2, Bd. 2, Heft 1 (1917).

20. SVERDRUP, H. U. and HOLTSMARK, J. "Über die Reibung an der Erdoberfläche........" *Geo. Inst. Leipzig*, Ser. 2, Bd. 2, Heft 2 (1917).

21. TAYLOR, G. I. "Phenomena connected with Turbulence in the Lower Atmosphere." *Roy. Soc. Lond. Proc.* A, Vol. 94 (1917).

22. TAYLOR, G. I. "Fog Conditions." *Aeronautical Journal*, Jan.—March, 1917, Vol. XXI, p. 75.

23. DOUGLAS, C. K. M. "The Lapse Line in its Relation to Cloud Formation." *J. Scott. Met. Soc.* 1917.

24. DOUGLAS, C. K. M. "The Upper Air, Some Impressions Gained by Flying." *J. Scott. Met. Soc.* 1918.

25. SCHMIDT, W. "Wirkungen des Luftaustausches auf das Klima und den täglichen Gang der Luft-temperatur in der Höhe." *Akad. Wiss. Wien*, 1918.

26. SVERDRUP, H. U. "Über den Energieverbrauch der Atmosphäre." *Geo. Inst. Leipzig*, Series 2, Bd. 2, Heft 4 (1918).

27. ÅNGSTRÖM, A. "On the Radiation and Temperature of Snow and the Convection of Air at its Surface." *Arkiv för Mat. Astr. och Fysik*, Bd. 13, No. 21 (1918).

28. RICHARDSON, L. F. "Atmospheric Stirring Measured by Precipitation." *Roy. Soc. Lond. Proc.* A, Vol. 96, 1919.

29. JEFFREYS, H. "Relation between Wind and Distribution of Pressure." *Roy. Soc. Proc.* 1919.

30. WHIPPLE, F. J. W. "Laws of Approach to the Geostrophic Wind." *Q. J. R. Met. Soc.*, Jan. 1920.

31. BRUNT, D. "Internal Friction in the Atmosphere." *Q. J. R. Met. Soc.*, April, 1920.

32. RICHARDSON, L. F. "Some Measurements of Atmospheric Turbulence." *Phil. Trans.* A, Vol. 221, pp. 1—28 (1920).

33. RICHARDSON, L. F. "The Supply of Energy from and to Atmospheric Eddies." *Roy. Soc. Proc.* A, 97, July, 1920.

34. SCHMIDT, W. "Über den täglichen Temperaturgang in den unteren Luftschichten." *Met. Zeit.* Heft 3/4, 1920.

35. SCHMIDT, W. "Worauf beruht der Unterschied zwischen See- und Landklima?"

## CH. 4/9. HETEROGENEITY*

### CH. 4/9/0. GENERAL OBSERVATIONS

Not only is the velocity usually turbulent but the temperature is often patchy. An ordinary pen thermograph in a Stevenson screen shows variations of about 1° C on a sunny afternoon. When motoring on a calm evening one can sometimes distinctly feel alterations of warm and cool air not corresponding to height. E. Barkow (*Met. Zeit.* 1915 März), using a thermometer of quick period, found variations of some tenths of a degree occurring at time intervals which could be interpreted as meaning that patches of linear dimension of 14 to 127 metres were moving past the instrument with the speed of the wind. Barkow's observations were made at Potsdam 41 metres above the ground on a wooded hill. The daily weather reports show that in a square on the ground of 200 km in the side the temperature at the screen on an anticyclonic morning may vary as much as 10° C.

For the upper air an estimate of diversity has been extracted from the aeroplane records published in the London Daily Weather Report for 1920, January 1 to October 30. Only those pairs are included which are simultaneous to an hour or less. The aspect of the data to which attention is invited is that the square root of the mean square of the difference of temperature between two stations is not at all proportional to their distance apart, as it would be if the stations lay in a straight line and the horizontal temperature gradients were uniform. The differences may be in part due to observational error. If we ignore this unknown error we get the impression that there are irregular variations in temperature, represented by a standard deviation of about 2° C, in distances comparable with the side of our co-ordinate chequer of $200 \times 200$ km. But, as the stations do not lie in a straight line, the statement cannot be tested rigorously.

* In what follows for brevity I have written "diverse" for "heterogeneous" and "diversity" for "heterogeneity."

| Height. Kilometres above M.S.L. | Number of pairs | Mean difference of temperature | |
|---|---|---|---|
| | | Arithmetic mean. Centigrade | Square root of mean square. Centigrade |
| Baldonnel *minus* either South Farnborough, Upavon or Andover. Distance about 400 kilometres. | | | |
| 3·55 | 12 | +0·8 | 3·7 |
| 3·05 | 16 | 0·2 | 3·1 |
| 2·44 | 15 | −0·3 | 2·6 |
| 1·83 | 17 | −0·3 | 2·5 |
| 1·22 | 17 | 0·0 | 2·9 |
| 0·61 | 17 | 0 2 | 3·1 |
| 0·30 | 14 | −0·1 | 2·0 |
| 0·0 | 16 | 1·1 | 3·9 |
| South Farnborough *minus* Andover. Distance 53 kilometres. | | | |
| 1·83 | 4 | 0·0 | 1·0 |
| 1·22 | 5 | −0·7 | 2·3 |
| 0·61 | 5 | −0·1 | 1·4 |
| 0·0 | 4 | −0·7 | 1·5 |

In telegraphic meteorology the smaller diversities are smoothed-out. The temperature is commonly measured by a thermometer with a bulb large enough to damp out rapid variations, and if the barometer is "pumping," owing to a gusty gale, the observer does his best to estimate the mean reading. In the scheme of this book it is proposed to replace the instantaneous and local values by means over times as long as 3 hours and spaces as great as 100 km horizontally or several km vertically. That this neglect of detail has important consequences we have already seen in the case of velocity. It now remains to enquire whether the diversity of density, temperature, pressure and moisture produces any statistical effects of like importance to eddy-viscosity.

### CH. 4/9/1. NOTATION AND ASSUMPTIONS

The theory of Osborne Reynolds concerning turbulence suggests a mathematical method. The bar-and-dash notation of H. Lamb (*Hydrodynamics*, IVth edn. § 369) allows the theory to be put into a compact form. We begin by supposing that the actual distribution is replaced by a smoothed one, which is indicated by putting a bar over the symbol. Thus for any quantity whatever $A$, we have $A = \bar{A} + A'$ where $A'$ is the deviation. ........................................................................................(1)

The subsequent algebra is based on the following suppositions, no one of which is strictly accurate, but all of which tend to become decent approximations if the diversities are sufficiently numerous and random. These suppositions would become exact if it were possible to choose a smoothing-interval which could be regarded as an infinitesimal for the smoothed distribution, and yet as infinitely large as compared with the diversities.

(i) A smoothed-value is unaltered by a second smoothing.

$$\bar{\bar{A}} = \bar{A}. \qquad \text{......(2)}$$

It follows from (1) and (2) that $\bar{A'} = 0$. ......(3)

(ii) The smoothed-value of the product of a deviation into a smoothed-value vanishes

$$\overline{\bar{A}B'} = 0. \qquad \text{......(4)}$$

(iii) The product of a number of smoothed-values is unaltered by smoothing.

$$\overline{\bar{A}\,.\,\bar{B}\,.\,\bar{C}} = \bar{A}\,.\,\bar{B}\,.\,\bar{C}. \qquad \text{......(5)}$$

(iv) The smoothed-value of a differential coefficient is equal to the differential coefficient of the smoothed-value

$$\overline{\frac{\partial A}{\partial B}} = \frac{\partial}{\partial B}\bar{A}. \qquad \text{......(6)}$$

The *mean of the product* of any two diversified quantities $A$ and $B$ is accordingly

$$\overline{AB} = \overline{(\bar{A}+A')(\bar{B}+B')} = \bar{A}\bar{B} + \overline{A'B'}. \qquad \text{......(7)}$$

And the *deviation of a product* takes the following alternative forms, however large the deviations of the factors may be

$$(AB)' = AB - \overline{AB} \qquad \text{......(8)}$$

$$= \bar{A}B' + A'\bar{B} + A'B' - \overline{A'B'} \qquad \text{......(9)}$$

$$= AB' - A'\bar{B} - \overline{A'B'} \text{ containing } A \text{ without bar or dash} \qquad \text{......(10)}$$

$$= A'B + \bar{A}B' - \overline{A'B'} \text{ containing } B \text{ without bar or dash.} \qquad \text{......(11)}$$

In other parts of this book a good deal of attention is given to integrations with respect to height across strata, and the normal variation with height is then taken into account. But any "normal" variation is essentially a smooth one, so that in those integrations we are concerned with the facts, which we here ignore, that $\bar{\bar{A}}$ is not exactly equal to $\bar{A}$ nor $\overline{\bar{A}\,.\,\bar{B}\,.\,\bar{C}}$ to $\bar{A}\,.\,\bar{B}\,.\,\bar{C}$.

Let us now apply these smoothing operations to each of the chief equations.

### CH. 4/9/2. SMOOTHING THE CHARACTERISTIC GAS EQUATION

Under all circumstances for dry air $p = b\rho\theta$ so that $\overline{p} = b\overline{\rho}\overline{\theta} + b\overline{\rho'\theta'}$. ............(1)

We might give to $b\overline{\rho'\theta'}$ the name "*pressure of heterogeneity*."

Suppose that the standard deviation of $\theta$ were 2°·8 C that is to say 1 % of its mean value, and suppose that the pressure had no local variations. Then the "pressure of heterogeneity" would come to $10^{-4}$ of the mean, that is to 0·1 millibar.

If the moisture is also diversified then $b'$ will not vanish and

$$\overline{p} = \overline{b}\,.\,\overline{\rho}\overline{\theta} + \overline{b}\,.\,\overline{\rho'\theta'} + \overline{\rho}\,.\,\overline{\theta'b'} + \overline{\theta}\,.\,\overline{b'\rho'} + \ldots. \quad \text{..................(2)}$$

The term $\overline{\theta'b'\rho'}$ has been omitted as being of a smaller order.

### CH. 4/9/3. SMOOTHING THE EQUATION OF CONTINUITY OF MASS

In the form given in Ch. 4/2 # 2 the equation is linear in the dependent variables, so that it transforms simply into

$$-\frac{\partial\overline{\rho}}{\partial t} = \frac{\partial\overline{m}_E}{\partial e} + \frac{\partial\overline{m}_N}{\partial n} - \frac{\overline{m}_N\,.\tan\phi}{a} + \frac{\partial\overline{m}_H}{\partial h} + \frac{2\overline{m}_H}{a}, \quad \text{..................(1)}$$

which is of exactly the same form in the smoothed variables as the original equation was in the unsmoothed. The neglect here of the variation $\tan\phi$ is as justifiable as the statement that $\overline{\overline{A}A'} = 0$; for $\tan\phi$ is already smooth.

But if the equation of continuity had been written in terms of velocities in place of momenta-per-volume, then smoothing would have produced a crop of new terms. That is another reason for preferring momenta-per-volume to velocities.

### CH. 4/9/4. SMOOTHING THE EQUATION FOR CONVEYANCE OF WATER

If we begin with the form (Ch. 4/3 # 5)

$$\frac{\partial\mu}{\partial t} = -v_E\frac{\partial\mu}{\partial e} - v_N\frac{\partial\mu}{\partial n} - v_H\frac{\partial\mu}{\partial h} + \frac{D\mu}{Dt}, \quad \text{........................(1)}$$

then there results

$$\left(\frac{\partial}{\partial t} + \overline{v}_E\frac{\partial}{\partial e} + \overline{v}_N\frac{\partial}{\partial n} + \overline{v}_H\frac{\partial}{\partial h}\right)\overline{\mu} = \overline{\left(\frac{D\mu}{Dt}\right)} - \overline{v'_E\frac{\partial\mu'}{\partial e}} - \overline{v'_N\frac{\partial\mu'}{\partial n}} - \overline{v'_H\frac{\partial\mu'}{\partial h}}. \quad \text{........(2)}$$

The first member is a rate of change following in some sense the smoothed motion, but is not quite the same as $\frac{\overline{D}}{\overline{D}t}\,.\,\overline{\mu}$ defined in Ch. 4/8/0 # 1. There are three new terms on the right.

On the other hand if we begin with the equation for $\partial w/\partial t$ which taken from Ch. 4/3 # 6, and written in Cartesians, runs

$$\frac{\partial (\rho\mu)}{\partial t} + \frac{\partial (\mu m_X)}{\partial x} + \frac{\partial (\mu m_Y)}{\partial y} + \frac{\partial (\mu m_H)}{\partial h} = \rho \frac{D\mu}{Dt}, \quad \text{.................(3)}$$

we obtain on smoothing

$$\frac{\partial \bar{w}}{\partial t} + \frac{\partial (\bar{\mu}\bar{m}_X)}{\partial x} + \frac{\partial (\bar{\mu}\bar{m}_Y)}{\partial y} + \frac{\partial (\bar{\mu}\bar{m}_H)}{\partial h}$$

$$+ \frac{\partial}{\partial x}(\overline{\mu' m'_X}) + \frac{\partial}{\partial y}(\overline{\mu' m'_Y}) + \frac{\partial}{\partial h}(\overline{\mu' m'_H}) = \left(\overline{\rho \frac{D\mu}{Dt}}\right). \quad \text{......(4)}$$

Here there are again three new terms, which in this case have a close formal resemblance to the eddy-stress terms in the smoothed dynamical equation, $\mu$ replacing each of the components of $\mathbf{v}$. The new terms are the divergence of a flux which has components $\overline{\mu' m'_X}$, $\overline{\mu' m'_Y}$, $\overline{\mu' m'_H}$; and that is the flux relative to a point which moves so that, relative to the point, $\bar{m}_X = 0$, $\bar{m}_Y = 0$, $\bar{m}_H = 0$. This point corresponds to the "definite" portion of a turbulent fluid as it was defined in Ch. 4/8/0 # 1.

But according to the quite different view of Ch. 4/8 we have been accustomed to express the vertical flux as $-c\partial\mu/\partial h$. ..........................................................(5)

In this statement, as in so many others not in the present Ch. 4/9, there is an implied bar over the $\mu$.

We must also admit that there may possibly be different conductivities in different directions, say

$$c_{XX}, \quad c_{YY}, \quad c_{HH}.$$

It follows, from (5), as thus modified, that

$$c_{XX}\frac{\partial \bar{\mu}}{\partial x} = -\overline{\mu' m'_X}; \quad c_{YY}\frac{\partial \bar{\mu}}{\partial y} = -\overline{\mu' m'_Y}; \quad c_{HH}\frac{\partial \bar{\mu}}{\partial h} = -\overline{\mu' m'_H}. \quad \text{...........(6)}$$

There is a close analogy between these equations and

(viscosity) × (space-rate of mean-shear) = (eddy-shearing-stress).

### CH. 4/9/5. SMOOTHING THE DYNAMICAL EQUATIONS

For the present purpose Cartesian co-ordinates are preferable to polar ones. We get the equations in Cartesians from Ch. 4/4 # 3, 4 by making the polar-coordinate-radius infinite and by replacing $E$ by $X$ and $N$ by $Y$. Only the terms containing products of dependent variables, namely $\frac{\partial}{\partial x}(m_X v_X)$ and the like, produce on being smoothed any additional terms. As Reynolds showed, these additions express the body-force produced by a system of eddy-stress. But, if the density is diverse, the components of stress are not quite of the form $-\rho \, . \, \overline{v'_X v'_X}$ etc., which Reynolds first gave for an incompressible fluid, and which the present writer has elsewhere applied to the atmosphere. On referring to Ch. 4/4 # 3, 4, 5 we see that the exact form of the additional terms is

$$\frac{\partial}{\partial x}\overline{(m'_X v'_X)}+\frac{\partial}{\partial y}\overline{(m'_X v'_Y)}+\frac{\partial}{\partial h}\overline{(m'_X v'_H)} \text{ in the } X \text{ equation,}$$

$$\frac{\partial}{\partial x}\overline{(m'_Y v'_X)}+\frac{\partial}{\partial y}\overline{(m'_Y v'_Y)}+\frac{\partial}{\partial h}\overline{(m'_Y v'_H)} \text{ in the } Y \text{ equation,}$$

$$\frac{\partial}{\partial x}\overline{(m'_H v'_X)}+\frac{\partial}{\partial y}\overline{(m'_H v'_Y)}+\frac{\partial}{\partial h}\overline{(m'_H v'_H)} \text{ in the } H \text{ equation.}$$

Thus in place of the stress $-\rho\,.\,\overline{v'_X v'_X}$ we have strictly $-\overline{m'_X v'_X}$. Furthermore the stress in the $x$ direction on the plane normal to the $y$-axis is $-\overline{m'_X v'_Y}$ which may differ from $-\overline{m'_Y v'_X}$, the stress in the $y$ direction on the plane normal to the $x$-axis. There are thus nine components of eddy-stress in contrast to the six found when the density is not diverse. Silberstein* gives a proof to show that any stress arising from causes which conform to d'Alembert's principle will have only six different components. The question then arises: does smoothing the equation, which expresses d'Alembert's principle, cause it to throw off extra terms which invalidate the aforesaid proof? We have here the suggestion of an analogy to the stresses in magnetised bodies. But the subject will not be pursued further now except to note that the eddy-stresses could be reduced in number from nine to six if we rearranged the dynamical equations, before smoothing them, so as to have in the first $\frac{\partial}{\partial n}(m_E v_N)$ and in the second $\frac{\partial}{\partial e}(m_E v_N)$, and so as to make corresponding changes elsewhere. But then the smoothed equations would have an ugly lack of symmetry.

### CH. 4/9/6. DETACHED CLOUDS AND LOCAL SHOWERS

The scheme proposed in Ch. 4/6 for forecasting cloud was that there would be general condensation, if the mean value of the water-content of a stratum exceeded its saturation value at the mean temperature.

If air is cooled *by expansion* the result of diversity is that the wetter parts of the volume become cloudy before the volume is on the average saturated, and after it has become on the average saturated the drier parts may still remain clear. Diversity thus converts a sudden transition into a gradual one. The more diverse the atmosphere the more gradual the change.

If the air is cooled *by radiation*, the parts containing more water than the average absorb and emit more radiation, so that the question is more complicated than that of a simple expansion.

But certainly the most familiar example of diversity is that connected with cumuli, and here the adiabatic cooling as well as the moisture is diversified. It is known that for cumuli to be formed there must be (i) in the upper air a lapse rate not too stable, (ii) a supply of heat and moisture below. The problem of the statistical mechanics of cumuli deserves further attention.

* *Vectorial Mechanics* (Macmillan & Co.).

## CH. 4/9/7. THE PRODUCTION OF DIVERSITY

The measure of diversity which fits best with our mathematical habits is half the smoothed square of the deviation. For if the deviating quantity were the velocity the aforesaid measure would be the eddying-kinetic-energy-per-mass. We have already a theory of its production and dissipation in connection with the Criterion of turbulence*. Our present task is to find what happens when $v$ is replaced by $\mu$, $\tau$ or other deviating quantities.

Let us consider the diversity of $\mu$, the water per mass. That is to say, we seek to formulate the changes in $\overline{\{(\mu')^2\}}$ the smoothed square of its deviation. The equation for the conveyance of water (Ch. 4/3 # 5) may be written, after multiplication by $\rho$,

$$\rho\frac{\partial\mu}{\partial t} = -m_X\frac{\partial\mu}{\partial x} - m_Y\frac{\partial\mu}{\partial y} - m_H\frac{\partial\mu}{\partial h} + \rho\frac{D\mu}{Dt}, \quad\text{.....................(1)}$$

where $D\mu/Dt$ is zero except in so far as precipitation and molecular diffusion come in.

Now take the deviation of this equation term by term using the formula for the deviation of a product (Ch. 4/9/1 # 10) in part of which the actual $\rho$ and $m$ appear without either bar or dash.

$$\begin{aligned}\rho\frac{\partial\mu'}{\partial t} + \rho'\frac{\partial\bar\mu}{\partial t} - \overline{\rho'\frac{\partial\mu'}{\partial t}} &\\ = -m_X\frac{\partial\mu'}{\partial x} - m_Y\frac{\partial\mu'}{\partial y} - m_H\frac{\partial\mu'}{\partial h}&\\ -m'_X\frac{\partial\bar\mu}{\partial x} - m'_Y\frac{\partial\bar\mu}{\partial y} - m'_H\frac{\partial\bar\mu}{\partial h}&\\ +\overline{m'_X\frac{\partial\mu'}{\partial x}} + \overline{m'_Y\frac{\partial\mu'}{\partial y}} + \overline{m'_H\frac{\partial\mu'}{\partial h}} + \left(\rho\frac{D\mu}{Dt}\right)'. &\quad\text{.....................(2)}\end{aligned}$$

Now multiply by $\mu'$ and bring the terms in the complete $m_X$, $m_Y$, $m_H$ to the first member. Then

$$\begin{aligned}\left(\rho\frac{\partial}{\partial t} + m_X\frac{\partial}{\partial x} + m_Y\frac{\partial}{\partial y} + m_H\frac{\partial}{\partial h}\right)\{\tfrac{1}{2}(\mu')^2\} + \rho'\mu'\frac{\partial\bar\mu}{\partial t} - \mu'.\overline{\left(\rho'\frac{\partial\mu'}{\partial t}\right)}&\\ = -\mu'm'_X\frac{\partial\bar\mu}{\partial x} - \mu'm'_Y\frac{\partial\bar\mu}{\partial y} - \mu'm'_H\frac{\partial\bar\mu}{\partial h}&\\ -\mu'\times(\text{a smoothed quantity}) + \mu'\left(\rho\frac{D\mu}{Dt}\right)'. &\quad\text{..............(3)}\end{aligned}$$

On smoothing this equation various terms disappear, but it is necessary to split $m$ and $\rho$ into their means and deviations. Thus there results

$$\begin{aligned}\left(\bar\rho\frac{\partial}{\partial t} + \bar m_X\frac{\partial}{\partial x} + \bar m_Y\frac{\partial}{\partial y} + \bar m_H\frac{\partial}{\partial h}\right)\{\tfrac{1}{2}\overline{(\mu')^2}\}&\\ +\tfrac{1}{2}\left(\overline{\rho'\frac{\partial\mu'^2}{\partial t}} + \overline{m'_X\frac{\partial\mu'^2}{\partial x}} + \overline{m'_Y\frac{\partial\mu'^2}{\partial y}} + \overline{m'_H\frac{\partial\mu'^2}{\partial h}}\right) + \overline{\rho'\mu'}.\frac{\partial\bar\mu}{\partial t}&\\ = -\overline{\mu'm'_X}.\frac{\partial\bar\mu}{\partial x} - \overline{\mu'm'_Y}.\frac{\partial\bar\mu}{\partial y} - \overline{\mu'm'_H}.\frac{\partial\bar\mu}{\partial h} + \overline{\mu'\left(\rho\frac{D\mu}{Dt}\right)'}. &\quad\text{..................(4)}\end{aligned}$$

* O. Reynolds, *Phil. Trans.* A, 186, p. 123 (1894); L. F. Richardson, *Roy. Soc. Proc.* A, 97 (1920), p. 354.

The first term resembles that which we seek, for the term is $\bar{\rho}$ multiplied by the time-rate of the smoothed square of the deviation of $\mu$, following some sort of mean motion. To see what sort of mean motion, let the co-ordinate axes move so that

$$\overline{m}_X = 0, \quad \overline{m}_Y = 0, \quad \overline{m}_H = 0.$$

This supposition is permissible because the equation from which we began was true for any motion of the axes. But if $\overline{\mathbf{m}} = 0$ then no mass is crossing the co-ordinate planes. That is the same sort of mean motion as the one considered in Ch. 4/8/0 # 1. Let us put a bar over a capital $D$ to denote change following this kind of mean motion.

We have already (Ch. 4/9/4 # 6) arrived at an interpretation of $\overline{\mu' m'_H}$ as the upward flux of water, equal to $-c_{HH}\frac{\partial \bar{\mu}}{\partial h}$. Inserting this and similar terms in (4) there results

$$\bar{\rho}\,\frac{\bar{D}}{\bar{D}t}\{\tfrac{1}{2}\overline{(\mu')^2}\} + \overline{\mu'\rho'}\,.\,\frac{\partial \bar{\mu}}{\partial t} + \overline{\rho'\,\frac{\partial}{\partial t}\,\tfrac{1}{2}\mu'^2}$$

$$= c_{XX}\left(\frac{\partial \bar{\mu}}{\partial x}\right)^2 + c_{YY}\left(\frac{\partial \bar{\mu}}{\partial y}\right)^2 + c_{HH}\left(\frac{\partial \bar{\mu}}{\partial h}\right)^2$$

$$-\tfrac{1}{2}\left\{\overline{m'_X\frac{\partial \mu'^2}{\partial x}} + \overline{m'_Y\frac{\partial \mu'^2}{\partial y}} + \overline{m'_H\frac{\partial \mu'^2}{\partial h}}\right\} + \overline{\mu'\left(\rho\,\frac{D\mu}{Dt}\right)'}. \quad \text{..............(5)}$$

There is no obvious reason why $m'_H$ should be correlated with $\frac{\partial\,(\mu')^2}{\partial h}$, nor why $\rho'$ should be correlated appreciably with $\mu'$ or with $\frac{\partial}{\partial t}\,(\mu')^2$. If these correlations and the corresponding ones in $m'_X$, $m'_Y$ were to vanish, then

$$\bar{\rho}\,\frac{\bar{D}}{\bar{D}t}\{\tfrac{1}{2}\overline{(\mu')^2}\} = c_{XX}\left(\frac{\partial \bar{\mu}}{\partial x}\right)^2 + c_{YY}\left(\frac{\partial \bar{\mu}}{\partial y}\right)^2 + c_{HH}\left(\frac{\partial \bar{\mu}}{\partial h}\right)^2 + \overline{\mu'\left(\rho\,\frac{D\mu}{Dt}\right)'}. \quad \text{...........(6)}$$

This equation has a close formal resemblance to the one originally worked out by O. Reynolds to express the activity of the eddy stresses. We may leave the last term over for consideration in the next section. The equation signifies then, since $c$ is always observed to be positive, that unless the mean distribution of $\mu$ is originally uniform, any turbulence tends to increase the diversity. This one can believe, without the aid of mathematics, after watching the process of stirring together water and lime-juice. But we have obtained mathematically a numerical measure, which could conceivably be tested by observations such as Barkow's*.

Obviously there would be a precisely similar theory in which potential temperature $\tau$ replaced $\mu$ as far as (5). But the supposed vanishing of the correlations which cause the simplification to (6) would need more careful scrutiny.

* E. Barkow, "Über die thermische Struktur des Windes." *Met. Zeit.* 1915, März.

## Ch. 4/9/8. THE DISSIPATION OF DIVERSITY

This is carried out presumably by the molecular agitation. The term $\overline{\mu'\left(\rho\,\frac{D\mu}{Dt}\right)'}$ in the equation at the end of the last section has been put there as a reminder of the existence of dissipation. It now remains to examine the term in detail. Since the molecular velocities are enormously greater than the relative molar velocities of two portions separated by a distance equal to the mean-free-path, we may now ignore the molar velocity and treat the molecular dissipation as if it were proceeding in still air. For the effect of eddies is represented by the other terms in the aforesaid equation.

The usual statement* is that the vapour density $w$ tends to become uniform according to the equation

$$\frac{\partial w}{\partial t}=\kappa\left(\frac{\partial^2}{\partial x^2}+\frac{\partial^2}{\partial y^2}+\frac{\partial^2}{\partial h^2}\right)w, \qquad\text{.............................(1)}$$

where $\kappa$ is the molecular diffusivity and equal to $0{\cdot}2\ \text{cm}^2\,\text{sec}^{-1}$ for the interdiffusion of water vapour and air.

Smooth equation (1) and subtract the smoothed form from the original. So we get an equation in the deviation

$$\frac{\partial w'}{\partial t}=\kappa\left(\frac{\partial^2}{\partial x^2}+\frac{\partial^2}{\partial y^2}+\frac{\partial^2}{\partial h^2}\right)w'. \qquad\text{.............................(2)}$$

Multiply by $w'$

$$\frac{1}{2}\frac{\partial}{\partial t}\,w'^2=\kappa\,.\,w'\left(\frac{\partial^2}{\partial x^2}+\frac{\partial^2}{\partial y^2}+\frac{\partial^2}{\partial h^2}\right)w'=\kappa\,.\,w'\nabla^2 w'. \qquad\text{.................(3)}$$

Then because, by the rule for differentiating a product,

$$\frac{\partial}{\partial x}\left(w'\,\frac{\partial w'}{\partial x}\right)=w'\,\frac{\partial^2 w'}{\partial x^2}+\left(\frac{\partial w'}{\partial x}\right)^2, \qquad\text{.............................(4)}$$

on rearranging it follows that

$$w'\,\frac{\partial^2 w'}{\partial x^2}=-\left(\frac{\partial w'}{\partial x}\right)^2+\frac{1}{2}\,\frac{\partial^2}{\partial x^2}\,w'^2. \qquad\text{.............................(5)}$$

Inserting (5) together with similar expressions in the other two co-ordinates into (3) the latter becomes

$$\frac{1}{2}\frac{\partial}{\partial t}\,w'^2=\kappa\,.\left\{\tfrac{1}{2}\nabla^2 w'^2-\left(\frac{\partial w'}{\partial x}\right)^2-\left(\frac{\partial w'}{\partial y}\right)^2-\left(\frac{\partial w'}{\partial h}\right)^2\right\}. \qquad\text{.................(6)}$$

Now smooth this equation in order to obtain one in $\overline{(w'^2)}$ the mean square of the diversity. It reads

$$\frac{\partial}{\partial t}\,\overline{(w'^2)}=\kappa\,.\left[\nabla^2\,\overline{(w'^2)}-2\,\overline{\left\{\left(\frac{\partial w'}{\partial x}\right)^2+\left(\frac{\partial w'}{\partial y}\right)^2+\left(\frac{\partial w'}{\partial h}\right)^2\right\}}\right]. \qquad\text{...........(7)}$$

* Jeans' *Dynamical Theory of Gases*, 2nd edn., eqn (868).

The term $\nabla^2(\overline{w'^2})$ on the right represents a diffusion of mean-square-diversity from regions where it is large to those where it is small. The coefficient of this diffusion is $\kappa$ just the same as for the diffusion of $w$. But, as the rate of space-variation of the smoothed quantity $(\overline{w'^2})$ is usually very much smaller than the rate of space-variation of the unsmoothed $w$, the curious diffusion of $(\overline{w'^2})$ is likely to be of minor importance. We must look for the main effect in the second term on the right of (7) for in it the space-variation is taken before the smoothing. This term is composed of squares. Its sign is such as to correspond always to a decrease of the diversity. It may be contrasted with a rather similar term in Ch. 4/9/7 # 5, which has however the opposite sign and which contains the space rates of a smoothed quantity $\bar{\mu}$ in place of the deviation $w'$. The term in Ch. 4/9/7 # 5 expresses the production of diversity, when an atmosphere, non-uniform in its smoothed distribution, is stirred; the present term expresses the dissipation of the diversity so produced. The two equations could be reduced to comparable quantities, either $\mu$ or $w$, if density were independent of time and place. For then we should have, for the extra term in Ch. 4/9/7 # 5,

$$\overline{\mu'\left(\rho\frac{D\mu}{Dt}\right)'} = \tfrac{1}{2}\rho\frac{D}{Dt}(\overline{w'^2})$$

$$= \frac{\kappa}{\rho}\left[\tfrac{1}{2}\nabla^2(\overline{w'^2}) - \overline{\left\{\left(\frac{\partial w'}{\partial x}\right)^2 + \left(\frac{\partial w'}{\partial y}\right)^2 + \left(\frac{\partial w'}{\partial h}\right)^2\right\}}\right] \quad \text{..........(8)}$$

This expression, although instructive, must be regarded as only a stage on the way to a working theory, for we have as yet no theoretical way of finding

$$\frac{\partial w'}{\partial x}, \quad \frac{\partial w'}{\partial y}, \quad \frac{\partial w'}{\partial h}.$$

Let us now examine the hypotheses. If the temperature distribution is variegated, will it be the vapour density which tends to become uniform? The ordinary theory of distillation proceeds from the assumption that it is not $w$ but the vapour pressure $p_w$ which tends to uniformity horizontally. In the vertical, according to Dalton's law, the tendency of diffusion alone is not towards uniformity of either $w$ or $p_w$ but to a state such that $\partial p_w/\partial h = -gw$. Then again the diffusion coefficient $\kappa$ which we have assumed to be a constant, really increases with temperature, and if temperature is diverse that increase should be taken into account. But to bring in all these corrections would make the equations very elaborate and would perhaps obscure their main features.

The dissipation of temperature must follow very similar lines, $\theta$ replacing $w$, and the molecular thermal conductivity replacing $\kappa$.

## Ch. 4/10. BENEATH THE EARTH'S SURFACE

### Ch. 4/10/0. GENERAL

The atmosphere and the upper layers of the soil or sea form together a united system. This is evident since the first metre of ground has a thermal capacity comparable with 1/10 that of the entire atmospheric column standing upon it, and since buried thermometers show that its changes of temperature are considerable. Similar considerations apply to the sea, and to the capacity of the soil for water.

As it will not do to neglect the changes in the land or sea, two courses are open:

(i) The variables expressing the temperature in the sea or land might, at least in imagination, be eliminated. The result of the elimination would be to yield a complicated boundary-condition for the atmosphere.

(ii) A forecast for the land and sea might be attempted concurrently with that for the air. Let this be the ideal which we here set before us. One reason for this choice is that a forecast for the soil would itself be of value to agriculturalists.

The quality of the forecast might range from a mere use of the normal variables for the time of day and year, to a thorough treatment by finite differences. Possibly a combination of the two may eventuate.

We have already regarded the earth's surface in Ch. 4/8/3, 4, 5, 6, from above, now let us look at it from below. Ultimately these two points of view must be combined.

In this connection there are three principal varieties of surface: the sea, bare earth, and earth covered by vegetation*. It will be convenient to consider these separately and afterwards to attempt to form average constants for our horizontal chequers of about 200 km square, by reference to the relative amounts of the three kinds of surface in each chequer; due regard being paid to the season and to the customary times of ploughing, harrowing, sowing and the like.

The changes in the soil may be described by two differential equations, one for the conduction of heat, the other for the transference of water. It is intended to treat these equations by finite arithmetical differences. The soil must accordingly be divided into conventional strata. Let $z$ be the depth reckoned positive downwards from the surface of the soil or sea. It will be remembered that $h$ the height is reckoned positive upwards, and always from mean sea-level. At what depths $z$ shall we make the divisions between the conventional strata? Well on referring to Rambaut's† observations of temperature in the Oxford gravel, it is seen that at a depth of one or two metres the temperature depends simply on the time of year. The more rapid oscillations, associated with the passage of cyclones, scarcely penetrate to these depths. In other words it will suffice to consider a layer of soil having a thermal capacity comparable with that of the atmosphere.

* And next in interest perhaps snow and rock.

† A. A. Rambaut, *Phil. Trans.* A, Vol. 195, 1901.

Again, in the first few centimetres below the surface the down-grade of temperature is often steep so that a detailed treatment is required. Whereas, at a depth of a metre it would be wasteful to keep account of the temperatures at heights differing by so little as a centimetre. Accordingly it is proposed to change the independent variable from the depth $z$ to some function of $z$, which is called $j$, and which, when divided equally, gives thin conventional strata near the surface, thick ones lower down. The choice of the particular function is arbitrary. A convenient form would be

$$j = \log_e (z+1). \qquad (1)$$

Giving to $j$ the values 0, 1, 2, 3, 4, 5 in succession we get the following depths in centimetres as the boundaries of the conventional strata: 0; 1·72; 6·39; 19·1; 53·6; 147·5. To transform the differential equations we must substitute

$$\frac{1}{(z+1)} \frac{d}{dj} \text{ for } \frac{d}{dz}, \qquad (2)$$

and

$$\frac{1}{(z+1)^2} \left\{ \frac{d^2}{dj^2} - \frac{d}{dj} \right\} \text{ for } \frac{d^2}{dz^2}. \qquad (3)$$

Now the equations for the motion of water and heat are interlinked by various terms, so that it is convenient to use the same conventional strata for both. It is accordingly intended to make the substitutions (2) and (3) in all the equations for the soil. For simplicity however the independent $z$ is written in what follows.

## CH. 4/10/1. THE SEA

A layer of sea 2·4 metres deep has a thermal capacity equalling that of the whole of a dry atmosphere standing upon it.

The temperature of the sea surface is much steadier than that of the land. We might, for some purposes, assume the sea surface to have the mean temperature observed at the given place and date in previous years, by consulting, for example, the monthly Charts of the Atlantic Ocean*, published by the Meteorological Office. The error which we should thereby commit may be represented by a standard deviation. For areas on the Atlantic Ocean measuring 2° in longitude by 1° in latitude this standard deviation would be about 1° C in the eastern North Atlantic, while on the western side, where the gulf-stream is narrow but not quite fixed in its course, the standard deviation might amount to 5° C. Hann (*Meteorologie*, III edn. p. 65) has computed the daily variation of temperature in the sea and in the air immediately over it. The average difference between the two, at any given time of day, rarely exceeds 1° C. So that the temperature of the air in contact with the eastern North Atlantic, is now predictable to 2° C.

* Helland-Hansen und Nansen, *Temperatur-Schwankungen des Nordatlantischen Ozeans und in der Atmosphäre*, Kristiania, Jacob Dybwad, pp. 50—84. English Edition published by Smithsonian Institution, 1920.

If a more exact value of the surface temperature had to be predicted, the system of prediction would have to take into account:

(i) The long-wave radiation which is given off or absorbed, on account of the great opacity of water, only by the uppermost centimetre*.

(ii) The solar radiation which penetrates to much greater depths. O. Krümmel† gives a table of absorption coefficients ranging from 0·01 per metre for a wave length of $450\mu\mu$ to 0·30 per metre for a wave length of $660\mu\mu$.

(iii) The turbulence in the sea which depends on the down-grade of "potential density." The problem here is more complicated than that of the atmosphere on account of the variation of salinity with depth. Presumably the shearing of horizontal velocity must also be a prime cause of turbulence in the sea as it is in the atmosphere; and the shearing will in turn depend on the wind.

(iv) The turbulence and temperature of the air which convey heat to or from the surface. We have already discussed this in Ch. 4/8 and will not go into it further, except to note that the interaction of sea and air must be treated as one problem.

(v) The ocean currents, their dynamics, and the heat which they convey.

Presumably these five principles could be put into a scheme of prediction such as has been worked out for the atmosphere in this book. It may come to that, but let us hope that something simpler will suffice.

**Interaction of Sea and Air.** If the sea is colder than the air at deck-level, the turbulence in the air becomes small and very little interchange of heat goes on. A low eddy-viscosity under these circumstances was observed by G. I. Taylor during the Scotia cruise‡. Prof. Helland-Hansen—to whom I am indebted for much of the information in this section—tells me that when the sea is colder than the air at deck-level there may sometimes be a decided change of relative humidity in the first few metres above the water, indicating a protecting film so thin as that. The warmer air tends, if anything, to diminish turbulence in the colder sea and thus to cut off the flow of heat on that side also.

On the other hand if the sea is warmer than the air at deck-level, as it tends to be in winter, there is much turbulence in the air, which tends to have no up-grade of potential-temperature nor of water-per-mass. This state of affairs is illustrated by observations in the North Atlantic trade wind and by the eddy-viscosity deduced therefrom by Sverdrup‡. The sea, being cooled above, tends also to become turbulent, if the down-grade of salinity permits.

Thus the sea surface acts as a leaky valve, which allows heat readily to flow upwards, but hinders its descent.

More observations of turbulence over the sea and within it are much needed. The smoke method§ might be suitable.

* Winkelmann, *Handb. der Physik*, 2 Aufl. Bd. III. p. 342. † *Handbuch der Oceanographie*, I. 389.

‡ *Vide* p. 72 above. § *Phil. Trans.* A, Vol. 221, pp. 5—26 (1920).

## CH. 4/10/2. THE BARE SOIL

In winter great areas of arable land are bare. As in the case of the sea, we require to forecast the temperature and humidity of the air in immediate contact with the surface of the soil.

A wealth of observational material has been brought together by Warington in his *Physical Properties of Soil* (Oxford Press) from which I have drawn freely.

**The Motion of Water in Soil.** The evaporation from bare land has been measured by the Rothamsted drain gauges* with the remarkable result that in the six winter months, October to March, the evaporation is practically identical with that from a water surface, as measured by Greaves*, near London. Provided that we may assume the atmospheric conditions above Mr Greaves' gauge to have been the same as at Rothamstead, we may say: during that portion of the year in which any considerable fraction of the land is bare, the bare surface is so wet as to saturate the air in contact with it.

This rough generalization might suffice, but it seems desirable also to develop the general equation applicable to summer as well as winter. In this equation it will be convenient to regard the soil as a continuous medium, that is to say the "infinitesimal" differences of the coordinates must be large compared with the soil particles and yet small enough to give a good representation of the variation of moisture with position. The theory of percolation in saturated soils has already been put in mathematical form by Boussinesq†. For unsaturated soil, such as is often found near the surface, I have not found the equation anywhere. It may be developed from the ideas of Briggs, according to whom the water in the unsaturated soil may be typified by a waist-shaped piece partly filling the crevice between two spherical soil particles. If the amount of water in the waist decreases, the curvature of its surface becomes more strongly concave and the negative pressure in the water is thereby increased. If the water in all the crevices is continuous with itself, the pressure will tend to become everywhere equal, in the absence of gravity. Denote the mass of water per volume of soil by $w$. From the point of view of our large infinitesimals, the pressure in the water will be a single valued continuous function of position, and will depend on $w$. Denote the pressure in the water by $\Psi(w)$. The form of the function $\Psi$ can be determined, for any particular soil, by experiments similar to those of Loughridge. He put air-dried soil into vertical metal pipes closed at their lower ends by muslin. Water was supplied through the muslin and rose by capillarity. From the mode of its entry this water was probably all continuous. When the steady state was established, samples taken at different heights were weighed, dried and reweighed. This gave $w$. At the same time $\Psi(w)$ is equal to $g$ times the height of the sample above the free

* Warington, *loc. cit.* p. 109.

† *Journal de Mathématiques*, Paris, 1904, pp. 1 and 363. See also L. F. Richardson, *Proc. Roy. Soc. Dublin*, 1908, May. F. H. King, *Irrigation and Drainage*, Macmillan, 1882. J. M. K. Pennick, "Over de Bewegung van Grundwater," *De Ingenieur*, 29 July, 1905.

water-level outside the tube, as the density of water is unity. In the more general case, when there is not equilibrium, $\frac{\partial \Psi(w)}{\partial z} - g$ will be the unbalanced pressure gradient producing a flow upwards. Since the flow is in very narrow channels it is non-turbulent. It is therefore proportional to the unbalanced pressure gradient, and varies inversely as the viscosity. The flow also depends on the dimensions of the channels. These are bounded in part by the water-air surfaces, with the result that the conductance of the channels diminishes rapidly with diminishing $w$. Denote the conductance of the channels connecting the opposite faces of a centimetre cube of soil by ϣ, which is a Coptic letter pronounced "shai." Then the flow of water upwards is, in $cm^3 \, sec^{-1}$ per horizontal $cm^2$,

$$\text{ϣ} \left\{ \frac{\partial \Psi}{\partial z} - g \right\}. \qquad \text{.....................................(1)}$$

So the rate at which water accumulates to any point by creeping is

$$\frac{\partial w}{\partial t} = \frac{\partial}{\partial z} \left[ \text{ϣ} \left\{ \frac{\partial \Psi}{\partial z} - g \right\} \right]. \qquad \text{.....................................(2)}$$

The conductivity ϣ could, I think, be determined experimentally as a function of $w$ by means of apparatus similar to that which gives the pressure $\Psi$. For let a uniform slow current of water be established, either down the tube by dropping water on the top, or up the tube by promoting evaporation at the top. Let the current be measured and be maintained constant until a steady state has become established throughout the tube; and then let $w$ be determined at a series of heights by drying samples. From the distribution of $w$, that of the pressure $\Psi(w)$ could be found, since the form of $\Psi$ is known from the previous experiment. Taking the gradient of the pressure $\Psi(w)$ and inserting it along with the constant flow in (1) we should have the conductivity ϣ for a series of values of $w$.

$\Psi$ and ϣ depend on the temperature in known ways; if, as usually happens, the soil is not uniform, they also depend on the depth. The possibility of the existence of isolated water must not be forgotten, although no simple way of treating it mathematically may be to hand. Nor must the loss of water by surface drainage, root-holes and fissures be overlooked.

In the unsaturated soil, simultaneous with the creeping of liquid water, there goes on a distillation of vapour. In summer there is often a thin coat of dry soil on the surface; and the distillation through this may dispose of quantities of water which cannot be neglected. Where liquid water is adhering to the soil particles the vapour density will be saturated for the temperature and for water surfaces of the existing curvature. The curvature in turn depends on $w$, so that we may write for the vapour density $F(w, \theta)$. Soil which feels dry to the hand may still contain considerable quantities of occluded, adsorbed or "hygroscopic" water. From this cause $w$ may amount to as much as 0·05 grm $cm^{-3}$ or more*. Now the adsorption of gases by solids has been the subject of many investigations†, which have shown that an equilibrium state is

* Warington, *loc. cit.* p. 60. † Winkelmann, *Handb. der Physik*, 2 Aufl. Bd. II. p. 1524.

reached in which the mass adsorbed depends on the temperature and on the density of the gas. So we may write here: vapour density $= F(w, \theta)$, thus prolonging the function $F$ from the region of liquid water into that of adsorbed water. For such different substances as charcoal *, glass * and peat dust † it has been found that $F$, in the region of adsorption, is more or less proportional to the mass adsorbed, and decreases rapidly with increasing temperature.

The rate of motion of vapour will be proportional to the gradient of density $\frac{\partial F}{\partial z}$ and also to the porosity which we may denote by ϫ, a Coptic letter pronounced "janja." The porosity will be diminished if the passages are partly choked with water, so that it should be written ϫ $(w)$, a function of $w$. It may be defined so that the mass of water distilling upwards per unit of time and of horizontal area is

$$ϫ \frac{\partial F}{\partial z}. \qquad \text{.......................................(3)}$$

And the rate of accumulation of water by distillation is therefore

$$\frac{\partial w}{\partial t} = \frac{\partial}{\partial z}\left[ϫ \frac{\partial F}{\partial z}\right]. \qquad \text{.......................................(4)}$$

I have attempted to measure the porosity ϫ of peat dust to water vapour in the following way. A little water was put into the bottom of a test-tube. Above the water-level a partition of brass wire gauze and linen was fixed in the tube. The dust was packed in the tube above the partition. The tube was stood upright in a desiccator, and was weighed at intervals during several weeks, until the rate of loss became constant. The experiment was repeated with a column of peat dust of a different length. It became apparent that a length of 5 cms of peat dust allowed the water vapour to diffuse almost as freely as in a tube containing only still air. The diffusivity in still air is well known. To have obtained a good measurement special attention would have had to be paid to the constancy of the resistance offered by the open end of the tube.

Adding the accumulation by creeping, given by (2), to that by distillation, given by (4), we obtain the total rate of accumulation of water in an unsaturated soil, as follows:

$$\frac{\partial w}{\partial t} = \frac{\partial}{\partial z}\left[ϣ \left\{\frac{\partial ϒ}{\partial z} - g\right\} + ϫ \frac{\partial F}{\partial z}\right]. \qquad \text{.........................(5)}$$

If we were to add the two terms representing the horizontal components of motion, equation (5) would then apply to saturated soils as well as to unsaturated ones; for when saturation occurs ϫ vanishes, ϣ is the porosity to water, and the pressure ϒ is no longer a function of $w$ but depends on position instead.

Natural soils are rarely uniform so that in (5) ϒ, ϣ, $F$, ϫ will depend on height. Temperature reduces the capillary tension, and therefore ϒ, by 0·002 of itself per 1° C. The porosity ϣ varies rapidly with the temperature in a known manner, because it is proportional to the reciprocal of the viscosity.

* Winkelmann, *loc. cit.* † Unpublished experiments by the author.

The flow of water across the surface $z=0$ depends on precipitation and evaporation, and can be calculated at the initial instant, at which $w$ is given, according to the methods of Ch. 4/6, Ch. 4/8/5, Ch. 4/9/6. Then the approximate distribution of $w$ after $\delta t$ can be found from (5).

**The Motion of Heat in Soil.** Turning now to the temperature equation, numerous researches* have shown that it is, at least approximately, of the form

$$\frac{\partial \theta}{\partial t} = s \frac{\partial^2 \theta}{\partial z^2}, \quad \text{.........................................(6)}$$

where the diffusivity $s$ is of the order of 0·008 cm² sec⁻¹. The treatment of this equation by arithmetical differences has been illustrated by the present writer‡. Callendar* and Pott† have found that the diffusivity $s$ is much greater for wet soil than for dry, so that we must regard $s$ as a function of $w$. If the soil is stratified $s$ will also depend upon $z$, or rather it will be necessary to regard the conductivity and thermal capacity separately as functions of $z$, so that, in place of (6) we have

$$\frac{\partial \theta}{\partial t} = \frac{1}{u} \frac{\partial}{\partial z} \left( k \frac{\partial \theta}{\partial z} \right), \quad \text{.........................................(7)}$$

where $u$ is the thermal capacity of unit volume and $k$ is the conductivity.

We may improve upon equation (7) by taking account of the disappearance of heat wherever water is evaporating. The rate of condensation is given by (4). Multiplying the rate-of-condensation by the latent-heat-of-evaporation and dividing the result by the heat-capacity-of-unit-volume-of-the-soil, we get a measure of the rate of temperature due to condensation, in the form

$$\frac{4{\cdot}18 \times 10^7 \{598 - 0{\cdot}60\,(\theta - 273^\circ)\}}{u} \frac{\partial}{\partial z} \left\{ \varkappa \frac{\partial F}{\partial z} \right\}, \quad \text{..................(8)}$$

where $u$ is the number of ergs required to raise the temperature of one cubic centimetre of soil by one degree. For sand, for example, if the grains are of pure quartz and occupy half the volume of the soil, then $u$ will be

$$\tfrac{1}{2}\,(\text{density}) \times (\text{specific heat of quartz}) = 0{\cdot}23 \times 4{\cdot}18 \times 10^7,$$

when the sand is quite dry; and, when water is present $u = (0{\cdot}23 + w) \times 4{\cdot}18 \times 10^7$, with an upper limit of $u = 0{\cdot}73 \times 4{\cdot}18 \times 10^7$, when the sand is saturated. On the other hand, if the soil contains fine powders which attract water, as peat dust does, both the specific and the latent heat of the water may be expected to differ from their normal values.

Callendar§ has found that heat is also carried through soil by percolating water. If, neglecting any rapid motion through root-holes and fissures, we assume that the water is moving slowly between the solid particles, then the temperature of the water at any point will be the same as that of the particles, so that the symbol $\theta$ will serve

* *Encyc. Britt.* XI edn. Vol. 6, pp. 893 to 894; A. A. Rambaut, *Radliffe Observations*, Vol. LI. Oxford; Hann, *Lehrbuch der Meteorologie*, 3rd edn. pp. 48 to 55.

† Warington, *loc. cit.* pp. 161 to 171.

‡ *Phil. Trans. Roy. Soc. London*, A, Vol. 210, p. 313.

§ *Encyc. Britt.* XI edn. Vol. 6, pp. 893 to 894.

for both. The upward flow of water is given by (1) in grm $\sec^{-1}\,\mathrm{cm}^{-2}$. Multiplying this flow by $\theta$ and by the specific heat of water, which is $4{\cdot}2\times10^7$ ergs $\mathrm{gram}^{-1}$ $\mathrm{degree}^{-1}$ we get the upward flow of heat in so far as it depends on the motion of water. Differentiating the flow of heat with respect to $z$, and dividing the result by the thermal capacity of unit volume of the soil, we get the rate of rise of temperature, due to the flow of water, in the form:

$$\frac{1}{u}\frac{\partial}{\partial z}\left[\theta\times4{\cdot}2\times10^7\times\mathfrak{y}\left\{\frac{\partial\Psi}{\partial z}-g\right\}\right]. \qquad \text{......................(9)}$$

Collecting now the three parts of the temperature-rise due severally to conduction, condensation and convection, we have from (7), (8), and (9)

$$\frac{\partial\theta}{\partial t}=\frac{1}{u}\frac{\partial}{\partial z}\left(k\frac{\partial\theta}{\partial z}\right)+\frac{\text{latent ht. of evapn.}}{u}\frac{\partial}{\partial z}\left[\varkappa\frac{\partial F}{\partial z}\right]$$
$$+\frac{4{\cdot}2\times10^7}{u}\frac{\partial}{\partial z}\left[\theta\,.\,\mathfrak{y}\left\{\frac{\partial\Psi}{\partial z}-g\right\}\right]. \qquad \text{......(10)}$$

This is the complete temperature equation. The latent heat of fusion of ice in the soil may be regarded as a very large increase in the thermal capacity in the immediate neighbourhood of the freezing point.

The flow of heat across the earth-air surface, $z=0$, depends on precipitation, evaporation and radiation, and can be calculated at the initial instant at which $\theta$ is given, according to the methods of Ch. 4/6, 4/7, 4/8/4, 4/9/6. The approximate distribution of temperature after $\delta t$ can then be found from (10).

Now as to the treatment of equations (5) and (10) by **finite differences.** The combinations of $\theta$ and $w$ which occur in these equations, mostly require $\theta$ to be tabulated at the same depths as $w$. The last term in (10) is an exception to this statement, but as we cannot have it both ways, this exception has been ignored. Again, in order to calculate the radiation and the evaporation from bare soil, it is desirable that $\theta$ and $w$ should be tabulated actually at the surface, rather than as mean values for the upper layer which is 1·7 centimetres thick. These principles have been embodied in the corresponding forms in Ch. 9.

### CH. 4/10/3. EARTH COVERED BY VEGETATION

Leaves, when present, exert a paramount influence on the interchanges of moisture and heat. They absorb the sunshine and screen the soil beneath. Being very freely exposed to the air they very rapidly communicate the absorbed energy to the air, either by raising its temperature or by evaporating water into it. The amounts evaporated by crops are considerable; barley for instance has been estimated to give off the equivalent of one millimetre of rain per day during its growth. A portion of rain, and the greater part of dew, is caught on foliage and evaporated there without ever reaching the soil. Leaves and stems exert a retarding friction on the air, so that within a forest it is very doubtful whether we can still estimate the "eddy diffusion" of entropy and moisture, from observations of the change of wind velocity with height, as G. I. Taylor and W. Schmidt have done in the free air (p. 76 above).

For a numerical treatment I have referred to a series of papers by Brown and Escombe*, and Brown and Wilson*. Their observations are built around the following framework of theory. The air in the intercellular spaces of the leaf is supposed to be saturated in equilibrium with pure water at the temperature of the leaf. The vapour diffuses out through the stomata at a rate proportional to difference of vapour density between inside and outside. The transpiration also depends on the size, shape and number of the stomata. By analogy with electric conduction, the rate of transpiration may be said to be inversely as the "resistance" of the stomata. The resistance consists of two parts in series. The larger part is due to the air in the intercellular spaces and in the constricted passages of the stomata; this part is unaffected by wind. The remainder of the resistance is due to the air immediately outside the stomata and is reduced by wind. Thus the wind can only alter the transpiration in a limited range. The openings of the stomata are sensitive to sunlight. The careful experiments set out on pages 106 to 109 of Brown and Escombe's paper show that a reduction of sunshine to one-half of its intensity may increase the resistance by perhaps 10 °/。 in the case of *Helianthus annus*. The initial sunshine averaged about 0·4 cal $cm^{-2}$ $min^{-1}$.

The temperature of the leaf is above or below that of the surrounding air according as the radiation received by the leaf is greater or less than the energy absorbed by the water in evaporating. The difference between these two sets of energy, divided by the emissivity of the leaf for heat, gives the temperature difference between the leaf and the air. The emissivity here is the total rate of loss of heat to the surrounding air by conduction, convection and radiation jointly, for 1° C temperature difference. Brown and Wilson measured the emissivity and found that it was a linear function of the wind velocity; thus for *Helianthus multiflorus*

$$\text{emissivity in cal cm}^{-2}\text{ min}^{-1}\text{ (degree C)}^{-1} = 0{\cdot}015 + 0{\cdot}010 \times \text{(velocity in metres sec}^{-1}\text{)}.$$

The emissivity is so great that although, in the absence of any loss of energy, sunshine would raise the temperature of a particular leaf at the rate of 35° C per minute, yet the temperatures of the leaves observed by Brown and Escombe† never differed by more than 2° C from that of the air, and were much nearer to the dry-bulb temperature than to that of the wet-bulb.

To put this in a formula: Let the rate of loss of water from a leaf be denoted by ♂, then

$$♂ = K\{F(\theta_{\text{leaf}}) - w_{\text{air}}\}. \quad \text{.................................(1)}$$

Here $K$ is the conductance of the stomatal openings and $F(\theta)$ is the saturated vapour density at $\theta$.

* *Roy. Soc. Lond. Proc.* B, Jan. 1905.

† *Loc. cit.* Tables VIII. and X.

In (1) we must substitute

$$\theta_{\text{leaf}} = \theta_{\text{air}} + \frac{(\text{radiation}) \times (\text{absorptivity}) - \not{c} \times (\text{latent heat of evapn.})}{\text{emissivity}}, \quad \ldots(2)$$

and then ȼ appears on both sides of the equation.

However, to a first approximation, we may take the dry bulb temperature as equal to the leaf temperature so that

$$\not{c} = K\{F(\theta_{\text{air}}) - w_{\text{air}}\}. \ldots\ldots\ldots\ldots\ldots\ldots\ldots\ldots\ldots(3)$$

The equations (1), (2), (3) apply to a single leaf. We require the corresponding expressions for a mass of foliage. The chief difference will be the large increase in the conductance $K$. We can estimate $K$ for a crop from the following considerations. The total water transpired by a crop* is from 250 to 700 times its dry weight. For example for barley it is at the average of about $10^{-6}$ gram per $\text{cm}^2$ of land surface per second. The average increment of vapour density, which would just saturate the air at the temperature of the dry bulb, could be determined by keeping a recording psychrometer in a barley field. Taking this increment provisionally as of the order of $5 \times 10^{-6}$ gram $\text{cm}^{-3}$, it follows that $K$ is of the order of

$$10^{-6} \div 5 \times 10^{-6} = \tfrac{1}{5}\ \text{cm}^3\ \text{sec}^{-1} \text{ per cm}^2 \text{ of land surface.}$$

In Brown and Escombe's experiments the leaves were supplied with abundant water. If the amount of water in the soil at the level of the roots falls below a certain small amount, the leaves lose their turgescence and transpiration diminishes. This critical amount of water has been determined by Heinrich †. It is, for example, about $w = 0{\cdot}02$ for a coarse sandy soil, $w = 0{\cdot}1$ for a sandy loam. Now $w$ is already one of our dependent variables, so its effect on transpiration can be taken into account.

**A Conventional Film of Vegetation.** In order to bring (1), (2), (3) into connection with the other equations of the atmosphere, we must take the temperature and vapour density of the air surrounding the leaves as dependent variables. Unfortunately it is not possible to estimate these quantities, even approximately, from the known instantaneous mean values for the stratum which extends up to $h = 2$ kilometres; because the estimate would be spoilt by the steep and changing temperature gradients near the surface. For example, at Eskdalemuir, E. H. Chapman found gradients of $\pm 2^\circ$ C per 10 metres near the earth. It is therefore necessary to introduce another conventional layer extending from the earth to the top of the foliage. It is called hereafter the "vegetation film" and its lower and upper limits will be denoted respectively by the subscripts $G$ and $L$. This layer or film might be treated on the same general plan as are the conventional strata, with the distinction that the air must be considered as having its opacity and its thermal capacity increased by the leaves mixed with it, and that the friction does not vanish when $\frac{\partial v}{\partial h}$ vanishes.

* Various observers quoted by Russell in *Soil Conditions and Plant Growth* (Longmans, Green & Co.), p. 28.

† Warington, *loc. cit.* p. 64.

But such an elaborate treatment is hardly necessary. For the air in the vegetation is dragged along by friction with the air above. In other words, the frictional terms in the dynamical equations are so large that the other terms may be neglected. To the contrary it might be objected that on a clear night there are often rivers of cold air running downhill near the surface. But these are after all local currents of small velocity, so that, in dealing with an area as large as $200 \times 200$ km$^2$ no great error would be committed if their average were taken as zero.

In Ch. 8 we shall again discuss the conventional treatment of the film.

### CH. 4/10/4. SNOW AND ICE

The chief peculiarities of a snow surface are perhaps its large reflection of solar radiation and its contrasting full radiation of long-waves. By these properties Dr G. C. Simpson has explained the low temperature of the antarctic summer*. The thermal conductivity of snow is remarkable in being low. Jansson found that when expressed in G-cal cm$^{-1}$ sec$^{-1}$ degree-C$^{-1}$ it was equal to

$$0{\cdot}00005 + 0{\cdot}0019\rho_1 + 0{\cdot}006\rho_1^4,$$

where $\rho_1$ is the density of the snow. This figure is quoted from an interesting paper on the radiation and temperature of snow by Anders Ångström†.

* "British Antarctic Expedition 1910—1913," *Meteorology*, Vol. I. p. 89.

† Stockholm *Arkiv för Mat. Ast. och Fysik*, Bd. 13, No. 21.

# CHAPTER V

## FINDING THE VERTICAL VELOCITY

### CH. 5/0. PRELIMINARY

SOME observations of vertical velocity have been obtained by J. S. Dines*. He took the difference between the rate of ascent of a balloon, as observed in free air by two theodolites, and its rate of ascent in a closed shed.

However, this method is not yet usual and in particular there are no observations of vertical velocity for 1910 May 20 d over the region studied in the example of Ch. 9. Even if we knew the complete initial distribution of vertical velocity, together with the distributions of $m_E$, $m_N$, $p$, $\rho$, $\sigma$, $\mu$ etc. we could hardly compute the new distribution of vertical velocity after an interval $\delta t$ from the expression $v_H + \frac{\partial v_H}{\partial t}\delta t$, because the vertical acceleration $\frac{\partial v_H}{\partial t}$ would have to be found from a tiny difference between two large terms in the vertical dynamical equation. If progress is to be possible it can only be by eliminating the vertical velocity.

There are various theories which arrive at the vertical velocity by treating the air as if it were not shearing, or which either neglect, or else fail to eliminate, some of the time changes at fixed points. For such reasons the otherwise interesting discussions by W. H. Dines† and by M. Berek‡ will not serve for the purposes of numerical prediction. In the theory here presented $\partial p/\partial t$, $\partial \rho/\partial t$, $\partial \theta/\partial t$ are all eliminated, the horizontal velocities are quite general, and the air may be either clear or cloudy. The equation for the vertical velocity at which we thus arrive (No. 12 below) is complicated and I have not succeeded in simplifying it much. But its deduction has been carefully checked and it yields intelligible results when applied to the problem of Ch. 9.

Consider a vertical prism of air. From the known horizontal winds we can find the amount of air entering or leaving the prism at any level. The excess of that entering over that leaving—the net amount entering at any level—spreads upwards or downwards so as to satisfy the hydrostatic equation. Note that the pressures due to the vertical acceleration $\frac{\partial v_H}{\partial t}$ are here neglected. Now the density $\rho$ may be expressed as a function of the pressure, of the entropy-per-unit-mass, $\sigma$, and of the water-per-unit-mass-of-atmosphere, $\mu$. Also a fixed mass of air, in moving, carries its entropy and moisture with it (except in so far as radiation, stirring and precipitation come in), so that the changes in these quantities at any level, depend partly on the

* Advisory Committee on Aeronautics, *Fourth Report on Wind Structure.*

† "Statical Changes of Pressure and Temperature...," *Q. J. R. Met. Soc.* Jan. 1912.

‡ "Die Bestimmung der Vertikal-komponente...," *Leipzig Geophys. Inst.* Spezial II. 6 (1919).

15—2

known horizontal velocities, partly on the unknown vertical velocity. From these relations the vertical velocity can be determined.

As the vertical velocity occurs also in other equations, for instance the horizontal dynamical equations, and as we have to remove it from all of them, we may conveniently take the elimination in two stages, first finding the vertical velocity and then substituting it throughout. The two stages have the further advantage that they permit a comparison of the computed vertical velocity with any observed values which there may be.

## CH. 5/1. DEDUCTION OF A GENERAL EQUATION

Analytically we suppose that $m_E$, $m_N$, $\rho$, $p$, $\mu$, $\sigma$ are known functions of position at an arbitrary fixed instant, to which all the equations of this Ch. 5 relate; but that the changes of these quantities in time are unknown. Required to find $v_H$ or $m_H$ as a function of position. For the present all variables are regarded as varying continuously; the treatment of conventional strata will be considered later. In writing down the following equations, the terms involving the unknowns $m_H$ and $\frac{\partial}{\partial t}$ have been kept in the first members, consequently the second members are known quantities. The increments $D\sigma$ and $D\mu$ represent the changes in $\sigma$ and $\mu$ which a definite portion of air experiences as it moves. Apart from precipitation, stirring and radiation, these changes would both be zero. We suppose, in Ch. 5, that $D\sigma$ and $D\mu$ have already been determined by the methods of Ch. 4/6, Ch. 4/9/6, Ch. 4/7, Ch. 4/8. Accordingly $D\sigma$ and $D\mu$ appear as known quantities and are placed in the second members of the equations.

We begin with the equation (Ch. 4/5/2 # 1) for the conveyance of heat

$$\frac{\partial \sigma}{\partial t} + v_H \frac{\partial \sigma}{\partial h} = -v_E \frac{\partial \sigma}{\partial e} - v_N \frac{\partial \sigma}{\partial n} + \frac{D\sigma}{Dt}, \quad \text{.........................(1)}$$

the equation (Ch. 4/3 # 5) for the conveyance of water

$$\frac{\partial \mu}{\partial t} + v_H \frac{\partial \mu}{\partial h} = -v_E \frac{\partial \mu}{\partial e} - v_N \frac{\partial \mu}{\partial n} + \frac{D\mu}{Dt}, \quad \text{.........................(2)}$$

and the equation (Ch. 4/2 # 2) for the continuity of mass

$$\frac{\partial \rho}{\partial t} + \frac{\partial m_H}{\partial h} + \frac{2m_H}{a} = -\operatorname{div}_{EN} m, \quad \text{.........................(3)}$$

where $\operatorname{div}_{EN} m$ is a contraction for an oft-recurring expression, given by

$$\operatorname{div}_{EN} m = \frac{\partial m_E}{\partial e} + \frac{\partial m_N}{\partial n} - m_N \frac{\tan\phi}{a}. \quad \text{.........................(4)}$$

To obtain an equation soluble for $v_H$ we must first eliminate $\frac{\partial \sigma}{\partial t}, \frac{\partial \mu}{\partial t}, \frac{\partial \rho}{\partial t}$. Two of these can be removed in the following way. The entropy-per-mass $\sigma$ can be expressed as a function of $\mu$, $\rho$, and of one other variable, which it will be convenient later to have taken as $p$. Let $a_\mu$, $a_\rho$, $a_p$ be defined to be such that

$$d\sigma = a_\mu \, . \, d\mu + a_\rho \, . \, d\rho + a_p \, . \, dp. \quad \text{.........................(5)}$$

For clear air the values of $\alpha_\mu$, $\alpha_\rho$, $\alpha_p$ have been given in Ch. 4/5/1. In general they are functions of $\mu$, $\rho$, $p$ and therefore may vary with position and time. Now multiply (2) by $\alpha_\mu$, and (3) by $\alpha_\rho$, and subtract both products from (1). The result is

$$\alpha_p \frac{\partial p}{\partial t} + v_H\left(\alpha_p \frac{\partial p}{\partial h} + \alpha_\rho \frac{\partial \rho}{\partial h}\right) - \alpha_\rho\left(\frac{\partial m_H}{\partial h} + \frac{2m_H}{a}\right)$$

$$= -v_E\left(\frac{\partial \sigma}{\partial e} - \alpha_\mu \frac{\partial \mu}{\partial e}\right) - v_N\left(\frac{\partial \sigma}{\partial n} - \alpha_\mu \frac{\partial \mu}{\partial n}\right) + \alpha_\rho \,.\, \mathrm{div}_{EN} m + \frac{D\sigma}{Dt} - \alpha_\mu \frac{D\mu}{Dt}\,. \quad \text{......(6)}$$

Next $\partial p/\partial t$ in (6) must be replaced by known quantities. This can be done by bringing in the hydrostatic equation

$$p = \int_h^{h_0} g\rho\, dh, \quad \text{..................................(7)}$$

where $h_0$ is a height so great that the pressure at it may be neglected. Take the time-rate of (7). The lower limit of integration is independent of time because we require $\partial p/\partial t$ at a height which is fixed. Thus

$$\frac{\partial p}{\partial t} = \int_h^{h_0} g \frac{\partial \rho}{\partial t}\, dh. \quad \text{..................................(8)}$$

Substitute in (8) the value of $\partial \rho/\partial t$ from (3) then

$$\frac{\partial p}{\partial t} = -\int_h^{h_0} g\left\{\mathrm{div}_{EN} m + \frac{\partial m_H}{\partial h} + \frac{2m_H}{a}\right\} dh. \quad \text{....................(9)}$$

Note that equations (8) and (9) are exceptions to the rule that the second members are completely known. Next substitute in (6) the value of $\partial p/\partial t$ given by (9), and in so doing separate the integral into unknown and known parts, and arrange them respectively in the first and second members. Thus we arrive at

$$-\alpha_p \int_h^{h_0} g\left\{\frac{\partial m_H}{\partial h} + \frac{2m_H}{a}\right\} dh + v_H\left(\alpha_p \frac{\partial p}{\partial h} + \alpha_\rho \frac{\partial \rho}{\partial h}\right) - \alpha_\rho\left(\frac{\partial m_H}{\partial h} + \frac{2m_H}{a}\right)$$

$$= +\alpha_p \int_h^{h_0} g \,.\, \mathrm{div}_{EN}\, m \,.\, dh + \alpha_\rho \,.\, \mathrm{div}_{EN}\, m$$

$$-v_E\left(\frac{\partial \sigma}{\partial e} - \alpha_\mu \frac{\partial \mu}{\partial e}\right) - v_N\left(\frac{\partial \sigma}{\partial n} - \alpha_\mu \frac{\partial \mu}{\partial n}\right)$$

$$+ \frac{D\sigma}{Dt} - \alpha_\mu \frac{D\mu}{Dt} \quad \text{..........................................................(10)}$$

The equation (10) involves only the vertical velocity and quantities expressing the instantaneous distribution of other meteorological elements, a distribution which is supposed to be known.

But (10) can be simplified. For in the first member

$$v_H \alpha_\rho \frac{\partial \rho}{\partial h} - \alpha_\rho \frac{\partial m_H}{\partial h} = -\rho \alpha_\rho \frac{\partial v_H}{\partial h}\,. \quad \text{........................(11)}$$

And again for $\partial p/\partial h$ we can put $-g\rho$.

Then (10) reduces to

$$-\alpha_p \int_h^{h_0} g \left\{ \frac{\partial m_H}{\partial h} + \frac{2m_H}{a} \right\} dh - \alpha_p \,.\, g \,.\, m_H - \alpha_\rho \rho \left( \frac{\partial v_H}{\partial h} + \frac{2v_H}{a} \right) =$$

$$= \alpha_p \int_h^{h_0} g \,.\, \mathrm{div}_{EN}\, m \,.\, dh + \alpha_\rho \,.\, \mathrm{div}_{EN}\, m$$

$$- v_E \left( \frac{\partial \sigma}{\partial e} - \alpha_\mu \frac{\partial \mu}{\partial e} \right) - v_N \left( \frac{\partial \sigma}{\partial n} - \alpha_\mu \frac{\partial \mu}{\partial n} \right)$$

$$+ \frac{D\sigma}{Dt} - \alpha_\mu \frac{D\mu}{Dt} \,. \qquad \ldots\ldots\ldots\ldots(12)$$

So far the equation is very free from assumptions, almost the only one being that the pressures due to vertical acceleration are negligible in comparison with the pressure due to gravity. In particular in deducing (12) we have not assumed $g$ to be independent of height.

## CH. 5/2. SIMPLIFICATION BY APPROXIMATION

As (12) is inconveniently complicated, let us simplify it by making the following two approximations which will probably not affect the accuracy by one per cent.

$g$ is independent of height, ..........(13)

the terms in the reciprocal of the radius of the earth are negligible. ......(14)

There is then a cancelling, in the first member of (12), of

$$-\alpha_p \int_h^{h_0} g \frac{\partial m_H}{\partial h} dh \quad \text{with} \quad -\alpha_p \,.\, g \,.\, m_H,$$

so that the first member of (12) reduces to

$$-\alpha_\rho \,.\, \rho \,.\, \partial v_H / \partial h. \qquad \ldots\ldots\ldots\ldots(15)$$

Let us next divide through by $-\alpha_\rho \,.\, \rho$ so as to leave $\partial v_H/\partial h$ by itself. In so doing let us express $\alpha_p$ and $\alpha_\rho$ in terms of the more familiar quantities $\gamma_v$ and $\gamma_p$, the thermal capacities, in erg units, of unit mass at constant volume and pressure. By Ch. 4/5/1 # 11, 12, for clear* air

$$\alpha_p = \frac{\gamma_v}{p}; \quad \alpha_\rho = -\frac{\gamma_p}{\rho} \,. \qquad \ldots\ldots\ldots\ldots(16), (17)$$

With these substitutions equation (12) reduces to

$$\frac{\partial v_H}{\partial h} = \frac{\gamma_v}{\gamma_p \,.\, p} \int_h^{h_0} g \,.\, \mathrm{div}_{EN}\, m \,.\, dh - \frac{1}{\rho}\, \mathrm{div}_{EN}\, m$$

$$- \frac{1}{\gamma_p} \left( v_E \frac{\partial \sigma}{\partial e} + v_N \frac{\partial \sigma}{\partial n} \right) + \frac{\alpha_\mu}{\gamma_p} \left( v_E \frac{\partial \mu}{\partial e} + v_N \frac{\partial \mu}{\partial n} \right)$$

$$+ \frac{1}{\gamma_p} \left( \frac{D\sigma}{Dt} - \alpha_\mu \frac{D\mu}{Dt} \right) \qquad \ldots\ldots\ldots\ldots(18)$$

in which $\mathrm{div}_{EN}\, m$ is given by (4).

* A temporary restriction for illustrative purposes. In Ch. 8 the unrestricted $\alpha_\rho$, $\alpha_p$, $\alpha_\mu$ are used.

It may be said that (18) would look neater if the integral were removed by differentiation. But in so doing we should also remove the valuable statement, which is implied in the definiteness of the integral, namely that the pressure becomes negligible at a great height. As a matter of fact the differential form was first derived directly from $\partial p/\partial h = -g\rho$ and was employed throughout the example of Ch. 9; but a constant of integration kept on appearing inconveniently in places where it could not be determined. This "hysterical manifestation" was eventually traced to the suppression of the limits of integration which are now explicit in equation (18).

## CH. 5/3. METHOD OF SOLVING THE EQUATION

The approximate solution may be obtained by arithmetical steps. The explicit integral in (18) is begun at the top, where it vanishes, and is carried downwards to each conventional level in turn. The equation is then integrated for $v_H$ beginning at the bottom and working upwards, because at the ground $v_H$ is known from the horizontal winds and the slope of the surface according to the equation

$$(v_H)_G = \left(v_E \frac{dh}{de} + v_N \frac{dh}{dn}\right)_G, \qquad \text{......(19)}$$

which Prof. V. Bjerknes has used to prepare maps of the vertical velocity at the surface (see his *Dynamical Meteorology and Hydrography*, Part II).

As equations (18) and (19) are linear in $v_H$ and $\partial v_H/\partial h$, the sum of any number of solutions is itself the solution for the sum of the corresponding distributions of $m_E$, $m_N$, $p$, $\sigma$, $\mu$.

## CH. 5/4. ILLUSTRATIVE SPECIAL CASES

**Case (i).** *No horizontal velocity anywhere, no radiation, precipitation or stirring.* Then Ch. 5/2 # 18 reduces to $\partial v_H/\partial h = 0$.

But we may derive this result independently. For the pressure on any definite portion of moving air, being simply the weight of air in a column of unit cross-section above it, remains constant. Therefore, as the motion is adiabatic, the density is constant for a moving portion. Thus the atmosphere rises or falls like a rigid body, so that $\partial v_H/\partial h = 0$. If we bring in also the boundary condition (19) we see that $v_H$ is zero everywhere.

**Case (ii).** *No horizontal velocity, no precipitation or stirring, but radiation in progress.*

Then Ch. 5/2 # 18 reduces to

$$\frac{\partial v_H}{\partial h} = \frac{1}{\gamma_p} \frac{D\sigma}{Dt}. \qquad \text{......(1)}$$

The change in entropy $D\sigma$ occurs under the constant pressure due to the air above. Therefore $D\sigma = \frac{1}{\theta}\gamma^p D\theta$ and (18) implies that

$$\frac{\partial v_H}{\partial h} = \frac{1}{\theta} \frac{D\theta}{D}. \qquad \text{......(2)}$$

But we may deduce this independently. For consider a portion $\delta h$ of a vertical column. If it changes in temperature by $D\theta$ while the pressure is kept constant, then it will expand by a fraction $D\theta/\theta$ of its original length $\delta h$. The displacement of the upper end of the short column relative to the lower is therefore $\delta h \,.\, D\theta/\theta$. If the displacement takes place in a time $Dt$, the velocity of the upper end of $\delta h$ relative to its lower end is

$$\delta h \,.\, \frac{1}{\theta}\frac{D\theta}{Dt}.$$

But this relative velocity is $\delta v_H$. Equating and dividing by $\delta h$ we get

$$\frac{\delta v_H}{\delta h} = \frac{1}{\theta}\frac{D\theta}{Dt},$$

which agrees with (2).

**Case (iii).** *Horizontal winds exist, but are such as to cause no divergence of horizontal momentum per volume, that is to say* $\mathrm{div}_{EN} m = 0$ *at all levels. The air is dry and there is no stirring or radiation.*

Then Ch. 5/2 # 18 reduces to

$$\frac{\partial v_H}{\partial h} = -\frac{1}{\gamma_p}\left(v_E\frac{\partial\sigma}{\partial e} + v_N\frac{\partial\sigma}{\partial n}\right).$$

Suppose that the wind is coming from a region where the entropy-per-mass is greater towards one where it is less. Then the quantity in the bracket will be negative, as is easily seen by considering the special case $v_N = 0$ and remembering that, contrary to sailors' usage, $v_E$ is positive when the wind blows to the east. In the case imagined $\partial v_H/\partial h$ will be positive. The arrival of air of greater entropy-per-mass causes an increase of upward velocity with height. If the motion happens to be at constant pressure then we may say: the arrival of warmer air causes a swelling. An independent check is lacking.

**Case (iv).** *At a level* $h_i$ *air is being abstracted from the column. Moisture is absent. The entropy-per-mass has no horizontal variation and does not change with time following the motion.*

The general equation Ch. 5/2 # 18 then reduces to

$$\frac{\partial v_H}{\partial h} = \frac{\gamma_v}{\gamma_p \,.\, p}\int_h^{h_0} g \,.\, \mathrm{div}_{EN} m \,.\, dh - \frac{1}{\rho}\,\mathrm{div}_{EN} m. \quad\text{.....................(1)}$$

**Case (iv A).** Let us suppose, for mathematical simplicity, that $\mathrm{div}_{EN} m$ *is zero everywhere except in a very short range* lying between $h_i - \delta h$, and $h_i + \delta h$, and let us represent the total rate of abstraction of mass from a column of unit cross-section by $A$, where

$$A = \int_{h_i - \delta h}^{h_i + \delta h} \mathrm{div}_{EN} m \,.\, dh. \quad\text{..........................(2)}$$

Then below the level of abstraction the integral in (1) includes the range in (2) so that

$$\frac{\partial v_H}{\partial h} = \frac{\gamma_v}{\gamma_p} \,.\, \frac{gA}{p}. \quad\text{...............................(3)}$$

Above the level of abstraction the integral in (1) does not include that in (2) and so

$$\frac{\partial v_H}{\partial h} = 0. \qquad \ldots\ldots(4)$$

Let the ground be flat and at a height $h_G$. Then

$$v_H = 0 \quad \text{at} \quad h = h_G. \qquad \ldots\ldots(5)$$

Now integrating (3) upwards from the ground it follows that, below the level of abstraction $h_i$

$$\frac{v_H}{A} = \int_{h_G}^{h} \frac{\gamma_v}{\gamma_p}\, g\, \frac{dh}{p}. \qquad \ldots\ldots(6)$$

Next, when this second integration is carried upwards across the level $h_i$, there is an abrupt change in $v_H$ arising from the last term in (1), thus

$$-\int_{h_i - \delta h}^{h_i + \delta h} \frac{1}{\rho}\, \text{div}_{EN}\, m = -\frac{A}{\rho_i}. \qquad \ldots\ldots(7)$$

So from (6) and (7) it follows that, just above the level of abstraction,

$$\frac{v_H}{A} = \int_{h_G}^{h_i} \frac{\gamma_v}{\gamma_p}\, g\, \frac{dh}{p} - \frac{1}{\rho_i}. \qquad \ldots\ldots(8)$$

And (4) shows us that the vertical velocity remains constant for all greater heights.

The integral in (6) and (8) has been computed from the observed* mean pressures over Europe, with the following results:

| Height above M.S.L. kilometres | $\frac{\gamma_v}{\gamma_p} g \int_0^h \frac{dh}{p}$ cm$^3$ grm$^{-1}$ | Height above M.S.L. kilometres | $\frac{\gamma_v}{\gamma_p} g \int_0^h \frac{dh}{p}$ cm$^3$ grm$^{-1}$ |
|---|---|---|---|
| 20·5 | 8540 | 9·5 | 1293 |
| 19·5 | 7220 | 8·5 | 1064 |
| 18·5 | 6180 | 7·5 | 866 |
| 17·5 | 5240 | 6·5 | 694 |
| 16·5 | 4430 | 5·5 | 545 |
| 15·5 | 3750 | 4·5 | 415 |
| 14·5 | 3170 | 3·5 | 301 |
| 13·5 | 2670 | 2·5 | 201 |
| 12·5 | 2240 | 1·5 | 112 |
| 11·5 | 1870 | 0·5 | 35 |
| 10·5 | 1560 | [0·0 | 0] |

Note $g$ has here been taken as 980 and $\gamma_v/\gamma_p$ as 1/1·405. The latter is of doubtful validity in a moisture-containing atmosphere.

* W. H. Dines, "Characteristics of the Free Atmosphere," Meteor. Office, London, *Geophys. Mem.* No. 13, Table X.

The curves in the adjoining figure have been drawn by using this table in equations (6) and (8).

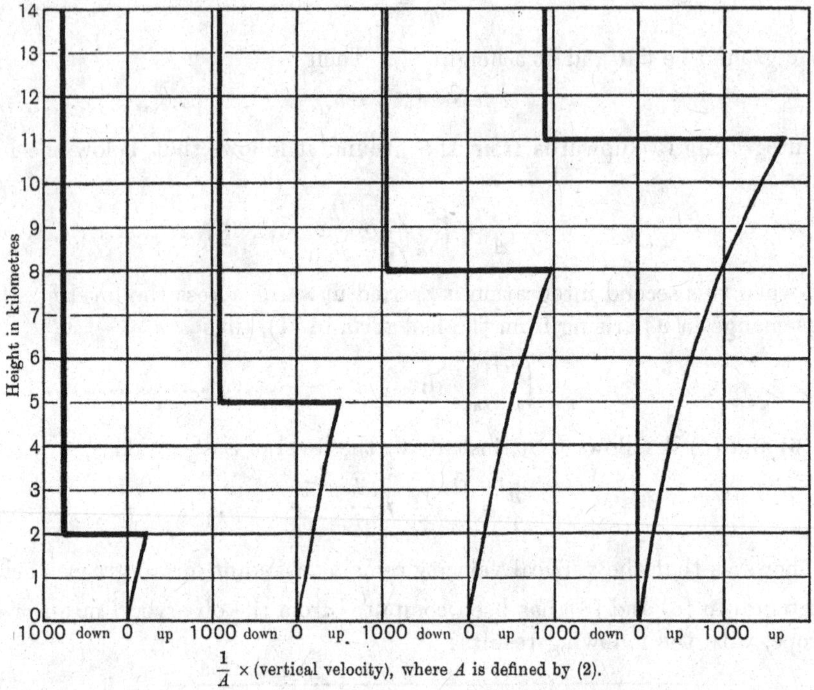


**Four distributions of vertical velocity produced by the horizontal removal of mass at equal speeds but at different heights, namely at 2, 5, 8 and 11 km.**

Since $\text{div}_{EN} m$ and $v_H$ enter equation (1) linearly, we may add together any fixed multiples of the abscissae of any of these curves in order to obtain a new distribution satisfying (1). For example if we multiply the velocities of the third curve by $-1$ and add them to those of the fourth curve we obtain the distribution due to the insertion of mass at 8 km and its withdrawal at an equal speed at 11 km. Below 8 km there is no vertical motion; between 8 km and 11 km there is an upward velocity increasing with height; above 11 km there is a small downward velocity. The latter is at first sight surprising, but it must be remembered that we are not discussing a steady state, but a sudden disturbance of the actual mean distribution, and that the air in the neighbourhood of 10 km would be in the course of replacement by colder air, on account of the difference between the actual and the adiabatic lapse-rates.

The integral in (6) may be put into a variety of equivalent forms which are sometimes useful. Thus by using the hydrostatic and characteristic equations

$$\frac{\gamma_v}{\gamma_p} g \int_{h_G}^{h} \frac{dh}{p} = -\frac{\gamma_v}{\gamma_p} \int_{p_G}^{p} \frac{d \log p}{\rho} = \frac{\gamma_v}{\gamma_p} b \int_{p_G}^{p} \theta d\left(\frac{1}{p}\right)$$

$$= \int_{\rho_G}^{\rho} d\left(\frac{1}{\rho}\right) + \int_{h_G}^{h} \frac{g}{\gamma_p} \left(\frac{\partial \sigma}{\partial p}\right)_{\text{vertically}} dh \quad \ldots\ldots\ldots\ldots\ldots\ldots(9)$$

By way of these it may be shown that the vertical velocity above a level of abstraction is downwards, unless the atmosphere has a degree of instability far exceeding anything that is observed. To illustrate this exception we may take the hypothetical case in which the density is the same at all heights. Then the second form in (9) integrates, and when inserted in (8) gives for the vertical velocity above the level of abstraction

$$\frac{v_H}{\rho A} = \frac{\gamma_v}{\gamma_p} \log \frac{p_G}{p} - 1, \quad \text{.................................(10)}$$

which is positive if $p$ is small enough. Thus in such an atmosphere the removal of mass at a sufficiently high level would cause the lower part to raise the part which is above the level of abstraction.

## CH. 5/5. FURTHER VARIETIES OF THE SIMPLIFIED GENERAL EQUATION.

In some applications it is desirable to replace $\text{div}_{EN}\, m$ by an expression in velocity. Now

$$\text{div}_{EN}\, m = \rho \left\{ \frac{\partial v_E}{\partial e} + \frac{\partial v_N}{\partial n} - \frac{\tan\phi}{a} v_N \right\} + v_E \frac{\partial \rho}{\partial e} + v_N \frac{\partial \rho}{\partial n}. \quad \text{..............(1)}$$

If this be substituted in Ch. 5/2 # 18 then because, when different samples of air are compared*,

$$d\sigma = \frac{\gamma_v}{p} dp - \frac{\gamma_p}{\rho} d\rho + a_\mu d\mu \quad \text{..............................(2)}$$

it follows that a term in Ch. 5/2 # 18 derived from $-\frac{1}{\rho} \text{div}_{EN}\, m$ combines with the two terms following it giving us

$$-\frac{d\sigma}{\gamma_p} - \frac{d\rho}{\rho} + \frac{a_\mu}{\gamma_p} d\mu = -\frac{\gamma_v}{\gamma_p} \cdot \frac{dp}{p}. \quad \text{.............................(3)}$$

And Ch. 5/2 # 18 becomes

$$\frac{\partial v_H}{\partial h} = \frac{\gamma_v}{\gamma_p \cdot p} \int_h^{h_0} g \left\{ \rho\, \text{div}_{EN}\, v + v_E \frac{\partial \rho}{\partial e} + v_N \frac{\partial \rho}{\partial n} \right\} dh - \text{div}_{EN}\, v$$

$$- \frac{\gamma_v}{\gamma_p \cdot p} \left( v_E \frac{\partial p}{\partial e} + v_N \frac{\partial p}{\partial n} \right) + \frac{1}{\gamma_p} \left( \frac{D\sigma}{Dt} - a_\mu \frac{D\mu}{Dt} \right). \quad \text{.........(4)}$$

But, by the rule for differentiating a product

$$-\frac{\partial}{\partial h} \left( v_E \frac{\partial p}{\partial e} \right) = -\frac{\partial v_E}{\partial h} \cdot \frac{\partial p}{\partial e} + v_E \cdot g \frac{\partial \rho}{\partial e}, \quad \text{..........................(5)}$$

and there is a similar expression in the northward coordinate and component. If in the integral in (4) we substitute for $g\, v_E \frac{\partial \rho}{\partial e}$ the value given by (5) we obtain a term $\int_h^{h_0} -\frac{\partial}{\partial h} \left( v_E \frac{\partial p}{\partial e} \right) dh$ which transforms simply, because at the upper limit $p$ vanishes everywhere, so that

$$\left( v_E \cdot \frac{\partial p}{\partial e} \right)_{h_0} = 0. \quad \text{..........................................(6)}$$

* See Ch. 4/5/1. Clear air is here assumed for illustration, but in Ch. 8 we revert to the more general form in $a_p$, $a_\rho$.

The value of the same integral at its lower limit cancels another term in (4). Thus we arrive at

$$\frac{\partial v_H}{\partial h} = \frac{\gamma_v}{\gamma_p \cdot p}\int_h^{h_0}\left\{g\rho \,.\, \mathrm{div}_{EN} v + \frac{\partial v_E}{\partial h}\cdot\frac{\partial p}{\partial e} + \frac{\partial v_N}{\partial h}\cdot\frac{\partial p}{\partial n}\right\} dh - \mathrm{div}_{EN} v + \frac{1}{\gamma_p}\left(\frac{D\sigma}{Dt} - \alpha_\mu \frac{D\mu}{Dt}\right), \quad \text{.........(7)}$$

where

$$\mathrm{div}_{EN} v = \frac{\partial v_E}{\partial e} + \frac{\partial v_N}{\partial n} - v_N \frac{\tan\phi}{a}. \quad \text{.........................(8)}$$

The above has been found to be the most convenient form of the equation when we have to do with the stratosphere, as in Ch. 6/6.

In places where we measure height by means of pressure it is convenient to take $p$ as the independent in (7) in place of $h$. In these differentiations with respect to $p$, it is implied that latitude and longitude are constant, whereas in $\alpha_p = \partial\sigma/\partial p$ it is $\rho$ and $\mu$ which are constant. Where there is any risk of confusion the symbols of the quantities which are constant during the differentiation are added as suffixes to the differential coefficient. Then (7) transforms into

$$\frac{\partial v_H}{\partial h} = \frac{\gamma_v}{\gamma_p \cdot p}\int_{p_0}^{p}\left\{\mathrm{div}_{EN} v - \left(\frac{\partial v_E}{\partial p}\right)_{\lambda,\phi} \times \left(\frac{\partial p}{\partial e}\right)_{h,\phi} - \left(\frac{\partial v_N}{\partial p}\right)_{\lambda,\phi} \times \left(\frac{\partial p}{\partial n}\right)_{h,\lambda}\right\} dp - \mathrm{div}_{EN} v + \frac{1}{\gamma_p}\left(\frac{D\sigma}{Dt} - \alpha_\mu \frac{D\mu}{Dt}\right). \quad \text{.........(9)}$$

## CH. 5/6. THE INFLUENCE OF EDDIES

**Smoothing the equation for vertical velocity.** This equation is linear in the three velocity components. So if the other variables $\rho$, $p$, $\alpha_\rho$, $\alpha_p$, $\alpha_\mu$, $\sigma$, $\mu$ are not heterogeneous, the process of smoothing, carried out as in Ch. 4/10, merely puts a bar over all the velocity components and over the capital $D$, which now denotes a differentiation following the mean motion.

If the other quantities are also diversified great complications will arise. But as became evident in Ch. 4/10, the percentage variations in velocity are usually so much greater than the percentage variations of the other quantities that for many purposes the latter are negligible.

**Vertical velocity connected with eddy-motion.** Wherever light warm air is displaced and pushed up by heavy cold air, the joint centre of mass falls, for gravity supplies $g/\gamma_p$ of the kinetic energy of the motion*. Again wherever a thermally stable atmosphere is stirred by eddies derived from the wind, there heat is carried down and the centre of mass rises. These motions of the centre of mass imply a small, but highly significant, mean vertical velocity.

In view of what has been said about smoothing, it follows that the equations already given in case (ii) in Ch. 5/4 for the vertical velocity due to radiation, will also apply to that due to eddy motion, if we place a bar over $v_H$ and over $D$. That is true provided $\rho$, $p$, $\alpha_\rho$, $\alpha_p$, $\alpha_\mu$, $\sigma$, $\mu$ are not diversified.

* W. H. Dines' theorem, *Q. J. R. Met. Soc.* 1913, July, p. 188, see also *Roy. Soc. Proc.* A, Vol. 97, 1920.

# CHAPTER VI

## SPECIAL TREATMENT FOR THE STRATOSPHERE

### CH. 6/0. INTRODUCTION

THE equations developed in Ch. 4, between the integrated quantities represented by the capital letters $P$, $R$, $M_E$, $M_N$, hold good, with the approximations indicated, for any one of the conventional strata. But in the stratum, which has its base at 11·8 km and extends upwards to at least 40 km, the ratio of pressure is so great that the aforesaid approximations are all open to criticism and must be reexamined. This examination is one of the principal aims in Ch. 6. Another aim is to find a way of extrapolating observations made by balloons, which seldom penetrate into the upper tenth of the mass of the atmosphere, so as to obtain $P$, $R$, $M_E$, $M_N$ which are integrals up to the top. The final aim is to choose a set of quantities, either $P$, $R$, $M_E$, $M_N$ or some other equivalent ones, and to find a corresponding set of equations so that, when the quantities are given at one instant, the equations will give their time rates. This problem is for the most part carried over to Ch. 8. But its general features have already been described in Ch. 4/0.

The quantities $P$, $R$, $M_E$ etc. are the definite integrals of $p$, $\rho$, $m_E$ etc. with respect to height, taken so as to include the whole thickness of the stratum. In such integrals, values of quantities at the lower limit will be denoted by the subscript 2, because the mean pressure there is 2 decibars; thus we have $p_2$, $h_2$, $m_{E2}$ etc. Similarly the subscript 0 will be used to denote the upper limit. But while $h_0$ resembles infinity in that it denotes a height so great that its exact value does not concern us, it must be understood to lie somewhere between 50 and 100 kilometres above sea-level, at a pressure between 1 millibar and 0·001 millibar. By this convention we free ourselves from the necessity for entering into difficult questions concerning that outer atmosphere* which is ionized, which may be escaping, and in which the variation of gravity, and the term $\frac{2m_E}{a}$ in the equation of continuity of mass, would cause mathematical difficulties. By this convention also we assume that whatever the rare gas above $h_0$ may do, it has no influence on the surface weather.

The vertical variation of gravity has been neglected. It may however be mentioned that the theory could be carried through with integrals with respect to gravity-potential instead of with respect to height, $\mathscr{M}_E = \int m_E \, d\psi$ replacing $M_E = \int m_E \, dh$ and so on; but it would then be necessary to follow this less familiar system throughout the whole subject.

* Stoermer, *Terrestrial Magnetism*, March 1915; Jeans, *Dynamical Theory of Gases*, § 375; S. Chapman and A. E. Milne, *Q. J. R. Met. Soc.* 1920, Oct. Would not the "empyrean" form a good name for this highly ionized atmosphere?

It has been assumed in this chapter that vertical isothermy extends right down to 11·8 km above M.S.L. Observations sometimes reveal disturbances at greater heights in temperate latitudes, and in particular in the tropics the normal level of the tropopause is about 16 km. These are arguments for raising the boundary of the conventional stratum. There would be no difficulty about so doing, except the increasing scarcity of observations.

## CH. 6/1. INTEGRALS OF PRESSURE AND DENSITY

It is here assumed that

$$\frac{\partial \theta}{\partial h} = 0 \qquad \text{.......... (1)}$$

at all times and places within the stratum.

The air is dry, so that in

$$p = b\rho\theta \qquad \text{.......... (2)}$$

$b$ has the fixed* value $2{\cdot}87 \times 10^6$ cm$^2$ sec$^{-2}$ per degree centigrade.

The hydrostatic equation,

$$\frac{\partial p}{\partial h} = -g\rho = -g\frac{p}{b\theta}, \qquad \text{.......... (3)}$$

may be written

$$\frac{\partial \log p}{\partial h} = -\frac{g}{b\theta}, \qquad \text{.......... (4)}$$

that is to say "the logarithm of the pressure is a linear function of height"; it follows on integrating that

$$p = \text{const} \times e^{-\frac{g}{b\theta}h}. \qquad \text{.......... (5)}$$

So that

$$\int p\, dh = -\frac{b\theta}{g} p + \text{arbitrary const.} \qquad \text{.......... (6)}$$

And, putting in the limits of integration,

$$P = +\frac{b\theta}{g} p_2. \qquad \text{.......... (7)}$$

Again

$$R = \int_{h_2}^{h_0} \rho\, dh = \frac{p_2}{g}. \qquad \text{.......... (8)}$$

Thus $P$ and $R$ have been expressed in terms of the pressure and temperature at 11·8 kilometres, a height not infrequently reached by recording balloons.

* G. Gouy concludes that there is no stratification of the different components by molecular weight where the pressure exceeds that of a Crookes' tube [that is about 0.01 millibar]. *Comptes Rendus*, clviii. p. 664 quoted in *Nature*, 1914, Apr. 23. According to Chapman and Milne (*Q. J. R. Met. Soc.* 1920, Oct.) $b$ might increase to $2{\cdot}92 \times 10^6$ cm$^2$ sec$^{-2}$ degree$^{-1}$ at a height of 60 km where the pressure is 0.12 millibar.

## CH. 6/2. THE CONTINUITY OF MASS IN THE STRATOSPHERE

Since all the terms in the equation of continuity of mass are linear, the equation transforms without approximation, and without any reference to the manner of variation of wind with height. The transformation has already been effected in Ch. 4/2, and the result, when appropriate suffixes are inserted, reads

$$-\frac{\partial R_{20}}{\partial t}=\frac{\partial M_{E20}}{\partial e}+\frac{\partial M_{N20}}{\partial n}-\frac{\tan\phi\,.\,M_{N20}}{a}-m_{H2}+\frac{2M_H}{a}. \quad \text{..............(1)}$$

We assume here, as elsewhere, that there is no escape of air from the top of the atmosphere, so that

$$m_{H0}=0. \quad \text{......................................(2)}$$

Now from Ch. 6/1 # 8

$$\frac{\partial R_{20}}{\partial t}=\frac{1}{g}\frac{\partial p_2}{\partial t}, \quad \text{......................................(3)}$$

so that equation (1) gives us pressure-changes at the base of the stratum. $M_E$, $M_N$, for insertion in (1), are main tabulated variables, while $m_{H2}$ and $M_H$ are given by the equation for the vertical velocity to be discussed in Ch. 6/6 below.

## CH. 6/3. EXTRAPOLATING OBSERVATIONS OF WIND

For this purpose, in latitudes not too near the equator, we can use a theory developed from "The Upper Air Calculus"* of Sir Napier Shaw. The dynamical equations may be written

$$-\frac{1}{\rho}\frac{\partial p}{\partial e}=-2\omega\sin\phi\,.\,v_N+v_E\frac{\partial v_E}{\partial e}+v_N\frac{\partial v_N}{\partial n}+\frac{\partial v_E}{\partial t}+\text{etc.}, \quad \text{..........(1)}$$

$$-\frac{1}{\rho}\frac{\partial p}{\partial n}=+2\omega\sin\phi\,.\,v_E+v_E\frac{\partial v_N}{\partial e}+v_N\frac{\partial v_N}{\partial n}+\frac{\partial v_N}{\partial t}+\text{etc.} \quad \text{..............(2)}$$

Now since $\theta$ is independent of $h$ and since $p=b\rho\theta$, it follows that

$$\frac{\partial}{\partial h}\left(\frac{1}{\rho}\frac{\partial p}{\partial e}\right)=\frac{\partial}{\partial h}\left(b\theta\frac{\partial\log p}{\partial e}\right)=b\theta\frac{\partial^2\log p}{\partial h\partial e}. \quad \text{......................(3)}$$

But the hydrostatic equation may be written

$$\frac{\partial\log p}{\partial h}=-\frac{g}{b\theta}. \quad \text{......................................(4)}$$

On inserting (4) in (3)

$$\frac{\partial}{\partial h}\left(\frac{1}{\rho}\frac{\partial p}{\partial e}\right)=g\frac{\partial\log\theta}{\partial e}, \quad \text{..................................(5)}$$

and in (5) $\partial n$ might replace $\partial e$.

Therefore on differentiating again

$$\frac{\partial^2}{\partial h^2}\left(\frac{1}{\rho}\frac{\partial p}{\partial e}\right)=0 \text{ and similarly } \frac{\partial^2}{\partial h^2}\left(\frac{1}{\rho}\frac{\partial p}{\partial n}\right)=0. \quad \text{..................(6)}$$

Thus both sides of equations (1) and (2) are linear functions of height.

* *Nature*, 1913 Sept. 18, also *J. Scott. Met. Soc.* 1913.

Near the equator it will be difficult to interpret this general proposition in any simple way, because neither the linear nor the quadratic terms are negligible on the right-hand sides of (1) and (2). But in European latitudes it has been shown* that all terms except those in $\sin\phi$ may, with tolerable approximation, be neglected. In the stratosphere, as there is no vertical convection, there can be no appreciable friction between horizontal layers, so that one of the causes which disturb the "geostrophic wind" near the ground is here absent. Assuming then this simple geostrophic wind, with neglect of curvature of path, there follows from (5) and (1)

$$\frac{\partial v_N}{\partial h} = + \frac{g}{2\omega \sin\phi} \frac{\partial \log\theta}{\partial e}, \quad \text{.............................(7)}$$

and similarly

$$\frac{\partial v_E}{\partial h} = - \frac{g}{2\omega \sin\phi} \frac{\partial \log\theta}{\partial n}. \quad \text{.............................(8)}$$

And so

$$\frac{\partial^2 v_N}{\partial h^2} = 0; \quad \frac{\partial^2 v_E}{\partial h^2} = 0; \quad \text{.....................................(9)}$$

thus in the *stratosphere in not too low latitudes the horizontal component velocities are linear functions of the height*†.

To test this theory I selected the two highest balloon flights included in V. Bjerknes' Synoptic Charts, hefts 1, 2, 3. The ascents were at Zürich on 1910 Feb. 3 and May 19. One must resolve the velocities on rectangular axes, for (9) does *not* imply that the resultant $\sqrt{v_E^2 + v_N^2}$ is a linear function of height. The figures on p. 129 show that, in the stratosphere, the component velocities can be fairly well fitted to straight lines. The fluctuations may be in part attributed to errors made in observing balloons. For at a height of 16 km and at zenith distance 60°, a standard error of 0°·1 in the measurement of the zenith distance produces a standard error of 2·6 metres/sec in the radial velocity of a balloon, as deduced from successive differences of position observed at intervals of 60 seconds of time when the heights are correct. On the other hand, if the fluctuations are really gusts, we must suppose them to be due either to stable waves or else to eddies turning about vertical axes; for if the axes were not vertical then a fall of temperature with height would result.

From the slopes of the lines corresponding to $\frac{\partial v_E}{\partial h}$ and $\frac{\partial v_N}{\partial h}$ on May 19th we may calculate the horizontal temperature gradients by equations (7) and (8); it is thus found that

$$\frac{\partial\theta}{\partial e} = -0{\cdot}48^\circ\,\text{C per 100 km}; \quad \frac{\partial\theta}{\partial n} = +0^\circ{\cdot}11\,\text{C per 100 km}.$$

Unfortunately the temperatures obtained by sounding balloons are not precise enough to test these numbers, for according to Gold (*Geophys. Mem.* No. 5, p. 66) they are to be suspected of errors of $\pm 2^\circ$ C. Wenger‡ seems to be of much the same opinion. W. H. Dines finds a probable error of $< 1^\circ$C for a single ascent at night, but more in sunlight.

* E. Gold, "Barometric Gradient and Wind Force," *Report* to the Director of the Meteorological Office.

† First published in *Q. J. R. Met. Soc.* 1920 January, p. 63.

‡ R. Wenger, "Ueber den Einfluss der Instrumentalfehler..." *Geophys. Inst. Leipzig*, Specialarb. II. 1 (1913).

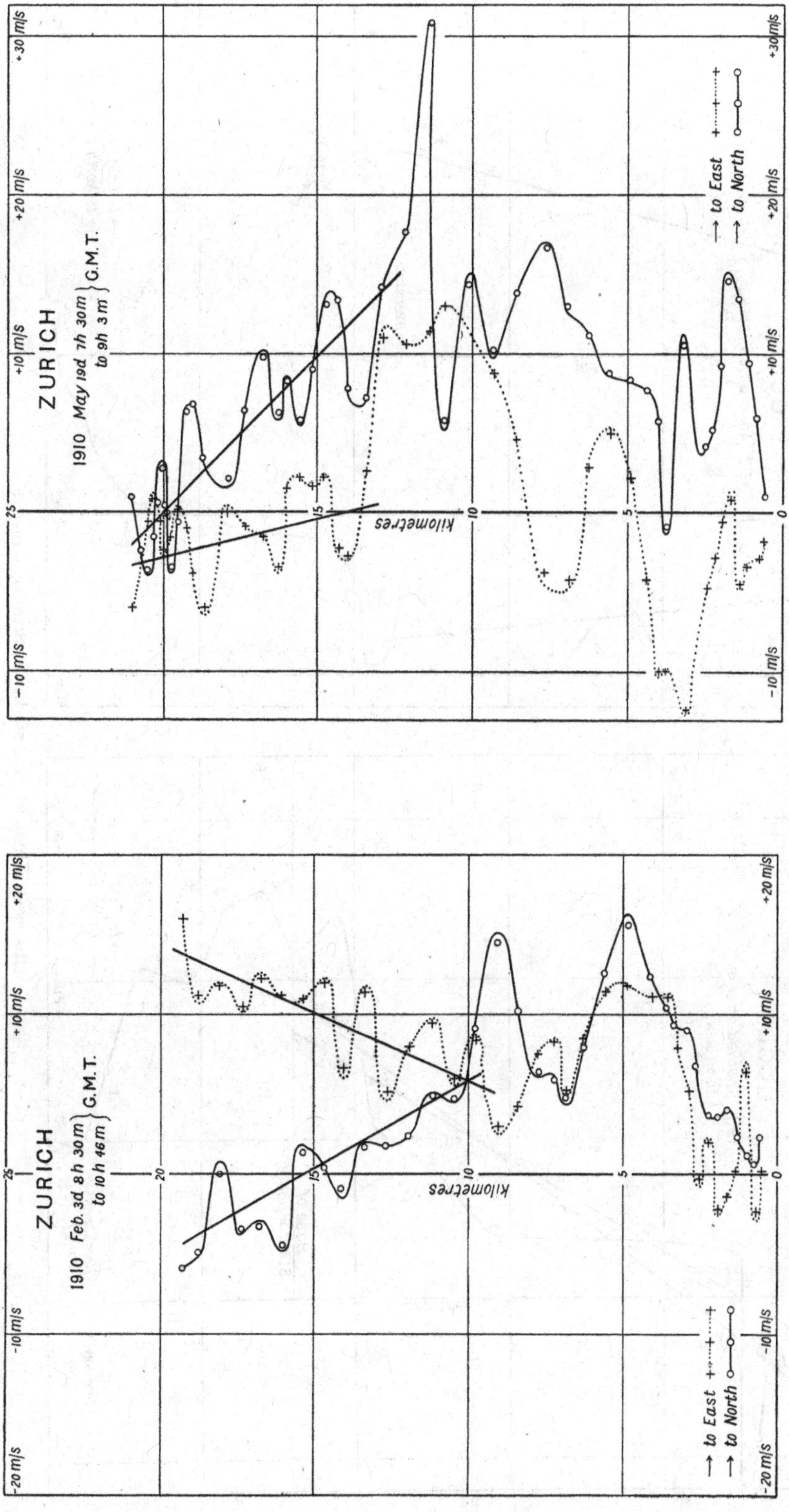


Observed Wind-Components in the Stratosphere.

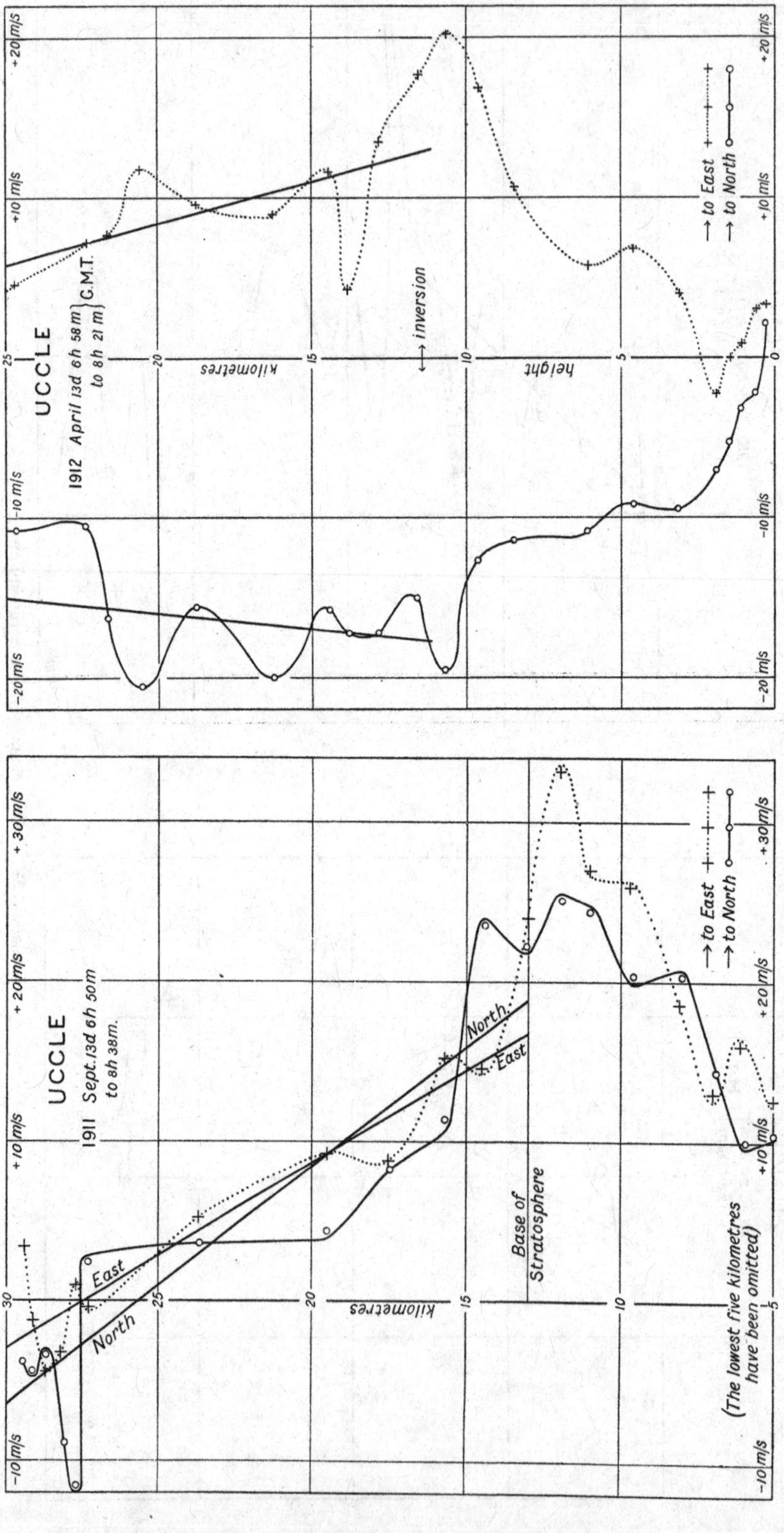


Observed Wind-Components in the Stratosphere.

As a further test of equations (9) one figure on p. 130 shows the highest theodolite-balloon record published in the international observations January to June 1912. It was made at Uccle 1912 Apr. 13. The remaining figure on p. 130 represents the highest ascent recorded in G. M. B. Dobson's paper* on "Winds and Temperature Gradients in the Stratosphere." This ascent was made at Uccle on 1911 Sept. 13. The straight lines fit moderately well. The selection of observations has been made by choosing the highest in order to eliminate any personal bias in favour of the theory to be tested.

For average conditions over England, Shaw† has calculated from a formula, with which (7) is really identical, that the wind velocity at 12 kilometres falls off by 9 per cent of itself in one kilometre. The formula however shows that the fall in each kilometre must be the same in absolute amount, not in percentage of the wind at that height, and so it follows that the velocity changes sign at about 23 kilometres and that at an infinite height the velocity would be infinite, if we might press the logic so far. Of course if the velocity became very great the square and product terms in the dynamical equations would cease to be negligible and the hypothesis of the simple geostrophic wind would be no longer tenable. That high velocities do sometimes occur at great heights has been shown by the Krakatoa glow stratum (33 metres/sec at 35 km) and by the bolide of 1909 Feb. 22‡ (70 metres/sec at 75 km). Even if the velocities $v_E$, $v_N$ did become infinite linearly, the total mass-transports $M_E$, $M_N$ would remain finite, because the density falls off exponentially.

**Height-integral of density times a linear function of height.** The momenta per unit volume $m_E$, $m_N$ are of the form $\rho(A+Bh)$, where $A$ and $B$ are independent of height. $M_E$, $M_N$ are therefore of the form $\int_{h_i}^{h_0} \rho(A+Bh)\,dh$. In what follows we shall frequently require the corresponding indefinite integral. To evaluate it, put $\rho = -\frac{1}{g}\frac{\partial p}{\partial h}$, integrate the coefficient of $B$ by parts and take $\int p\,dh$ from Ch. 6/1 # 6. There results

$$\int \rho(A+Bh)\,dh = -\frac{p}{g}\left[A+B\left(h+\frac{b\theta}{g}\right)\right] + \text{const.} \quad \text{..............(10)}$$

Now $\frac{b\theta}{g}$ is a length, and is equal to 6·45 kilometres when the stratosphere has the temperature of 220° A.

If the upper limit of the integral be $h_0$, the pressure there is very small and we have

$$\int_{h_i}^{h_0} \rho(A+Bh)\,dh = +\frac{p_i}{g}\left[A+B\left(h_i+\frac{b\theta}{g}\right)\right], \quad \text{................(11)}$$

where the subscript $i$ refers to any height in the stratosphere. But

$$\frac{p_i}{g} = \int_{h_i}^{h_0} \rho\,dh = \text{the whole mass above } h_i. \quad \text{...................(12)}$$

* *Q. J. R. Met. Soc.* Jan. 1920. † *J. Scott. Met. Soc.* 1913, p. 170.
‡ J. E. Clark, *Q. J. R. Met. Soc.* 1913.

And by taking the special case $A=0$, $B=1$ we see that $h_i+b\theta g^{-1}$ is the level of the centre of mass of the atmosphere above $h_i$. ..........................................(13)

Thus we may say: "*With reference to a column in the stratosphere extending from an arbitrary level to a region where pressure is negligible, the integral up the column, of the product of the density into any linear function of height, is equal to the pressure at the base of the column, multiplied by the value of the linear function at a height $b\theta/g$ above the base and divided by the acceleration of gravity.*" ..................(14)

When applied to velocity, statement (14) is equivalent to the following: The total eastward momentum above a small horizontal surface, at any level in the stratosphere in high latitudes, is equal to the total mass above that surface, multiplied by the eastward velocity at a height about 6·4 kilometres greater than the height of the surface.

An exactly similar statement holds for the northward momentum, but not, in general, for the resultant momentum.

Thus the initial values of $M_E$, $M_N$ can easily be computed from balloon observations which extend into the stratosphere far enough to allow the constant values of $\frac{\partial v_E}{\partial h}$, $\frac{\partial v_N}{\partial h}$ to be measured.

## Ch. 6/4. THE HORIZONTAL DYNAMICAL EQUATIONS IN THE STRATOSPHERE

In extrapolating the initial observations we have neglected all the small terms, but in tracing the course of the subsequent development it is both possible and desirable to retain them.

In Ch. 4/4, dealing with the dynamical equations, it is shown that a term such as $\int m_E v_N dh$ transforms approximately into $M_E M_N/R$. In the uppermost stratum the ranges of velocity and density are so great that one might expect the approximation to fail altogether. It will therefore be examined in detail. The argument proceeds thus: If the small terms vanish, the component velocities are linear functions of height. If the neglected terms are finite but small, a linear relation of component velocities to height may still be expected to hold approximately, as indeed the balloon-observations show that it does. This relation may therefore be used in calculating, to a first approximation, averages with respect to height.

Integrating by parts and using Ch. 6/3 # 9

$$\int m_E v_N dh = v_N \int m_E dh - \frac{\partial v_N}{\partial h} \iint m_E dh^2 + \text{const.} \qquad \text{.................(1)}$$

Now putting $(A+Bh)=v_E$ in Ch. 6/3 # 11 we get

$$\int^{h_0} m_E dh = -\frac{p}{g}\left[v_E + \frac{\partial v_E}{\partial h}\frac{b\theta}{g}\right]. \qquad \text{........................(2)}$$

And therefore

$$\int^{h_0}\!\!\int^{h_0} m_E dh^2 = -\frac{b\theta}{g}\int^{h_0} \rho v_E dh - \frac{b^2\theta^2}{g^2}\frac{\partial v_E}{\partial h}\int^{h_0} \rho dh$$

$$= +\frac{b\theta}{g}\cdot\frac{p}{g}\left[v_E + \frac{\partial v_E}{\partial h}\frac{b\theta}{g}\right] + \frac{p}{g}\frac{b^2\theta^2}{g^2}\frac{\partial v_E}{\partial h}. \qquad \text{.................(3)}$$

Collecting terms

$$\int m_E v_N dh = -\frac{p}{g}\left\{\left(v_E + \frac{b\theta}{g}\frac{\partial v_E}{\partial h}\right)\left(v_N + \frac{b\theta}{g}\frac{\partial v_N}{\partial h}\right) + \frac{b^2\theta^2}{g^2}\frac{\partial v_N}{\partial h}\frac{\partial v_E}{\partial h}\right\} + \text{const.} \qquad \text{...(4)}$$

Taking this between the limits $p_0$ and $p_2$

$$\int_{h_2}^{h_0} m_E v_N dh = \frac{M_E M_N}{R} + R\frac{b^2\theta^2}{g^2}\frac{\partial v_N}{\partial h}\frac{\partial v_E}{\partial h}. \qquad \text{.........................(5)}$$

Now $\frac{\partial v_N}{\partial h}$, $\frac{\partial v_E}{\partial h}$ are known initially, but not subsequently, as the stratosphere is represented by a single conventional stratum. But we may replace $\frac{\partial v_N}{\partial h}$, $\frac{\partial v_E}{\partial h}$ by means of Ch. 6/3 # 7, 8 and thus obtain

$$\int_{h_2}^{h_0} m_E v_N dh = \frac{M_E M_N}{R} - \frac{Rb^2}{(2\omega\sin\phi)^2}\frac{\partial\theta}{\partial e}\cdot\frac{\partial\theta}{\partial n}. \qquad \text{..................(6)}$$

The terms on the right are known both initially and subsequently and can be taken into account when dealing with the dynamical equation in $P$, $R$, $M_E$, $M_N$.

Inserting (6) in the dynamical equations of Ch. 4/4 and making other corresponding transformations, we obtain :

The dynamical equations for the uppermost conventional stratum*.

$$-\frac{\partial M_E}{\partial t} = \frac{\partial P}{\partial e} + \frac{\partial}{\partial e}\left(\frac{M_E^2}{R}\right) + \frac{\partial}{\partial n}\left(\frac{M_E M_N}{R}\right) - [m_E v_H]_2 - 2\omega\sin\phi M_N + 2\omega\cos\phi M_H + \frac{3}{a}\int_2^0 m_E v_H dh - \frac{2\tan\phi}{a}\left(\frac{M_E M_N}{R}\right)$$

$$+\frac{Rb^2}{(2\omega\sin\phi)^2}\left\{\frac{\partial}{\partial e}\left(\frac{\partial\theta}{\partial n}\right)^2 - \frac{\partial}{\partial n}\left(\frac{\partial\theta}{\partial e}\cdot\frac{\partial\theta}{\partial n}\right) + \frac{2\tan\phi}{a}\frac{\partial\theta}{\partial e}\cdot\frac{\partial\theta}{\partial n}\right\}. \qquad \text{..............................(7)}$$

Note that $\frac{\partial^2\theta}{\partial e\,\partial n}$ is not equal to $\frac{\partial^2\theta}{\partial n\partial e}$.

$$-\frac{\partial M_N}{\partial t} = -g_N R + \frac{\partial P}{\partial n} + \frac{\partial}{\partial e}\left(\frac{M_N M_E}{R}\right) + \frac{\partial}{\partial n}\left(\frac{M_N^2}{R}\right) - [m_N v_H]_2 + 2\omega\sin\phi M_E + \frac{3}{a}\int_2^0 m_N v_H dh + \frac{\tan\phi}{a}\left(\frac{M_E^2 - M_N^2}{R}\right)$$

$$+\frac{Rb^2}{(2\omega\sin\phi)^2}\left\{-\frac{\partial}{\partial e}\left(\frac{\partial\theta}{\partial e}\cdot\frac{\partial\theta}{\partial n}\right) + \frac{\partial}{\partial n}\left(\frac{\partial\theta}{\partial e}\right)^2 + \frac{\tan\phi}{a}\left[\left(\frac{\partial\theta}{\partial n}\right)^2 - \left(\frac{\partial\theta}{\partial e}\right)^2\right]\right\}. \qquad \text{.........(8)}$$

The terms in $v_H$, $m_H$, $M_H$ cannot be completely evaluated until the variation of $v_H$ with height has been determined for the particular time and place.

* These important equations are put in small print merely on account of their length.

## CH. 6/5. RADIATION IN THE STRATOSPHERE

### CH. 6/5/1. THE AVERAGE CONDITION

The independence of temperature on height was explained in 1909 by Gold and Humphreys as being a state of radiation-equilibrium undisturbed by mixing. This explanation has been further developed by Emden, by Friedmann and by Hergesell*.

In applying the process and constants developed in Ch. 4/7/1 to the example of Ch. 9, evidence is obtained that the lower side of the stratosphere has the greater flux of long-wave radiation passing through it. It is here the sum of the upward and downward fluxes which is meant. Accordingly it is difficult to see why the temperature in the stratosphere should not decrease a little upwards, unless the absorptivity for long waves is greater in the upper regions. The ratio of the sum-of-the-two-fluxes is as 613 at 11·7 km to 397 at the top. This is the same as the ratio of $(220)^4$ to $(197)^4$ so that temperatures in radiation equilibrium might be as 220° A at 11·7 km to 197° A at the top if absorptivity were uniform in height and wave length and sunshine were neglected. However that is not a large change. And a proper treatment of wave lengths might show it to be different.

### CH. 6/5/2. EVANESCENCE OF LOCAL DISTURBANCES

The data obtained also permit us to calculate the rate at which a small local disturbance of temperature would die away on account of radiation†. Suppose a thin horizontal layer of air $\delta p$ millibars thick is somehow raised $\delta\theta$ degrees centigrade above the temperature required for thermal equilibrium. A black surface at the temperature $\theta$ radiates $\mathfrak{O}\theta^4$ in all directions to one side. The increased emission due to the rise $\delta\theta$ is $4\mathfrak{O}\theta^3\delta\theta$ which is equal to

$$4{\cdot}85\delta\theta \text{ calorie cm}^{-2}\text{ day}^{-1} \text{ when } \theta = 220^\circ \text{ A}.$$

Now in Ch. 4/7/1#16 it is shown, from A. Ångström's observations, that a layer $\delta p$ millibars thick radiates by $0{\cdot}00302\delta p$ times the radiation for a black surface. This is for radiation scattered in all directions in the actual manner to one side. Doubling the figure, since the air radiates to both sides, we get for the loss

$$2\times 0{\cdot}00302\delta p \times 4{\cdot}85\delta\theta = 0{\cdot}0293\delta p\delta\theta\ \frac{\text{calorie}}{\text{cm}^2\text{ day}}.$$

But the thermal capacity of the layer, at constant pressure, is

$$0{\cdot}242\delta p\ \frac{\text{calorie}}{\text{cm}^2\text{ degree}}.$$

* E. Gold, *Roy. Soc. Lond. Proc.* A, Vol. 82, also Met. Office, London, *Geophys. Mem.* 5, p. 126; Humphreys, *Astrophysical Journal*, 1909; R. Emden, *Münch. Akad. Ber.* 1913, S. 55, said to be an important paper: I regret that I have not seen it; A. Friedmann, *Met. Zeit.* 1914 March; Hergesell (Lindenberg, XIII Bd.) reviewed in *Met. Zeit.* 1920 Aug.

† Compare E. Gold's analysis of surface observations, *Q. J. R. Met. Soc.* 1913 October.

Whence we conclude that "*if a thin horizontal layer in the stratosphere has its temperature slightly raised above that which would be in equilibrium with its surroundings, the disturbance will be decreased to $e^{-1}$ of itself, by radiation, in 8·3 days.*" ..............(1)

Now seeing that cyclones often sweep past a given place in a time short in comparison with 8·3 days, it is difficult to see how the uniformity of temperature with height in the stratosphere can be maintained by radiation alone.

One is led to the conclusion that the distribution of velocities must be such, that the adiabatic changes of temperature are the same at all levels.

This conclusion might be avoided if the stratosphere contained a larger proportion than the lower layers of some constituent which absorbs long wave radiation; for in calculating the absorptivity in Ch. 4/7/1 it was assumed that this quantity was the same at all heights provided the air were dry. The percentage of ozone has been observed by Pring*. He concludes that no very large increase in ozone content occurs between 5 kilometres and 20 kilometres. Also, according to A. Ångström†, spectroscopic observations have shown that in summer there is not enough ozone to have an appreciable effect. Is it possible that the presence of free ions increases the conversion of long wave radiation into heat?

Again, the inconsistency between (1) and a theory depending simply on radiation may conceivably disappear when enough is known about the selective absorption of dry air for certain wave lengths to enable calculations like those of Ch. 4/7/1 to be carried out strictly.

## CH. 6/6. VERTICAL VELOCITY IN THE STRATOSPHERE

This will now be investigated by means of the general equation Ch. 5/5#7. It is assumed that $\partial\theta/\partial h = 0$, but only at an initial instant, not necessarily subsequently. In other words

$$\frac{\partial^2\theta}{\partial h\,\partial t} \text{ is arbitrary.} \qquad \text{.........(1)}$$

As the air is dry the precipitation-term vanishes, the specific heats are strictly constant and $\gamma_p/\gamma_v = 1{\cdot}405$.

The second member of the general equation for vertical velocity must be expressed in terms of $\theta$, $p_2$, $M_{E20}$, $M_{N20}$, for these comprise all the information that is provided by the general process of computing. The discussion of what should be and what should not be provided involves a review of the entire subject and is treated in Ch. 8. I here anticipate one of the conclusions. Further, although it would be possible to do without $M_{E20}$, $M_{N20}$ by expressing $v_E$, $v_N$ and $\rho$ at all levels in terms of $\theta$ and $p_2$, by an extensive use of the hypothesis of the geostrophic wind; yet, remembering that this hypothesis is only an approximation, it will be much better to use it only to

* *Roy. Soc. Proc.* A, 90, p. 218. † *Smithsonian Miscell.* Vol. 65, p. 87.

give the variation of wind with height, and to introduce, for the motion of the centre of mass, $M_E$ and $M_N$ as given directly by observations, or by the general process of computing. For the centre of mass of the uppermost conventional stratum is at the level $h_2 + b\theta/g$, and at this level we have by Ch. 6/3 # 14 and Ch. 6/1 # 8

$$v_E = M_{E20}/R_{20} = gM_{E20}/p_2. \quad \text{......(2)}$$

Also, by the geostrophic-wind-theory, as expressed in Ch. 6/3 # 8

$$\frac{\partial v_E}{\partial h} = -\frac{g}{2\omega \sin\phi}\frac{\partial \log\theta}{\partial n}. \quad \text{......(3)}$$

Combining (2) with (3) we get $v_E$ at any level in the stratosphere, thus

$$v_E = \frac{M_{E20}}{R_{20}} - \frac{g}{2\omega\sin\phi}\frac{\partial\log\theta}{\partial n}\left(h - h_2 - \frac{b\theta}{g}\right). \quad \text{......(4)}$$

Similarly from Ch. 6/3 # 7

$$v_N = \frac{M_{N20}}{R_{20}} + \frac{g}{2\omega\sin\phi}\frac{\partial\log\theta}{\partial e}\left(h - h_2 - \frac{b\theta}{g}\right). \quad \text{......(5)}$$

We have also from Ch. 6/1 # 5

$$p = p_2 e^{-(h-h_2)g/b\theta}, \quad \text{......(6)}$$

$$\rho = \frac{p_2}{b\theta} e^{-(h-h_2)g/b\theta}. \quad \text{......(7)}$$

Let us now prepare to insert these expressions for the horizontal velocity, pressure and density in the general equation for the vertical velocity Ch. 5/5 # 7. In so doing it is best to watch for any groups of terms which are independent of, or proportional to, height, for these can be integrated as in Ch. 6/3 # 14. In the first place by Ch. 5/5 # 8

$$\operatorname{div}_{EN} v = \frac{\partial}{\partial e}\left(\frac{M_{E20}}{R_{20}}\right) + \left(\frac{\partial}{\partial n} - \frac{\tan\phi}{a}\right)\left(\frac{M_{N20}}{R_{20}}\right)$$

$$+ \frac{g}{2\omega\sin\phi}\left\{-\frac{\partial^2\log\theta}{\partial e\,\partial n} + \frac{\partial^2\log\theta}{\partial n\,\partial e}\right\}\left(h - h_2 - \frac{b\theta}{g}\right) - \frac{\tan\phi}{a}\frac{g}{2\omega\sin\phi}\cdot\frac{\partial\log\theta}{\partial e}\left(h - h_2 - \frac{b\theta}{g}\right).$$

$$+ \frac{\partial\log\theta}{\partial e}\left(h - h_2 - \frac{b\theta}{g}\right)\frac{g}{2\omega}\frac{\partial}{\partial n}\left(\frac{1}{\sin\phi}\right)$$

$$+ \frac{g}{2\omega\sin\phi}\frac{b}{\theta}\left\{\frac{\partial\theta}{\partial e}\cdot\frac{\partial\log\theta}{\partial n} - \frac{\partial\theta}{\partial n}\cdot\frac{\partial\log\theta}{\partial e}\right\}. \quad \text{......(8)}$$

This equation simplifies considerably. The last bracket vanishes, since $d\log\theta = d\theta/\theta$. Again in the third term in the second member

$$\frac{\partial}{\partial n}\left(\frac{1}{\sin\phi}\right) = \frac{\partial}{a\,\partial\phi}\left(\frac{1}{\sin\phi}\right) = -\frac{1}{a}\frac{\cot\phi}{\sin\phi}. \quad \text{......(9)}$$

The second term does not vanish because the differentials $\partial e$, $\partial n$, which are so brief and expressive, are not independent of one another. To evaluate the term we must bring in the original independents: longitude $\lambda$ and latitude $\phi$. Then

$$-\frac{\partial^2 \log\theta}{\partial e\partial n}+\frac{\partial^2 \log\theta}{\partial n\partial e}=\frac{1}{a}\left\{-\frac{\partial^2 \log\theta}{\cos\phi\,.\,\partial\lambda\,.\,\partial\phi}+\frac{\partial}{\partial\phi}\left(\frac{\partial\log\theta}{\cos\phi\,.\,\partial\lambda}\right)\right\}$$

$$=\frac{1}{a}\frac{\partial\log\theta}{\partial\lambda}\frac{\partial}{\partial\phi}\left(\frac{1}{\cos\phi}\right)$$

$$=+\frac{\partial\log\theta}{\partial e}\,.\tan\phi. \quad\text{.................................(10)}$$

Collecting terms in (8) we arrive at

$$\text{div}_{EN}\,v=\frac{\partial}{\partial e}\left(\frac{M_{E20}}{R_{20}}\right)+\frac{\partial}{\partial n}\left(\frac{M_{N20}}{R_{20}}\right)-\frac{\tan\phi}{a}\left(\frac{M_{N20}}{R_{20}}\right)$$

$$-\frac{g}{2\omega\sin\phi}\cdot\frac{\partial\log\theta}{\partial e}\cdot\left(\frac{\cot\phi}{a}\right)\left(h-h_2-\frac{b\theta}{g}\right).\quad\text{.........(11)}$$

Thus $\text{div}_{EN}v$ is a linear function of height. The part dependent on height does not involve the northward gradient of temperature.

Again, in the general equation Ch. 5/5 # 7 for the vertical velocity, there occurs an expression which transforms thus by the aid of (4) and (5)

$$\frac{\partial v_E}{\partial h}\cdot\frac{\partial p}{\partial e}+\frac{\partial v_N}{\partial h}\cdot\frac{\partial p}{\partial n}=\frac{g}{2\omega\sin\phi}\left\{-\frac{\partial\log\theta}{\partial n}\cdot\frac{\partial p}{\partial e}+\frac{\partial\log\theta}{\partial e}\cdot\frac{\partial p}{\partial n}\right\}$$

$$=\Gamma \text{ say, for short.}\quad\text{...................................(12)}$$

But by Ch. 6/3 # 5

$$\frac{\partial}{\partial h}\left(\frac{1}{\rho}\frac{\partial p}{\partial e}\right)=g\,\frac{\partial\log\theta}{\partial e},\ \text{which is independent of } h.\quad\text{..............(13)}$$

So that $\partial p/\partial e$ is $\rho$ times a linear function of height. By a similar argument the same may be said of $\partial p/\partial n$ and therefore also of $\Gamma$. It follows that we can write $\Gamma/\rho$ as the sum of its value at any basal level, say at $h_2$, plus a quantity proportional to height above $h_2$. This second term is obtained from (13) and from the similar expression in $\partial n$.

Thus

$$\frac{\Gamma}{\rho}=\frac{g}{2\omega\sin\phi}\left\{-\frac{\partial\log\theta}{\partial n}\cdot\frac{1}{\rho_2}\frac{\partial p_2}{\partial e}+\frac{\partial\log\theta}{\partial e}\cdot\frac{1}{\rho_2}\frac{\partial p_2}{\partial n}\right\}$$

$$+(h-h_2)\cdot\frac{g}{2\omega\sin\phi}\left\{-\frac{\partial\log\theta}{\partial n}g\,\frac{\partial\log\theta}{\partial e}+\frac{\partial\log\theta}{\partial e}g\,\frac{\partial\log\theta}{\partial n}\right\}.\quad\text{........(14)}$$

The part of (14) which depends on $h$ is seen to vanish. That vanishing may be traced back to its cause in the assumed connection with the geostrophic wind.

The preceding expressions (11) and (14) have been prepared for insertion in the general equation for the vertical velocity Ch. 5/5 #7. Consider first the definite integral in the latter. It is

$$\int_h^{h_0}\left\{g\rho \,.\, \mathrm{div}_{EN} v + \frac{\partial v_E}{\partial h}\cdot\frac{\partial p}{\partial e} + \frac{\partial v_N}{\partial h}\cdot\frac{\partial p}{\partial n}\right\} dh, \quad \text{.................(15)}$$

which may be written

$$\int_h^{h_0}\rho\{A_1 + B_1(h - h_2)\}\, dh, \quad \text{..........................(16)}$$

where $A_1$ and $B_1$ are independent of height and have the following values given by (11) and (14)

$$\frac{A_1}{g} = \frac{\partial}{\partial e}\left(\frac{M_{E20}}{R_{20}}\right) + \frac{\partial}{\partial n}\left(\frac{M_{N20}}{R_{20}}\right) - \frac{\tan\phi}{a}\left(\frac{M_{N20}}{R_{20}}\right) + \frac{b\theta}{2\omega\sin\phi}\cdot\frac{\partial\log\theta}{\partial e}\cdot\left(\frac{\cot\phi}{a}\right)$$

$$+\frac{1}{2\omega\sin\phi}\left\{-\frac{\partial\log\theta}{\partial n}\cdot\frac{1}{\rho_2}\frac{\partial p_2}{\partial e} + \frac{\partial\log\theta}{\partial e}\cdot\frac{1}{\rho_2}\frac{\partial p_2}{\partial n}\right\}, \quad \text{........(17)}$$

$$\frac{B_1}{g} = -\frac{g}{2\omega\sin\phi}\,\frac{\partial\log\theta}{\partial e}\left(\frac{\cot\phi}{a}\right). \quad \text{........................................(18)}$$

The integral (15) can therefore be written down by the rule Ch. 6/3 #14 in the form

$$\frac{p}{g}\left\{A_1 + B_1\left(h + \frac{b\theta}{g} - h_2\right)\right\}, \quad \text{..........................(19)}$$

so that the general equation for the vertical velocity, Ch. 5/5 #7, now runs

$$\frac{\partial v_H}{\partial h} = \frac{\gamma_v}{\gamma_p \,.\, g}\left\{A_1 + B_1\left(h + \frac{b\theta}{g} - h_2\right)\right\} - \mathrm{div}_{EN} v + \frac{1}{\gamma_p}\frac{D\sigma}{Dt}. \quad \text{...........(20)}$$

The terms in (20) can again be arranged so that

$$\frac{\partial v_H}{\partial h} = -\frac{1}{g}\{\Upsilon + (h - h_2)\Phi\} + \frac{1}{\gamma_p}\frac{D\sigma}{Dt}, \quad \text{.......................(21)}$$

where $\Upsilon$ and $\Phi$ are quantities independent of height, and have the following values in terms of the known quantities $M_{E20}$, $M_{N20}$, $p_2$, $\theta$:

$$\frac{\Upsilon}{g} = \left(1 - \frac{\gamma_v}{\gamma_p}\right)\left\{\frac{\partial}{\partial e}\left(\frac{M_{E20}}{R_{20}}\right) + \left(\frac{\partial}{\partial n} - \frac{\tan\phi}{a}\right)\left(\frac{M_{N20}}{R_{20}}\right)\right\}$$

$$+\frac{b\theta}{2\omega\sin\phi}\left[\frac{\gamma_v}{\gamma_p}\left(\frac{\partial\log\theta}{\partial n}\cdot\frac{\partial\log p_2}{\partial e} - \frac{\partial\log\theta}{\partial e}\cdot\frac{\partial\log p_2}{\partial n}\right) + \frac{\partial\log\theta}{\partial e}\left(\frac{\cot\phi}{a}\right)\right]. \quad \text{...(22)}$$

The dimension of $\Upsilon/g$ is the reciprocal of a time.

Also $$\frac{\Phi}{g} = -\left(1 - \frac{\gamma_v}{\gamma_p}\right)\frac{g}{2\omega\sin\phi}\cdot\frac{\partial\log\theta}{\partial e}\cdot\left(\frac{\cot\phi}{a}\right), \quad \text{..............(23)}$$

so that $\Phi/g$ is of dimensions: $(\text{time})^{-1}\,(\text{length})^{-1}$. These expressions $\Upsilon$ and $\Phi$ appear on the computing form. As the air is dry

$$1 - \frac{\gamma_v}{\gamma_p} = 1 - \frac{1}{1{\cdot}405} = 0{\cdot}288. \quad \text{.........................(24)}$$

We must next integrate (21) to find the vertical velocity. Before we can do that we need the relation of $D\sigma/Dt$ to height. As precipitation and stirring are absent $D\sigma$ arises wholly from radiation. The stratosphere is obviously very transparent to solar radiation and, if the calculations of Ch. 4/7/1 are reliable, the stratosphere only absorbs about $\frac{2}{5}$ of the diffuse long-wave radiation incident upon it. Under the circumstances the rise of temperature and the gain *of* entropy-per-mass will be roughly the same at all heights. Thus we assume that

$$\frac{1}{\gamma_p}\frac{D\sigma}{Dt} \text{ is independent of height.} \qquad \text{.......................(25)}$$

Then (21) integrates, yielding

$$v_H = -\frac{h}{g}\{\Upsilon + (\tfrac{1}{2}h - h_2)\,\Phi\} + \frac{1}{\gamma_p}\frac{D\sigma}{Dt}h + F, \qquad \text{..................(26)}$$

where the arbitrary "constant" of integration $F$ is a function of latitude and longitude determinable by the value of $v_H$ at $h_2$.

Thus eliminating $F$ by the introduction of $v_{H2}$ it follows that

$$v_H - v_{H2} = (h - h_2)\left\{\frac{1}{\gamma_p}\frac{D\sigma}{Dt} - \frac{\Upsilon}{g}\right\} - \tfrac{1}{2}(h - h_2)^2\,\frac{\Phi}{g}. \qquad \text{..............(27)}$$

Equation (27) shows among other things that *the vertical velocity in the stratosphere is a quadratic function of height.* It would become a linear function if $\Phi$ were zero, and (26) shows that $\Phi$ would be zero if the temperature had no east-west variation, or if the earth could be regarded as flat. See the conclusion of Ch. 6/7/2.

We shall subsequently require $M_{H20}$, **the total upward momentum of the air above the level** $h_2$. To find this multiply (27) by $\rho$ and integrate by parts. It follows that

$$M_{H20} = v_{H2}\frac{p_2}{g} + \frac{b\theta}{g}\cdot\frac{p_2}{g}\left\{\frac{1}{\gamma_p}\frac{D\sigma}{Dt} - \frac{\Upsilon}{g}\right\} - \left(\frac{b\theta}{g}\right)^2\cdot\frac{p_2}{g}\cdot\frac{\Phi}{g}. \qquad \text{...........(28)}$$

We are now in a position to evaluate **the terms** $\frac{3}{a}\int_2^0 m_E v_H\,dh$, $\frac{3}{a}\int_2^0 m_N v_H\,dh$, **which are required for the dynamical equations.** It will simplify the analysis to reckon all heights from $h_2 + b\theta/g$, the level of the centre of mass. Then, as $v_E$ and $v_H$ are respectively linear and quadratic functions of height, we may put

$$v_E = A + A'\left(h - h_2 - \frac{b\theta}{g}\right), \qquad \text{.....................................(29)}$$

$$v_H = B + B'\left(h - h_2 - \frac{b\theta}{g}\right) + B''\left(h - h_2 - \frac{b\theta}{g}\right)^2, \qquad \text{..................(30)}$$

where $A$, $A'$, $B$, $B'$, $B''$ are independent of height and need not be further specified.

$$\text{So } v_E v_H = AB + (A'B + AB')\left(h - h_2 - \frac{b\theta}{g}\right) + (A'B' + AB'')\left(h - h_2 - \frac{b\theta}{g}\right)^2$$
$$+ A'B''\left(h - h_2 - \frac{b\theta}{g}\right)^3. \qquad \text{..............(31)}$$

Next multiply (31) by $\rho$ and integrate the product "by parts" four times, remembering that by Ch. 6/6 #7

$$\int \rho\, dh + \text{const.} = -\frac{b\theta}{g}\rho = -\frac{p}{g}. \qquad \text{.......................(32)}$$

Then, on putting in the limits, the linear term disappears, leaving

$$\int_2^0 \rho v_E v_H dh = R_{20}\left\{AB + \left(\frac{b\theta}{g}\right)^2 (A'B' + AB'') + \left(\frac{b\theta}{g}\right)^3 2A'B''\right\}. \qquad \text{......(33)}$$

But, by similar integrations by parts,

$$\int_2^0 \rho v_E dh = R_{20} A = M_{E20}\,; \quad \int_2^0 \rho v_H dh = R_{20}\left\{B + B''\left(\frac{b\theta}{g}\right)^2\right\} = M_{H20}. \quad \text{...(34), (35)}$$

And on substituting these in the right-hand side of equation (33), it follows that

$$\frac{3}{a}\int_2^0 \rho v_E v_H dh = \frac{3}{a}\frac{M_{E20}\,.\,M_{H20}}{R_{20}} + \frac{3}{a}R_{20}\left(\frac{b\theta}{g}\right)^2 \frac{\partial v_E}{\partial h}\left\{\frac{\partial v_H}{\partial h} \text{ at } \left(h_2 + \text{twice } \frac{b\theta}{g}\right)\right\}, \text{...(36)}$$

which is in suitable form for computing. As $v_H$, $M_H$ are only found directly over the points on the map where pressure is tabulated, the mean of four surrounding values must be used here.

## CH. 6/7. DYNAMICAL CHANGES OF TEMPERATURE IN THE STRATOSPHERE

### CH. 6/7/1. MEAN HORIZONTAL TEMPERATURE GRADIENTS

Radiation-equilibrium does not explain why, when tropical are compared with polar regions or anticyclones with cyclones, the stratosphere is found in each case to be cooler where the air beneath it is warmer. Prof. V. Bjerknes* has published a dynamical explanation. Taking the case of the pole and equator he begins by assuming, as is more or less borne out by observation, that the lower part of the atmosphere is spinning about the polar axis more rapidly than the earth, while the upper part is spinning less rapidly than the lower part. From the distribution of velocity he finds the form of the isobaric surfaces, assuming that the motion is steady, and that it is uncomplicated by minor circulations. The vertical separation of a pair of adjacent isobaric surfaces varies from place to place proportionately to the volume-per-weight of the air between them. Bjerknes takes the volume-per-weight to be proportional to the absolute temperature. This last step is unconvincing as the weight depends not only on latitude but also on the eastward velocity relative to the earth. However a closer examination confirms Bjerknes' remarkable result, unless the variation of speed with latitude is unnaturally large. In any case it relates only to a steady motion along the parallels of latitude.

Bjerknes' theory of the temperature difference between cyclones and anticyclones is similar to the above, except that the rotation considered, instead of being around the polar axis, is around the vertical.

* V. Bjerknes, *Comptes Rendus*, Paris, t. 170, p. 604 (1920).

In the same publications Prof. Bjerknes explains the smaller height of the tropopause in cyclones by the fact that the angular velocity of the troposphere about a vertical axis is greater in cyclones than in anticyclones, while the angular velocity of the stratosphere is in both cases small. This explanation agrees with one given in other terms by Mr W. H. Dines*. It may easily be illustrated by floating oil on water in a glass vessel and stirring the lower layer.

But we must pass on; for in this book we are not concerned to explain mean distributions, but to forecast.

### CH. 6/7/2. DYNAMICAL INFLUENCE ON VERTICAL ISOTHERMY

It is indicated in Ch. 6/5 that radiation alone is probably insufficient to maintain the uniformity of temperature in a column. Let us enquire whether the compressions and rarefactions are so distributed as to produce the same adiabatic temperature changes at all heights. Sir Napier Shaw† has discussed this problem in a special case—that of the initial motion of a stratosphere disturbed from rest by the passage of a cyclone beneath it—and the conclusion is that the temperature change is the same at all points of a vertical. Let us now rework the problem more generally.

Indeed it must be solved as part of the general process of forecasting for the uppermost stratum. The case of the lower strata is different. In them we find $\partial R/\partial t$ from the equation of continuity of mass. Next by summing $g \,.\, \partial R/\partial t$ downwards from the top we get $\partial p/\partial t$ at the boundaries of the strata. Then by a process equivalent to interpolation a change of pressure is found for a mean level in the stratum so as to correspond to $\partial R/\partial t$, or to $\partial \rho/\partial t$ which follows from $\partial R/\partial t$. Having thus obtained $\partial p$ and $\partial \rho$ at the same level, the temperature change follows at once from $p = b\rho\theta$.

This process fails in the uppermost stratum for there $\partial R_{20}$ gives $\partial p_2$, and we have $\partial p_0 = 0$, but the range of pressure is far too great to allow us to interpolate between $\partial p_2$ and $\partial p_0$ a value for comparison with some mean density to be derived from $\partial R_{20}$. The failure is thus not due to isothermal conditions but to the application of finite differences to a quantity varying in an infinite ratio.

But we can proceed as follows. The equation for the conveyance of heat Ch. 4/5/2 # 1 is

$$\frac{\partial \sigma}{\partial t} + v_E \frac{\partial \sigma}{\partial e} + v_N \frac{\partial \sigma}{\partial n} + v_H \frac{\partial \sigma}{\partial h} = \frac{D\sigma}{Dt}, \quad \text{.......................(1)}$$

where $D\sigma$ is the change due to radiation which has been discussed in Ch. 6/5. As the instantaneous distribution of velocity components may be supposed, in view of the discussions in Ch. 6/3, Ch. 6/6, to be known at all heights, and as the same applies to the entropy-per-mass, equation (1) may be expected to give us temperature changes at all heights. In the following process it is assumed that

$$\partial\theta/\partial h = 0, \text{ all over the map at the instant considered,} \quad \text{............(2)}$$

* W. H. Dines, *Phil. Trans.* A, Vol. 211, p. 276.

† "The Perturbations of the Stratosphere," *Meteor. Office London Publications*, No. 202, 1909.

but not necessarily at neighbouring times before or after so that

$$\frac{\partial^2\theta}{\partial h\,\partial t} \text{ is arbitrary.} \qquad\qquad (3)$$

Begin with the expression for the entropy-per-mass Ch. 4/5/1 # 8 which may be written

$$d\sigma = -b\,.\,d\log p + \gamma_p\,.\,d\log\theta. \qquad\qquad (4)$$

Then by (2)

$$\frac{\partial\sigma}{\partial h} = -b\frac{\partial\log p}{\partial h}. \qquad\qquad (5)$$

But by the hydrostatic equation

$$\frac{\partial\log p}{\partial h} = -\frac{g}{b\theta}. \qquad\qquad (6)$$

So from (5) and (6)

$$\frac{\partial\sigma}{\partial h} = \frac{g}{\theta}. \qquad\qquad (7)$$

Thus *in the stratosphere the entropy-per-mass is a linear function of height increasing by* $g/\theta$ *per unit of height.* The statement applies to our hypothetical initial instant or to any other instant at which $\partial\theta/\partial h$ is shown by observation to vanish everywhere. In these circumstances $\partial\sigma/\partial e$, $\partial\sigma/\partial n$ are also linear functions of height although they increase at a different rate thus

$$\frac{\partial^2\sigma}{\partial h\,\partial e} = -\frac{g}{\theta^2}\frac{\partial\theta}{\partial e}; \qquad \frac{\partial^2\sigma}{\partial h\,\partial n} = -\frac{g}{\theta^2}\frac{\partial\theta}{\partial n}. \qquad\qquad (8)$$

It appears from (7) that temperature can be found more directly from $\partial\sigma/\partial h$ than from $\sigma$. Accordingly let us differentiate (1) with respect to height. The result is

$$\frac{\partial^2\sigma}{\partial h\,\partial t} + \frac{\partial v_E}{\partial h}\cdot\frac{\partial\sigma}{\partial e} + \frac{\partial v_N}{\partial h}\cdot\frac{\partial\sigma}{\partial n} + \frac{\partial v_H}{\partial h}\cdot\frac{\partial\sigma}{\partial h} + v_E\frac{\partial^2\sigma}{\partial e\,\partial h} + v_N\frac{\partial^2\sigma}{\partial n\,\partial h} + v_H\frac{\partial^2\sigma}{\partial h^2} = \frac{\partial}{\partial h}\left(\frac{D\sigma}{Dt}\right). \qquad (9)$$

Substitute in (9) expressions in $\theta$ for derivatives of $\sigma$, by making use of (4), (7), and (8). Then (9) becomes

$$\frac{\partial}{\partial t}\left(\frac{g}{\theta}\right) + \frac{\partial v_E}{\partial h}\left\{-\frac{b\,.\,d\log p}{\partial e} + \gamma_p\frac{\partial\log\theta}{\partial e}\right\} + \frac{\partial v_N}{\partial h}\left\{-\frac{b\,.\,d\log p}{\partial n} + \gamma_p\frac{\partial\log\theta}{\partial n}\right\}$$

$$+\frac{\partial v_H}{\partial h}\cdot\frac{g}{\theta} - \frac{g}{\theta^2}\left\{v_E\frac{\partial\theta}{\partial e} + v_N\frac{\partial\theta}{\partial n}\right\} = \frac{\partial}{\partial h}\left(\frac{D\sigma}{Dt}\right). \qquad (10)$$

It will be found that the theory is simplified if we differentiate (10) once more with respect to height, making use of (2) and of the hydrostatic equation in the form

$$\frac{\partial\log p}{\partial h} = -\frac{g}{b\theta}. \qquad\qquad (11)$$

Thus we obtain

$$-\frac{g}{\theta^2}\frac{\partial^2\theta}{\partial h\partial t}+\frac{\partial^2 v_E}{\partial h^2}\left\{-\frac{b\,.\,d\log p}{\partial e}+\gamma_p\frac{\partial\log\theta}{\partial e}\right\}+\frac{\partial^2 v_N}{\partial h^2}\left\{-\frac{b\,.\,d\log p}{\partial n}+\gamma_p\frac{\partial\log\theta}{\partial n}\right\}$$

$$+\frac{g}{\theta}\frac{\partial^2 v_H}{\partial h^2}-2\frac{g}{\theta^2}\left\{\frac{\partial v_E}{\partial h}\cdot\frac{\partial\theta}{\partial e}+\frac{\partial v_N}{\partial h}\cdot\frac{\partial\theta}{\partial n}\right\}=\frac{\partial^2}{\partial h^2}\left(\frac{D\sigma}{Dt}\right)\quad\text{.................(12)}$$

So far no assumption has been made which is not well-founded, with occasional exceptions in respect to (2), the isothermal state of the column.

To simplify (12) we must bring in the relation of horizontal velocity-components to height. We therefore now make the geostrophic assumption about this height-variation which is accordingly expressed by Ch. 6/3 # 7, 8, thus

$$\frac{\partial v_N}{\partial h}=+\frac{g}{2\omega\sin\phi}\frac{\partial\log\theta}{\partial e};\qquad\frac{\partial v_E}{\partial h}=-\frac{g}{2\omega\sin\phi}\frac{\partial\log\theta}{\partial n}.\quad\text{......(13)}$$

On introducing (13) the three terms of (12) in curly brackets disappear, and there is left after multiplication by $-\theta/g$

$$\frac{1}{\theta}\frac{\partial^2\theta}{\partial h\partial t}=\frac{\partial^2 v_H}{\partial h^2}-\frac{\theta}{g}\frac{\partial^2}{\partial h^2}\left(\frac{D\sigma}{Dt}\right).\quad\text{.........................(14)}$$

This relation between vertical velocity and temperature change is interesting. Now the theory of vertical velocity in the stratosphere, already given in Ch. 6/6, is based upon the use or avoidance of exactly the same assumptions as those used or avoided here, and in particular on (3). Differentiating the formula for $\partial v_H/\partial h$, namely Ch. 6/6 # 21 it follows that

$$\frac{\partial^2 v_H}{\partial h^2}=-\frac{\Phi}{g}+\frac{1}{\gamma_p}\frac{\partial}{\partial h}\left(\frac{D\sigma}{Dt}\right),\quad\text{..............................(15)}$$

where by Ch. 6/6 # 23

$$\frac{\Phi}{g}=-\left(1-\frac{\gamma_v}{\gamma_p}\right)\frac{g}{2\omega\sin\phi}\cdot\frac{\partial\log\theta}{\partial e}\cdot\frac{(\cot\phi)}{a},\quad\text{...........(15 A)}$$

so that $\Phi$ is independent of height. On eliminating $\partial^2 v_H/\partial h^2$ from (14) by means of (15 the result is

$$\frac{\partial}{\partial t}\left(\frac{\partial\log\theta}{\partial h}\right)=-\frac{\Phi}{g}+\left\{\frac{1}{\gamma_p}\frac{\partial}{\partial h}-\frac{\theta}{g}\frac{\partial^2}{\partial h^2}\right\}\left(\frac{D\sigma}{Dt}\right),\quad\text{.................(16)}$$

since as $\partial\theta/\partial h=0$ it follows that

$$\frac{\partial}{\partial t}\left(\frac{\partial\log\theta}{\partial h}\right)=\frac{1}{\theta}\frac{\partial^2\theta}{\partial h\partial t}.\quad\text{.................................(17)}$$

*The condition that the vertical distribution should not be departing from isothermy is that the second member of* (16) *should vanish.* We may distinguish the terms in it calling $-\Phi/g$ the adiabatic term and the rest the radiation term. In the latter $D\sigma/Dt$ has already been discussed in Ch. 6/5 and its smallness suggested that the adiabatic term must also be small, for otherwise $\partial^2\theta/(\partial h\partial t)$ would be noticeable in the observations.

For comparison with balloon records the adiabatic term may be put into a form which avoids horizontal temperature gradients, which have large observational errors. For, neglecting radiation, (13), (14), (15), (15 A) yield

$$\frac{1}{\theta}\frac{\partial^2\theta}{\partial h\partial t}=0{\cdot}288\,.\,\frac{\partial v_N}{\partial h}\frac{(\cot\phi)}{a}\,.\qquad\text{(18)}$$

Here 0·288 is the value of $1-\gamma_v/\gamma_p$ for dry air.

The presence of an infinite $\tan\phi$ at the pole is of purely geometrical origin. The term in $\cot\phi$, infinite at the equator, comes from $\frac{\partial}{\partial n}\left(\frac{1}{\sin\phi}\right)$ and so arises from the known failure at the equator of the geostrophic approximation. But let us consider middle latitudes.

Both $\partial\theta/\partial t$ and $\partial\theta/\partial e$ have their ups and downs, so that on the average of many balloon ascents both sides of equation (16), apart from the radiation term, are likely to vanish, without yielding any check on the theory. Individual sets of observations, sufficiently comprehensive to test (18) are scarce, but I have found a few. One relates to 1920 May 19, the day before that of the example of Ch. 9. Strassburg is 150 km NNW of Zürich and for the present purposes, on account of shortage of observations, we must regard them as one place. The theodolite observations yield the following up-grades of the north component of wind

| | | $\partial v_N/\partial h$ |
|---|---|---|
| May 19ᵈ 8ʰ | Zürich ... ... | $-2{\cdot}0\times10^{-3}\,\text{sec}^{-1}$ |
| May 20ᵈ 8ʰ | Strassburg ... ... | $-1{\cdot}9\times10^{-3}\,\text{sec}^{-1}$ |
| „ „ | Zürich ... .. | $-0{\cdot}6\times10^{-3}\,\text{sec}^{-1}$. |

The first of these is read from the diagram on p. 129. The last is very doubtful, being obtained by rejecting the highest reading. There is however an indication of persistent decrease of $v_N$ with height during this 24 hours, and we may put the mean at

$$\partial v_N/\partial h=-1{\cdot}5\times10^{-3}\,\text{sec}^{-1}.$$

At this latitude

$$a^{-1}(\tan\phi+\cot\phi)=3{\cdot}15\times10^{-9}\,\text{cm}^{-1}.$$

It follows from (18) that

$$\frac{1}{\theta}\frac{\partial^2\theta}{\partial h\partial t}=-1{\cdot}4\times10^{-2}\,\text{cm}^{-1}\,\text{sec}^{-1}.$$

The sign implies that the temperature would decrease more rapidly with time in the higher levels. Let us find the difference in temperature-increases during one day at two points separated vertically by one kilometre. The required quantity will, according to (16), be obtained by multiplying $-1{\cdot}4\times10^{-2}\,\text{cm}^{-1}\,\text{sec}^{-1}$ by the temperature, by 86,400 seconds, and by $10^5$ cms. It works out to

$$\frac{\partial^2\theta}{\partial h\partial t}=-2{\cdot}7^\circ\,\text{C per km per day}.$$

Now the balloon observations provide a check on this figure, for the record at Strassburg reaches to 15·5 km at 8 o'clock on both days. The differences between the two temperature curves are within the observational error. On both mornings the lowest temperature is at about 12 km and there is a large recovery above that. It may be said with some confidence that between 12 and 15·5 km the observations show that $\frac{\partial^2\theta}{\partial h\partial t}$ is less in absolute value than 1° C per km per day.

The table summarizes this comparison and adds others given thus briefly. They all are taken from the international balloon ascents* for the year 1910.

| Place | Time G.M.T. Year 1910 | $\frac{\partial\theta}{\partial h}$ °C/km | $0{\cdot}288\,\theta \cdot \frac{\partial v_N}{\partial h} \cdot \frac{(\tan\phi+\cot\phi)}{a}$ °C/km.day | $\frac{\partial^2\theta}{\partial h\partial t}$ °C/km.day |
|---|---|---|---|---|
| Zürich 12 to 14 k | Feb. $3^d\,9^h$ | | −2·3 | |
| Strassburg 12 to 14 k | $2^d\,8^h$ to $3^d\,8^h$ | 0·0 | | −4·0 |
| Zürich, Strassburg 10 to 20 k | May $19^d\,8^h$, $20^d\,8^h$ | | −2·7 | |
| Strassburg 12 to 15·5 k | $19^d\,8^h$ to $20^d\,8^h$ | +1·6 | | 0·0 |
| Lindenberg 14 to 17 k | May $19^d\,4^h$ | | +0·9 | |
| | $18^d\,8^h$ to $19^d\,4^h$ | +0·6 | | +1·7 |
| Vienna 14 to 18 k | May $19^d\,8^h$ | | −0·9 | |
| | $19^d\,4^h$ to $8^h$ | +0·6 | | −2·6 |
| The temperatures were taken from two thermometers of the same pattern, those by a different pattern being rejected. | | | | |
| Uccle 14 to 20 k | Aug. $10^d\,8^h$ | | +0·3 | |
| 14 to 16·5 k | $9^d\,8^h$ to $11^d\,8^h$ | +0·3 | | −0·2 |
| Hamburg 12 to 14 k | Aug. $9^d\,8^h$ | | +0·4 | |
| | $8^d\,8^h$ to $11^d\,8^h$ | 0·0 | | +0·2 |
| Lindenberg 10 to 15 k | Aug. $9^d\,8^h$ | | +1·4 | |
| | $8^d\,8^h$ to $9^d\,8^h$ | 0·1 | | 0·0 |

The numbers in the last two columns of the table should, according to equation (18), be equal in each row. Actually their standard deviations are of the same order, and there is a suggestion of a positive correlation between them. The observational errors are considerable. Before we can draw any sure conclusion there is a need for discussing

* *Beobachtungen mit bemannten, unbemannten Ballons,......*, 1910, herausgegeben von Prof. Dr H. Hergesell; Strassburg, Druck von M. du Mont Schauberg.

observations, more in number, further removed from the disturbed layer at the base of the stratosphere, and having the time-step correctly centred. But it looks as though equation (18) would be found to be inaccurate. As I have carefully checked its deduction, I suspect the error to reside in the hypotheses. It might be in the neglect of radiation, especially in the long-wave part. Now Mr W. H. Dines* has calculated the warming or cooling at different levels by long-wave radiation in various circumstances. Over England between 11·7 and 16·0 k on the average of his four diagrams

$$\frac{\partial^2\theta}{\partial h \partial t} = -0{\cdot}2^\circ \text{ C per km per day.}$$

That is a small number compared with some of those in the table, and so probably neglect of radiation is not the cause of the discrepancies, unless indeed the absorptivity per mass for long or short waves varies remarkably with height. Instead suspicion fastens on the hypothesis of geostrophic variation of wind with height, because in Ch. 2 a completer use of the geostrophic hypothesis was found to lead to pressure-changes having an unnatural sign †. In the stratosphere the correlation between observed and geostrophic is apparently going to be at least positive. However that may be, yet for purposes not connected with finding the divergence of wind, the geostrophic hypothesis appears to serve as well in the stratosphere as elsewhere. Of this Sir Napier Shaw gives one good example in his *Manual of Meteorology*, IV, p. 91, and the diagrams in the present chapter illustrate the same thing in a different way.

### CH. 6/7/3. TO FIND THE TIME-RATE OF TEMPERATURE

It is not clear just at what stage of the proceedings the geostrophic wind ceases to be a good approximation. For the immediate purpose of computing the temperature changes in the stratosphere, it seems best to take the permanence of vertical isothermy as an observed fact, and therefore to neglect $\Phi$ wherever it occurs, both in the equation for temperature change and in that for vertical velocity. As $\partial\theta/\partial t$ is taken as independent of height, we can find it at any one height. The height of the centre of mass $h_2 + b\theta/g$ commends itself as being the one that would be chosen on general principles, if there were no theory at all concerning variations with height.

The locus of the centres of mass of all vertical-sided columns, having their bases at a common level $h_2$, is not quite a level-surface, since $b\theta/g$ varies. However that does not appear to lead to any error in the following theory, in which the east and north differentiations are made strictly on the horizontal. Similarly time-changes must be taken at fixed heights.

Beginning again at the equation for the conveyance of heat Ch. 4/5/2 # 1 and expanding $d\sigma$ we have

$$\gamma_p\left(\frac{\partial}{\partial t} + v_E\frac{\partial}{\partial e} + v_N\frac{\partial}{\partial n} + v_H\frac{\partial}{\partial h}\right)\log\theta - b\left(\frac{\partial}{\partial t} + v_E\frac{\partial}{\partial e} + v_N\frac{\partial}{\partial n} + v_H\frac{\partial}{\partial h}\right)\log p = \frac{D\sigma}{Dt} \quad \ldots(1)$$

* W. H. Dines, *Q. J. R. Met. Soc.* 1920 April. In finding the temperature-change from the radiation he apparently ignores any vertical velocity produced by the radiation. See Ch. 5/4, Case ii.

† See p. 10.

One of the terms in $v_H$ disappears. In the other put $\partial \log p/\partial h = -g/b\theta$. Then take the equation to apply to the height $h_2 + b\theta/g$ so that by Ch. 6/3 # 14 and Ch. 6/1 # 8

$$v_E = M_{E20}/R_{20}, \quad v_N = M_{N20}/R_{20}. \quad \ldots\ldots(2)$$

Since we now neglect $\Phi$ we thereby take the vertical velocity to be, like the east and north velocities, a linear function of height and therefore

$$v_H = M_{H20}/R_{20}. \quad \ldots\ldots(3)$$

In simplifying the term in (1) which contains $\log p$ we must remember that the differentials of $\log p$ in latitude, longitude and time are to be taken at the *fixed* height $h_2 + b\theta/g$, so that $\theta$ in this expression alone must be regarded as fixed at its instantaneous local value, which we may distinguish by a dash. In all other connections $\theta$ is variable and undashed. Thus

$$\log p = \log p_2 + \frac{b\theta'}{g} \cdot \frac{\partial \log p}{\partial h} = \log p_2 - \frac{\theta'}{\theta}. \quad \ldots\ldots(4)$$

Whence

$$\frac{\partial \log p}{\partial t} = \frac{\partial \log p_2}{\partial t} + \frac{\theta'}{\theta^2}\frac{\partial \theta}{\partial t} = \frac{\partial \log p_2}{\partial t} + \frac{\partial \log \theta}{\partial t}. \quad \ldots\ldots(5)$$

And

$$\frac{\partial \log p}{\partial e} = \frac{\partial \log p_2}{\partial e} - \frac{\theta'}{\theta^2}\frac{\partial \theta}{\partial e} = \frac{\partial \log p_2}{\partial e} + \frac{\partial \log \theta}{\partial e}. \quad \ldots\ldots(6)$$

And similarly for the northward differentiation. These terms in $\log \theta$ combine with those in (1) which explicitly contain $\log \theta$ in such a way as to change the thermal capacity at constant pressure to one at constant volume, thus

$$\gamma_p - b = \gamma_v. \quad \ldots\ldots(7)$$

On inserting these relations in (1) and rearranging terms we get what is required, thus

$$\frac{\partial \theta}{\partial t} = -\frac{M_{E20}}{R_{20}}\frac{\partial \theta}{\partial e} - \frac{M_{N20}}{R_{20}}\frac{\partial \theta}{\partial n} + \frac{b\theta}{\gamma_v p_2}\left\{\frac{\partial}{\partial t} + \frac{M_{E20}}{R_{20}}\frac{\partial}{\partial e} + \frac{M_{N20}}{R_{20}}\frac{\partial}{\partial n}\right\} p_2 - \frac{g}{\gamma_v}\frac{M_{H20}}{R_{20}} + \frac{\theta}{\gamma_v}\frac{D\sigma}{Dt} \ldots(8)$$

This gives $\partial\theta/\partial t$ in terms of known quantities for $\partial p_2/\partial t$ is found from the accumulation of mass by Ch. 6/2 # 1, 3, and $M_{H20}$ is found from Ch. 6/6 # 28 in which $\Phi$ is to be neglected.

## CH. 6/8. SUMMARY

An effort has been made to treat all the air above 11·8 km as a single conventional stratum in the sense that the momenta, pressures, and densities in a column should each be represented by a single number. To do this, all quantities have to be integrated with respect to height. The integrals of $p$, $\rho$, $\sigma$ come out simply because the temperature is independent of height. But integrals involving velocity can only be obtained from the relation at any level of velocity to pressure. The strict treatment of this relation by analysis is too difficult, and so the geostrophic approximation has been introduced. This is probably good enough for transforming the dynamical equation, but when applied to finding the temperature change it yields results which are unlikely.

If on further consideration the single stratum has to be abandoned, another plan is ready. Divide the stratosphere into several conventional strata. For all of these except the uppermost the general processes of Ch. 4, Ch. 5 will apply. The mass of this one having been made small, any errors committed in treating it will be of little consequence near the earth where we live. Or again a conventional division may be made at a height of 20 km in order to benefit by the observed steadiness of pressure at this level, a steadiness to which Mr W. H. Dines has called attention*.

* "Characteristics of Free Atmosphere," p. 71, Meteor. Office, London, *Geophys. Mem.* No. 13.

# CHAPTER VII

## THE ARRANGEMENT OF POINTS AND INSTANTS

### Ch. 7/0. GENERAL

It will be convenient to have a brief distinctive name for the arrangement here to be discussed. For this purpose we may borrow from crystallography the term "lattice."

The approximate representation of a differential coefficient by a ratio of finite differences is notably more exact when the differences are centered*. If $A$, $B$ and $u$ be any three variables, the ideal arrangement would be such that:

(i) Wherever $B$ has to be equated to $\frac{\delta A}{\delta u}$, then $B$ should be tabulated at points on the $u$-scale half-way between the points where $A$ is tabulated.

(ii) Wherever two variables have to combine in an expression, not involving their differential coefficients, they should be tabulated at the same points and instants.

(iii) When $A$ is a function of $\frac{\delta A}{\delta u}$, special difficulties arise. See below under the arrangement of instants, Ch. 7/2.

Unfortunately it is not possible to satisfy conditions (i) and (ii), in their entirety, for the given differential equations. The best we can do is to satisfy (i) and (ii) for the largest terms, and to leave the rest to be centered by interpolation.

### Ch. 7/1. THE SIMPLEST ARRANGEMENT OF POINTS

($a$) To fit with $\frac{\partial p}{\partial h} = -g\rho$ it is convenient to tabulate $p$ at the heights where strata meet, so that, for example, $p_8 - p_6 = gR_{86}$, where $R_{86}$ is the mass per horizontal area of the stratum bounded by $h_8$ and $h_6$.

($b$) To fit with the equation of continuity of mass, $R$ should be tabulated at points intermediate between $M_E$ and $M_N$, when seen in plan; thus:

to ↑ north

$M_N$

$M_E$ $\quad R \quad$ $M_E$ ⟶ to east

$M_N$

* *Vide* W. F. Sheppard, *Proc. Lond. Math. Soc.* 1899 Dec.

(c) To fit with the two horizontal dynamical equations, when all terms are neglected except:

$$-\frac{\partial M_E}{\partial t} = \frac{\partial P}{\partial e} - 2\omega \sin \phi M_N$$

$$-\frac{\partial M_N}{\partial t} = \frac{\partial P}{\partial n} + 2\omega \sin \phi M_E$$

$M_E$ and $M_N$ should be tabulated at points intermediate between $P$ when seen in plan, thus:

| | | | | | | |
|---|---|---|---|---|---|---|
| | **P** | | | | **P** | |
| **P** | $M_E$ | **P** | | **P** | $M_N$ | **P** |
| | **P** | | | | **P** | |

(d) Fortunately condition (c) is consistent with both (b) and (a). The frontispiece shows the arrangement adopted. It satisfies (a), (b), (c). The points at which pressure and momenta are to be tabulated are indicated respectively by $P$ and $M$. The coordinate differences are 200 kilometres of arc in a north-south direction, and in longitude the intervals between 128 equally spaced meridians.

(e) To fit with the characteristic equation of moist air, and with the expression for the entropy, it is convenient to have $W$ the mass of water per unit area of a conventional stratum, $\bar{\sigma}$ the mean entropy-per-mass of the stratum, and $\theta$ the temperature, all tabulated at the same points as $R$ the mass per horizontal area of the stratum.

## CH. 7/2. THE ARRANGEMENT OF INSTANTS

To fit with $gR_{86} = p_8 - p_6$ it would be best to tabulate $R$ at the same instant as $p$. To fit with

$$\frac{\partial R}{\partial t} = \frac{\partial M_E}{\partial e} + \frac{\partial M_N}{\partial n} + \text{etc.}$$

$R$ should be tabulated at instants intermediate between those at which $M_E$, $M_N$ are tabulated. But there is no arrangement in time which will fit neatly with the above and with

$$-\frac{\partial M_E}{\partial t} = \frac{\partial P}{\partial e} - 2\omega \sin \phi M_N$$

$$-\frac{\partial M_N}{\partial t} = \frac{\partial P}{\partial n} + 2\omega \sin \phi M_E$$

without interpolations. For this reason it has seemed best to *tabulate all quantities at the same set of instants*. Then progress in time is made by what may be called the "step-over" method, in which, in order to secure proper centering, we multiply $\partial/\partial t$ of any quantity at the instant $t$, by twice $\delta t$, and add this product to the value of the quantity at $t - \delta t$, in order to find the new value at $t + \delta t$.

It is seen that this "step-over" method requires *two* initial distributions separated by an interval $\delta t$. Various ways of obtaining the second of these will now be described.

(i) The first step may be made by the inaccurate "advancing differences" and then we may return to the centered step-over progress. But the errors of the first step will persist.

(ii) Perhaps the best plan would be to make the first few steps with smaller but progressively increasing time-differences, of which only the first step would be uncentered. For instance from the distribution at $0^h$ we calculate, by an uncentered but short step, the distribution at $1\frac{1}{2}^h$. From the time rates deduced from the distribution at $1\frac{1}{2}^h$ we make a centered step of 3 hours from $0^h$ to $3^h$. Next, from the time rates at $3^h$ we make a centered step of 6 hours from $0^h$ to $6^h$. Then we begin the normal process, using the time rates at $6^h$ to step from $0^h$ to $12^h$.

(iii) Two initial distributions a few hours apart might be obtained by observation. But the errors of observation would be magnified in the difference. However, two sets of observations would form a valuable check on any purely deductive way of beginning, such as that described above.

(iv) There is another way of commencing, which is convenient and fairly exact when we are dealing with simple equations such as $\partial\theta/\partial t = \partial^2\theta/\partial z^2$, but which is probably unworkable in the present case, on account of the complexity of the system of equations. Still it may be worth mentioning here. We have, let us suppose, arithmetical values of all the dependent variables at $0^h$, given by observation. We assign algebraic symbols to values of these variables at $6^h$. The mean between the arithmetical value at $0^h$ and the symbol at $6^h$ is taken as the value at $3^h$, and this binomial expression is substituted for the variable in question wherever it occurs in the differential equations, except where it is differentiated with respect to the time. The differential equations then yield the rates of increase of all the dependent variables at $3^h$. The rate of increase of any variable at $3^h$ when multiplied by the time-step is put equal to the difference between its algebraic value at $6^h$ and its numerical value at $0^h$. Thus we obtain a set of algebraic simultaneous equations, which when solved yield the required arithmetical values at $6^h$. In the case of the weather the algebraic equations would be quadratic, because the differential equations contain some quadratic terms.

(v) There is another process which may be said to have a sort of inherent stability, because it allows the first step to be made in the simplest way possible, and yet gives high accuracy later on. If $\theta_0, \theta_1, \theta_2, \dots \theta_n$ be successive values of $\theta$ at equal intervals $\delta t$, the process is equivalent to using as many terms of the Maclaurin expansion as are available at each step, so that we put

$$\theta_1 = \theta_0 + \delta t \,.\, \frac{\partial\theta_0}{\partial t} \text{ to two terms,}$$

$$\theta_2 = \theta_0 + 2\delta t \,.\, \frac{\partial\theta_0}{\partial t} + \frac{(2\delta t)^2}{2!}\frac{\partial^2\theta_0}{\partial t^2} \text{ to three terms,}$$

..............................

$$\theta_n = \theta_0 + n\delta t \,.\, \frac{\partial\theta_0}{\partial t} + \frac{(n\delta t)^2}{2!}\,.\,\frac{\partial^2\theta_0}{\partial t^2} + \dots \text{ up to } \frac{(n\delta t)^n}{n!}\frac{\partial^n\theta_0}{\partial t^n}, \text{ that is } (n+1) \text{ terms.} \quad \dots(1)$$

Now what the differential equation gives us is not these successive differential coefficients of the initial value $\theta_0$, but instead the first differential coefficients of the successive values $\theta_0, \theta_1, \theta_2, \ldots \theta_n$. However we can find the former from the latter, because, by differentiating (1)

$$\frac{\partial}{\partial t}\theta_n = \frac{\partial}{\partial t}\theta_0 + n\delta t \cdot \frac{\partial^2 \theta_0}{\partial t^2} + \frac{(n\delta t)^2}{2!}\frac{\partial^3 \theta_0}{\partial t^3} + \text{as far as } \frac{(n\delta t)^n}{n!}\frac{\partial^{n+1}\theta_0}{\partial t^{n+1}}, \qquad \ldots\ldots\ldots(2)$$

and all the differential coefficients in the second member of this equation, except the highest one, can be expressed in terms of $\theta_n, \theta_{n-1}, \ldots \theta_2, \theta_1, \theta_0$. On making the substitutions it is found that the series begin thus:

$\theta_1 = \theta_0 + \delta t \cdot \dfrac{\partial \theta_0}{\partial t}$ an uncentered first step.

$\theta_2 = \theta_0 + 2\delta t \cdot \dfrac{\partial \theta_1}{\partial t}$ a simple step-over.

$\theta_3 = \frac{1}{4}\theta_0 + \frac{3}{4}\theta_1 + 2\frac{1}{4}\delta t \cdot \dfrac{\partial \theta_2}{\partial t}$ a step-over from an interpolated value $2\frac{1}{4}\delta t$ previous.

$$\theta_4 = -\tfrac{5}{27}\theta_0 + \tfrac{20}{27}\theta_1 + \tfrac{12}{27}\theta_2 + \tfrac{64}{27}\delta t \cdot \frac{\partial \theta_3}{\partial t}.$$

$$\theta_5 = \frac{1}{6912}\left\{-1823\,\theta_0 + 1785\,\theta_1 + 4950\,\theta_2 + 2000\,\theta_3 + 16875\,\delta t \cdot \frac{\partial \theta_4}{\partial t}\right\}.$$

These hold true quite independently of the nature of the differential equation which is to be solved.

These various methods will now be illustrated in a very simple case. Suppose that the differential equation is $\partial\theta/\partial t = -\theta$ and that the initial condition is that $\theta = 1$ when $t = 0$ so that the exact solution is $\theta = e^{-t}$. Let the normal interval between the tabulated values of $\theta$ be taken as $\delta t = 0{\cdot}2$.

| $t$ time | Exact $\theta = e^{-t}$ | $\theta$ by advancing time-steps of 0·2 throughout | $\theta$ by centered time-steps of 0·4, stepping over intermediate values, except at the beginning, where the following methods are employed: First step of 0·2 uncentered | Small initial steps; only first uncentered | Double initial data | Algebraic first step | $\theta$ by Maclaurin as far as available |
|---|---|---|---|---|---|---|---|
| | | Excess of approximate over exact | | | | | |
| 0 | 1·0000 | ·0000 | ·0000 | ·0000 | ·0000 | ·0000 | ·0000 |
| 0·025 | ·9753 | | | – ·0003 | | | |
| 0·05 | ·9512 | | | ·0000 | | | |
| 0·1 | ·9048 | | | ·0001 | | | |
| 0·2 | ·8187 | – ·0187 | – ·0187 | ·0003 | ·0000 | – ·0005 | – ·0187 |
| 0·4 | ·6703 | – ·0303 | ·0097 | ·0021 | ·0022 | ·0024 | ·0097 |
| 0·6 | ·5488 | – ·0368 | – ·0208 | ·0012 | ·0009 | ·0003 | – ·0048 |
| 0·8 | ·4493 | – ·0397 | ·0195 | ·0031 | ·0033 | ·0038 | ·0024 |
| 1·0 | ·3679 | – ·0402 | – ·0274 | ·0011 | ·0008 | ·0000 | – ·0012 |

On examining the last two rows of the above table it is seen that advancing time-steps have produced errors of about 10 per cent., while the errors of the four methods on the right are only about 0·5 per cent. There is a curious oscillation in time about the errors of the step-over method, the amplitude increasing as it goes. "Maclaurin as far as available" has also produced an oscillating error, but its amplitude decreases with time. If we consider not merely accuracy but also ease of performance, the most satisfactory process in this case must be judged to be the one which begins with a very small uncentered step and doubles the length of the step-over several times, in the manner described in (ii) above.

## CH. 7/3. STATISTICAL BOUNDARIES TO UNINHABITED REGIONS

On parts of the oceans, near the poles, and within the desert tracts of land there are no people to provide observations or to appreciate forecasts. For lack of observations some assumption has to be made about the behaviour of the atmosphere in those regions. This difficulty is not peculiar to finite-difference methods, but is common to all systems of forecasting. The favourite assumption is that the climatological values of the elements—say monthly means based on statistics of former years—are a sufficient representation. This assumption is very easily translated into our numerical process by writing the monthly means around the edge of the table of principal variables. The edge is thus maintained in being, and by contrast with the example of Ch. 2 the wasteful shrinkage of the table at each successive time-step no longer occurs. But of course errors are introduced if the monthly means at the edge do not represent what actually happens there.

## CH. 7/4. JOINTS IN THE LATTICE AT THE BORDERS OF SPARSELY INHABITED REGIONS

Again there are other portions of the globe, especially seas, where some rough sort of forecast might be possible and desirable if it could be carried out with a much opener network than that in use where population is dense. But the two networks must be united on the computing forms in such a way that air, represented by numbers, can flow across the joint. The figures (A) and (B)* show two kinds of joints between chessboard patterns. In Fig. A the number of stations per area on the map is reduced fourfold, in Fig. B the reduction is ninefold. An attempt has been made, in accordance with Ch. 7/1, to surround each $M$ point by four equally spaced $P$ points, and conversely. This has not succeeded in the case of the points marked by letters in parentheses $(P)$, $(M)$, $(P, M)$. The quantities thus indicated are required in making the prediction at other points, but they cannot be directly predicted by the scheme of Ch. 7/1 themselves. Instead they must be filled in by interpolation between their neighbours. Such interpolation is always possible. It is seen that the ninefold reduction gives a neater joint than the fourfold, in the sense that the latter involves more interpolations.

* p. 154.

M P M P M P (P) M P M P

Wilderness (M) P M P M

P M P M P M (P, M) P M P

(M) P M P M

M P M P M P (P) M P M P

(M) (P) (M) (M) (P) (M) (M) (P) M P M P M

(P, M) P M P (P, M) P M P (P, M) P M P M P M P

P M P M P M P M P M P M P M P M

M P M P M P M P M P M P M P M P

Cultivated

P M P M P M P M P M P M

Fig. A.

Joint reducing the density of stations in ratio 4 : 1.

---

M P M P M P M P

(P) M P M

Ocean

(M) P M P

P M P M P M P M

(M) P M P

(P) M P M

M (P) (M) P (M) (P) M (P) (M) P (M) (P) M P M P

P M P M P M P M P M P M P M P M

M P M P M P M P M P M P M P M P

Land

P M P M P M P M P M

Fig. B.

Joint reducing the density of stations in ratio 9 : 1.

See p. 153.

## CH. 7/5. THE POLAR CAPS

At the poles the terms in $\tan\phi$ become infinite and are balanced by an infinity of opposite sign in the terms in $\frac{\partial}{\cos\phi\partial\lambda}$. These infinities are artificial in the sense that they are not present in the weather, but arise solely from the polar co-ordinates by which we measure it.

If we had to deal only with local polar meteorology the most satisfactory plan would probably be to neglect the curvature of the earth and to arrange pressure and momentum points on a "chessboard" formed by straight lines intersecting at right angles. But there does not seem to be any way of making a smooth joint between the rectilinear chequers which suit the poles, and the chequers formed by meridians and parallels of latitude which suit the rest of the globe.

The crowding of meridians as the pole is approached has already been discussed in Ch. 3/5, Ch. 3/6, where it has been proposed to begin at the equator with 128 meridians, to omit alternate ones in latitude 63° or thereabouts, and again to halve the number at successive stages until only four are left close to the pole. The joint near latitude 63° could be made like Fig. A on p. 154. On drawing the scheme out it is found that in higher latitudes the joints follow one another so closely that their interpolations overlap. In other words, for latitudes between 70° and 90°, a special pattern needs to be devised together with a corresponding special transformation from infinitesimals to finite differences. This pattern, when it has been made, may have some influence on the number of meridians chosen for lower latitudes. The number 256 has been suggested as being a power of 2 and so making possible a series of joints like Fig. A. Alternatively a power of 3 such as $243 = 3^5$ would fit with a series of joints like Fig. B. Or we might have a product of a power of 2 into a power of 3 corresponding to some joints like Fig. A, some like Fig. B. But joints in which the number of meridians was reduced in a ratio as large as 1 : 5 would mark a violent change in the lattice, so that the number of meridians on the equator could not suitably contain 5 as a factor, still less 7 or any higher prime-number.

# CHAPTER VIII

## REVIEW OF OPERATIONS IN SEQUENCE

### Ch. 8/0. GENERAL

In writing the chapter on the fundamental equations the ideal was to obtain a description of atmospheric phenomena which should be in the first place correct, and which, secondly, might be used in prediction. Here in Chapter 8 the order of emphasis is reversed. The ideal is now to make a scheme first workable and secondly as exact as circumstances permit. After a new machine has emerged from the experimental stage, its workability is tested by the cost and the value of its product, by its satisfaction of human needs. But the present scheme has not yet emerged. The questions still are: does it conform to the nature of the external world? will the wheels go round at all? So the essence of workability is here taken to be that, when we have made a step forward in time, we should find ourselves provided with the data for making the next step. The initial data are arranged in a pattern which, by borrowing a term from crystallography, we may call a "space-lattice." Wherever in the lattice a pressure was given, there the numerical process must yield a pressure. And so for all the other meteorological elements. Such a numerical process will be referred to as a **"lattice-reproducing process."** All other processes have been rejected. There is an exception, as we have seen in Chapter 2, at the edge of the map covered by the lattice. In Chapter 7 two ways of dealing with this exception have been discussed under the heads of "statistical boundaries" and "joints in the lattice."

A notation is needed for partial differences, for whereas a partial differential coefficient such as $\delta p/\delta n$ is usually sufficiently clear, yet when we translate it into differences the $\delta p$ often becomes detached from the $\delta n$ so that it is necessary to write $\delta_N p$, the suffix indicating the coordinate which alone varies.

In Ch. 4/8/6 we noted the difficulties arising from having a lowest stratum of air as thick as 2 km. In Ch. 4/10/3 we came across the notion of a conventional film just thick enough to contain the vegetation. On further examination this film proves to be inconveniently thin, because its thermal capacity is in places negligibly small. If the lowest stratum is to be divided into two, the division might suitably be taken at a height of about 200 metres.

Those operations which have already been adequately described are in the present chapter but briefly mentioned in their proper sequence. More space is taken up in making good previous deficiencies, especially in regard to the entropy of moist air and the earth-air interface. The procedure to be described is certainly very complicated, but so are the atmospheric changes.

## CH. 8/1. INITIAL DATA

Suppose that, at the initial instant, we have numerical values of the dependent variables given by observation at points distributed in the following pattern with reference to the map which appears as frontispiece. Above every point marked $M$ on the map, suppose that the momenta-per-area, $M_E$ and $M_N$, are given for each of the five conventional strata. Above every point marked $P$ on the map suppose that the pressure is given at the ground and at the heights of exactly 2·0, 4·2, 7·2 and 11·8 kilometres above sea-level, that is to say at the dividing surfaces between the strata. Also, above the points marked $P$, suppose that the mass-of-water-per-horizontal-area, $W$, is given for the four lower strata; and that the temperature is given for the uppermost stratum only. Suppose further that, where the points marked $P$ fall on land, we are given the temperature and water-content of the soil at the surface and at a series of depths below. Suppose, lastly, that the sea temperature is given at the other points marked $P$. A sample of the data described above will be found in the large table in Ch. 9, but the information for the soil is there lacking.

## CH. 8/2. OPERATIONS CENTERED IN COLUMNS MARKED "$P$" ON THE MAP

### CH. 8/2/1. THE MASS-PER-AREA, $R$, OF EACH STRATUM

This is found from $p$ at the upper and lower limits of the stratum. Thus, for example,

$$R_{86} = \frac{p_8 - p_6}{g_{86}},$$

where $g_{86}$ is the mean acceleration of gravity for the stratum. The result is entered on Computing Form P I.

### CH. 8/2/2. THE FORCE-PER-HORIZONTAL LENGTH

This is $P = \int p\,dh$ and is found from the formula (Ch. 4/4 # 9), which for the stratum $h_8$ to $h_6$ becomes

$$P_{86} = \frac{(h_6 - h_8)(p_8 - p_6)}{\log_e p_8 - \log_e p_6}.$$

For the uppermost stratum this formula is replaced by

$$P_{20} = \frac{2{\cdot}870 \times 10^6 \theta_1}{g_{20}} \cdot p_2.$$

See Ch. 6/1 # 7. The result is entered on Computing Form P I.

### CH. 8/2/3. MEANS ACROSS A STRATUM

Where, in what follows, we require to know the mean values across the thickness of any stratum of the quantities $p$, $\rho$, $w$, $m_E$, $m_N$, they are usually taken respectively as $P/\delta h$, $R/\delta h$, $W/\delta h$, $M_E/\delta h$, $M_N/\delta h$, where $\delta h$ is the thickness of the stratum. Similarly, mean values for a stratum of $\mu$, $v_E$, $v_N$ are usually taken respectively as $W/R$, $M_E/R$, $M_N/R$. If some other process has to be employed in any case, it will be mentioned specially.

### CH. 8/2/4. SATURATION

For each stratum, except the uppermost, we find whether the air is saturated or not by means of Ch. 4/1 # 5, employing for $p$, $\rho$, $\mu$ the mean values which have just been indicated. From the appropriate characteristic equation (Ch. 4/1) are next found the mean temperature of the stratum and also if necessary $w_s$, the density of aqueous vapour saturated at this temperature. We may note that when the air is dry, the particular kind of mean temperature yielded by this process is

$$\int \theta\rho dh \div \int \rho dh.$$

The temperature is entered on Computing Form P I.

### CH. 8/2/5. UNIFORM CLOUD AND PRECIPITATION

We follow the scheme described in Ch. 4/6. For each stratum, in which the air is saturated, we multiply the density of saturated vapour $w_s$ by $\delta h$ so as to obtain $W_s$ the mass of vapour per horizontal area of the stratum. Then $W - W_s$ is the mass of condensed water per horizontal area. If $W - W_s$ exceeds 0·4 grm cm$^{-2}$, we deduct the excess from the stratum and transfer it to the ground as precipitation. See Ch. 4/6. The cloud is entered on Form P I, the precipitation on Forms P I and P XVIII. Local cloud and precipitation due to heterogeneity are treated separately, see Ch. 8/2/8 below.

### CH. 8/2/6. SHALL WE USE ENTROPY-PER-MASS $\sigma$, OR POTENTIAL TEMPERATURE $\tau$?

In the classical thermodynamics entropy occupies a central position, whereas potential temperature is rarely mentioned. In meteorological theories these positions are almost reversed.

Let us adopt the definition of entropy favoured by G. H. Bryan as being most suitable in connection with irreversible internal processes such as the smoothing out of heterogeneity:—"If from any cause whatever, the unavailable energy of a system with reference to an auxiliary medium of temperature $\theta_i$ undergoes any (positive or negative) increase and if this increase be divided by the temperature $\theta_i$ the quotient is called the increase of the entropy of the system*."

It should be noted that the choice of $\theta_i$ makes no difference whatever to the numerical value of the entropy provided that $\theta_i$ is lower than any other of the temperatures concerned. If $\theta_i$ is not lower, the interpretation is unphysical.

Let us define $\tau$ as "that temperature which the air would attain on being brought to equilibrium at a standard pressure $p_i$ without loss or gain of either heat or moisture." The numerical value of $\tau$ then depends upon the conventional $p_i$.

**The entropy when condensation occurs. Error in Ch. 4/5/1 # 13 and in the Hertz diagram.**

Even if the sky be cloudless, evaporation or condensation is usually in progress at the foliage. From general thermodynamics it is evident that a quantity $d\epsilon$ of radiant

* Bryan, *Thermodynamics*, § 71. B. G. Teubner, Leipzig, 1907.

energy absorbed by a leaf at $\theta_A$ and given out again to the air at almost the same temperature will increase the entropy of the atmosphere by $d\epsilon/\theta_A$ irrespective of whether the energy goes in warming or in separating water molecules during the process of evaporation. Therefore when entropy is stirred upwards by eddies it may be considered as going as the sum of two fluxes, one depending on sensible heat, the other on the water-vapour, and the latter flux must involve the latent heat. But we have shown that the flux of entropy, except for a possible modification depending on irreversibility, is

$$\frac{\xi}{g}\frac{\partial\sigma}{\partial p}.$$

Therefore $d\sigma$ must involve the latent heat $\Upsilon$, even if the air is clear. Now Hertz's well-known formula

$$d\sigma = \gamma_p\, d\log_e\theta - b d\log_e p \quad \ldots\ldots\ldots\ldots\ldots\ldots\ldots\ldots(1)$$

for the entropy of clear moist air does not involve the latent heat $\Upsilon$ and is therefore inapplicable for finding the eddy-flux of entropy even in clear air. On examining the manner in which Hertz derives this formula it is seen that he considers the addition of heat but not the addition of moisture, so that the formula gives $d\sigma$ correctly for two samples of clear air provided that both have the same $\mu$, but not otherwise. Now in the vertical $\mu$ is usually variable, so that it is only exceptionally that

$$d\sigma = \gamma_p d\log_e\theta - bd\log p.$$

I did not understand this at the time when Ch. 4/5/1 was printed off and so unfortunately the value of $a_\mu$ given by Ch. 4/5/1 # 13 depends only on the variations of $\gamma_p$ and $b$ with $\mu$ and neglects the more important effect of the latent heat.

To find the true difference of entropy we must trace in imagination some process of adding water-substance reversibly, and the question then arises: what energy and entropy are to be ascribed to unit mass of the incoming substance? As there is an arbitrary constant of integration in the entropy, we must ask what would be the effect of an increase in this constant for the incoming water. Approximations are not here permissible, for the constant might be made indefinitely large. Now it will be found, on examining the system of prediction, that all the calculations with $\sigma$ and $\mu$ lead up to finding $D\sigma/Dt$ and $D\mu/Dt$, and that the sole use to which these are put is in the equation for the vertical velocity, which they enter together in the form

$$D\sigma/Dt - a_\mu D\mu/Dt.$$

Now $a_\mu$ is here defined by the statement that

$$d\sigma = a_p\,.\,dp + a_\rho\,.\,d\rho + a_\mu\,.\,d\mu \quad \ldots\ldots\ldots\ldots\ldots\ldots\ldots\ldots(2)$$

for any arbitrary variations of $p$, $\rho$, $\mu$ so that $a_\mu = \left(\frac{\partial\sigma}{\partial\mu}\right)_{p,\,\rho\ \mathrm{const.}}$. An increase in the arbitrary constant which occurs in the entropy of water would increase $a_\mu$ by the same amount everywhere and would therefore leave $(D\sigma/Dt - a_\mu D\mu/Dt)$ unchanged since this expression is equal to $(a_p Dp/Dt + a_\rho D\rho/Dt)$ and the latter has nothing to do with the arbitrary constant in the entropy of water.

However it is convenient for many purposes to take account of the entropy of water, for otherwise the latent heats would lack any material embodiment. And if we deal with the entropy at all, we must reckon its variations with temperature.

The most natural way of reckoning the entropy of the water-substance would be to take it as zero at the absolute zero of temperature. It was formerly supposed that the presence of $\theta$ in the denominator of the integral which gives the entropy would make the integral have an infinity where $\theta = 0$; but the measurements of Nernst, Lindemann, Koref and others* have shown that the specific heats tend to zero at $\theta = 0$ in such a way that the entropy remains finite there.

However the most natural reckoning may not be the most expeditious, in view of the existence of the Hertz and Neuhoff diagrams of adiabatics of moist air.

If the earth's atmosphere were so hot that water-vapour never condensed in it, we could entirely ignore the latent heat of evaporation of water even as we now ignore those of oxygen and nitrogen. This thought suggests that we must recognise any energy which can be extracted from water substance by cooling it to the lowest temperature occurring in the atmosphere, say to 180° A, but that with regard to any energy or entropy remaining at that temperature we may either ignore it altogether or else give to it an arbitrary fixed value per mass of ice.

Now although Hertz's calculation of the entropy difference of two samples of air saturated with ice refers, as in the case of clear air, only to the putting in of heat and not of moisture, yet it may be used to compare samples having different quantities of moisture. For let all the samples be expanded adiabatically until their temperatures fall to 180° A. Then practically all the water will be condensed. Now we have seen that we may if we please ignore any energy remaining in ice at 180° A. Therefore, at that temperature, the ice-content may be taken to have no effect on the entropy of the sample. That is to say the difference of entropy per mass of two samples of atmosphere, both at 180° A, is taken to be simply

$$d\sigma = -b \, . \, d \log p, \quad \ldots\ldots\ldots\ldots\ldots\ldots \; \ldots\ldots\ldots\ldots (3)$$

even if they contain different amounts of ice. And then, by the addition of heat according to Hertz's process we may compare any samples whatever in the snow stage. Hertz states that the isopleths for the snow stage satisfy the equation, in our notation

$$\sigma = \text{const.} = \gamma_p \log \theta - b \log p + \frac{(\Upsilon + \Upsilon_s)}{\theta} \mu, \quad \ldots\ldots\ldots\ldots\ldots\ldots (4)$$

where $\Upsilon_s$ is the latent heat of fusion. At 180° A the water per mass $\mu$ would be practically zero, so that $d\sigma$ is simply $-b \, . \, d \log p$ as above. As the isopleths of $\sigma$ are drawn at intervals of 0·0025 calorie/degree C, it follows that $d\sigma$ at any temperature and pressure in the snow stage may be read off easily, without any need to trace the air down to 180° A, which lies beyond the limits of the diagram.

When we wish to find the difference of entropy-per-mass of two samples of air,

* Quoted by J. H. Jeans in his "Report on Radiation and the Quantum Theory," published by the Physical Society, London, 1914.

in a form approximately valid under all meteorological conditions, we may therefore proceed as follows. By the aid of Hertz's diagram trace the expansion of each sample when no heat or water is lost. Its course will in general be broken by the jump peculiar to the hail stage. On arriving at the snow stage read off the entropy difference at the rate of 0·0025 (kg-calorie) $(\text{kg})^{-1}$. $(\text{degree C})^{-1}$ per interval between consecutive adiabatic lines. In erg units the same interval is $0{\cdot}1045 \times 10^6$ erg $\text{grm}^{-1}$ $\text{degree}^{-1}$.

If instead we had read off $d\sigma$ on the lines in the same diagram representing the behaviour of unsaturated air, we should have obtained a quite different result. These lines in fact appear to have the right slope, but, unless $\mu$ is the same for all, a false spacing. They might with advantage have been made asymptotes, at a very low temperature, to the adiabatic lines of the snow stage.

Only if the limits of the diagram be reached before the air is saturated is there a need for calculation. In that case it is best to use the value of $a_\mu$ given below to correct the entropy of saturated air to that of the given unsaturated air.

In whatever way $d\sigma$ be reckoned it is essential that $a_\mu$, defined as $(\partial\sigma/\partial\mu)_{p,\,\rho\,\text{const.}}$, should be reckoned correspondingly. So to find $a_\mu$ we should take a pair of points on the Hertz diagram corresponding to the same $p$ and $\rho$ but to slightly different $\mu$. Owing to an approximation which Hertz makes, these points coincide. In other words, Hertz's diagram makes

$$(\partial\sigma/\partial\mu)_{p,\rho} = (\partial\sigma/\partial\mu)_{p,\theta}. \quad\text{.............................}(5)$$

But on tracing the samples down the adiabatic, they become saturated at different pressures, and so arrive at the snow stage with different values of $\sigma$. In this way the following values of $a_\mu$ have been computed for *unsaturated air*:

| temperature °A | pressure mb | $\mu$ $10^{-3}\times$ | $a_\mu$ $10^6$ ergs/degree |
|---|---|---|---|
| 300 | 1000 | 6 | 96 |
| 300 | 1000 | 18 | 84 |
| 280 | 700 | 5 | 98 |
| 272 | 600 | 3 | 101 |
| 263 | 450 | 3 | 117 |
| 253 | 500 | 1·5 | 120 roughly |

.........(6)

For *air saturated with water* the effect of adding more water is, according to the Hertz diagram, merely to increase the hail stage, and on this account

$$a_\mu = 17 \times 10^6 \text{ ergs/degree.} \quad\text{.............................}(7)$$

For *air saturated with ice* the addition of more ice is apparently to leave the entropy unchanged. That is because Hertz neglects the thermal capacity of the ice. Neuhoff's computations appear to be more accurate in this respect. But if great accuracy is to be attempted, it will no longer be possible to consider, as Hertz does, that $p$ and $\theta$ suffice to define the density, irrespective of $\mu$.

In the actual troposphere the entropy-per-mass is so very nearly independent of height—much more nearly so than would be supposed if we took the observed temperatures to apply to dry air—that quite small changes in procedure are enough to change the sign of the computed $\partial_H\sigma$.

In reckoning the entropy of water from 180° A, we include the latent heat of fusion of ice, divided by 273° A, so that the eddy flux of entropy in the form $\frac{\xi}{g}\frac{\partial\sigma}{\partial p}$ would also include the same. Now a flux of entropy found in this way at the level $h_8$ will often have to be compared with the flux at the foliage in order to find the accumulation in the air between 0 and 2 km. If we were to treat the latter flux as the radiant energy divided by the temperature of the foliage, the water substance leaving the foliage would be supposed to carry with it the latent heat of evaporation, but not that of fusion. The proper course is evidently to reckon the ground as supplying, *viâ* the plant stems, water having a certain entropy-per-mass which includes the effect of the latent heat of fusion. In general, wherever water-substance occurs, we must credit it with an entropy-per-mass reckoned from the same temperature, in this case ideally from 180° A, but with an approximation introduced by Hertz.

**The behaviour of entropy-per-mass and of potential temperature in regard to turbulence.** One of the chief questions is whether, when condensation occurs, either $\tau$ or $\sigma$ satisfies the three conditions Ch. 4/8/0 # 2, 3, 9 which any quantity $\chi$ must satisfy if it is to diffuse according to the equation

$$\frac{\bar{D}\chi}{\bar{D}t}=\frac{1}{\rho}\frac{\partial}{\partial h}\left(c\frac{\partial\chi}{\partial h}\right). \qquad \text{(8)}$$

The **first of these conditions**, Ch. 4/8/0 # 2, is equivalent to, and perhaps finds a clearer expression in, the "mixing-rule" of W. Schmidt*, which states that samples having masses $m_1$, $m_2$ and $\chi_1$, $\chi_2$ must give a mixture having $\chi_3$ such that

$$\chi_3=\frac{\chi_1 m_1+\chi_2 m_2}{m_1+m_2}. \qquad \text{(9)†}$$

The answer to this question becomes fairly clear if we adopt v. Bezold's‡ view of the process of mixture. Let us consider only processes which occur without loss or gain of heat or moisture. Bezold replaces in imagination the natural process by two artificial ones in sequence. In the first artificial process there is supposed to be neither evaporation nor condensation. So that as the specific heat $\gamma_p$ is nearly a constant, the temperature $\theta_3$ of this mixture is given approximately by the mixing rule (9) above. And a similar rule will be followed, without approximation, for $\mu$ the total water substance per mass of atmosphere and for $\nu$ the mass of water-vapour per mass of atmosphere, so that $\mu-\nu$ represents the liquid or solid. So that if subscripts 1 and 2 denote the component, and subscript 3 the mixture, the mixing rule (9) may be written

$$\frac{\mu_3-\mu_1}{\mu_2-\mu_3}=\frac{\nu_3-\nu_1}{\nu_2-\nu_3}=\frac{\theta_3-\theta_1}{\theta_2-\theta_3}=\frac{m_2}{m_1}. \qquad \text{(10)}$$

* Wm Schmidt, "Der Massenaustausch...," *Sitzb. Akad. Wiss. Wien, Mathem.-n. Klasse*, Abt. II a, 126 Band, 6 Heft, § 4. 1917.

† By an exception to the convention holding in the rest of this book, subscripts in Ch. 8/2/6 do not denote heights.

‡ *Sitzber. Akad. Wiss.* Berlin, 1890, pp. 355—390. English translation in Abbé's *Mechanics of the Earth's Atmosphere* (Smithsonian Institution, 1891).

This first process is not reversible because the mixture cannot be separated by reversing the motions. Therefore although no heat enters from outside, the entropy must increase.

The second artificial process is one of adiabatic condensation or evaporation of the intimate mixture of moist air and liquid or solid particles. During it $\mu$ is fixed, but $\nu$ and $\theta$ change in such a way that

$$\gamma_p d\theta = -\Upsilon d\nu, \qquad (11)$$

where $\Upsilon$ is the latent heat of evaporation or of sublimation. This change proceeds until either the air is saturated or $\mu - \nu$ is reduced to zero. Bezold gives a pretty diagrammatic representation of it in the $\mu$, $\theta$ plane, which however does not concern us here. Owing to the intimacy of the mixture this second process must be almost perfectly reversible, and therefore during it both entropy-per-mass and potential temperature are unchanged.

The question arises whether in the natural adiabatic process of mixing, the changes in $\tau$ and $\sigma$ would be the sum of their changes in Bezold's two consecutive adiabatic processes. If we provisionally assume that this is so, then we can deduce the following important proposition. Take at one height a sample of cloud having properties $\tau_1$, $\sigma_1$, $\mu_1$, $\nu_1$ and at another height a sample of clear air having properties $\tau_2$, $\sigma_2$, $\mu_2$, $\nu_2$. Let eddies move them adiabatically to the same level. In these journeys $\tau_1$, $\tau_2$, $\sigma_1$, $\sigma_2$, $\mu_1$, $\mu_2$ are all unchanged but $\nu_1$, $\nu_2$ probably alter. Now let the samples mix and come to equilibrium as regards evaporation. Then from what has been said it is evident that the mixing rule (9) applies to $\tau$ and to $\mu$ but not to $\sigma$ nor to $\nu$.

The **second condition** to be satisfied by $\chi$, if the diffusion equation (1) is to hold, is that $\chi$ should not be changed by delay. See Ch. 4/8/0 # 3. This applies obviously to $\sigma$ or to $\tau$ provided that the processes are really adiabatic or that we account separately for their non-adiabatic parts, as we do here in the case of radiation.

The **third condition** to be satisfied by the diffusing quantity $\chi$ is that the upward flux of ($\chi \times$ mass) should vanish when $\partial\chi/\partial h$ vanishes. .................... Ch. 4/8/0 # 9. This is satisfied by either $\sigma$, $\tau$ or $\mu$ since they are not changed when a portion of air is raised or lowered without loss of heat or of moisture; and when these losses occur they are taken into account separately.

As all three conditions are satisfied for $\tau$, the flux of ($\tau \times$ mass) is seen from Ch. 4/8/0 # 17 to be $-c\partial\tau/\partial h$ or equivalently $\frac{\xi}{g}\frac{\partial\tau}{\partial p}$. But now we come to a peculiarity. Suppose that a portion of cloud and a portion of dry air are in contact at the same temperature and pressure. Then it is impossible to say which has the greater potential temperature until the standard pressure $p_i$ has been fixed. For if $p_i$ be larger than $p$, then in changing from $p$ to $p_i$ some cloud will evaporate and the dry air will have the higher $\tau$. Whereas if $p_i$ be smaller than $p$, then in changing from $p$ to $p_i$ more water will condense in the cloud, so that the dry air will now have the lower $\tau$. Thus the flux of ($\tau \times$ mass) may be reversed in direction by a mere change in the conventional pressure $p_i$. But the fluxes of latent or of sensible heat are not thus fantastical, and the explanation of the behaviour of $\tau$ must be sought in its relation to heat. It is therefore evident that $\frac{\xi}{g}\frac{\partial \tau}{\partial p}$ does not in general measure the flux of sensible as distinct from latent heat, as it does at least approximately in the special case of clear air. The difference of entropy-per-mass between the cloud and clear air is free from the kind of artificiality that complicates $d\tau$.

Thus to sum up: if we use potential temperature we must attend to any local and temporal variations of $p_i$, and to the relation of the flux of heat to that of ($\tau \times$ mass), but irreversible mixing does not matter. If we use entropy-per-mass we must add on the gains due to irreversibility, but the flux of entropy at a surface is simply the heat entering per area per time divided by the temperature, and to compute it we need not know the evaporation which the heat produces.

**Increases of entropy by irreversible internal processes** such as the smoothing out of eddies or of patchiness by molecular diffusion. It has been noted on p. 40 that irreversible mixing of cold and warm air increases the total entropy but not the mass-mean of potential-temperature if the air is dry. For this reason on p. 69 the diffusion equation appears in a simpler form in potential temperature than in entropy. The latter equation Ch. 4/8/0 # 20 may, for clear air, be expanded to read

$$\frac{\bar{D}\sigma}{\bar{D}t} = \frac{\partial}{\partial p}\left(\xi \frac{\partial \sigma}{\partial p}\right) + \frac{\xi}{\gamma_p}\left(\frac{\partial \sigma}{\partial p}\right)^2. \qquad \text{......(12)}$$

This equation is, for dry air, of the same form in $\sigma$ as it would be in the potential temperature $\tau$ except for the presence of the last term. Now in a paper on the "Supply of energy to and from atmospheric eddies*" a term of just this form $\xi/\gamma_p \, . \, (\partial\sigma/\partial p)^2$ is found to correspond to the irreversible part of $\bar{D}\sigma/\bar{D}t$. The square seems appropriate as the irreversibility is not likely to depend on the sign of $\partial\sigma/\partial p$. However the argument so far is limited to the case of dry air, for in the paper on the "Supply of energy..." the coefficient $a_\mu$ is simply neglected in

$$d\sigma = a_p \, . \, dp + a_\rho \, . \, d\rho + a_\mu \, . \, d\mu.$$

In the general case the terms in $\bar{D}\sigma/\bar{D}t$ which express irreversibility may perhaps be found in all cases to have even indices.

It may be asked whether this irreversible increase of entropy has any importance. It is worth attending to if there is any practical difference between uniform air and thermally patchy air. Undoubtedly that is so, but we must take account of the difference consistently throughout our computations. For instance in using the hydrostatic equation to find the change of pressure with height in dry air, if the temperature of the air is specified by giving the mass-mean of its potential-temperature, that apparently would suffice, whereas if instead the temperature is specified by giving the total entropy, then we must also have given a measure of the heterogeneity.

Another increase in entropy is due to the dissipation of eddying kinetic energy by molecular viscosity.

* *Roy. Soc. Proc.* A, vol. 97, p. 366, equation (6·4). But note that in the second member $(\partial\sigma/\partial p)$ should be squared.

From what has been said it is seen that there ought to be some correction to the vertical velocity equation because in it entropy is used as a measure of density by way of

$$d\sigma = \alpha_\rho \, . \, d\rho + \alpha_p \, . \, dp + \alpha_\mu \, . \, d\mu.$$

The correction which would be applied to $\alpha_\rho$ would depend on the thermal heterogeneity of the air and would vanish if the air were not patchy.

All these matters connected with irreversibility are of rather secondary importance, and as they are still somewhat obscure I have thought it best to omit the irreversible terms until they can be fully investigated.

**Plan Adopted.**

It appears to be more convenient to use entropy-per-mass for treating the eddy diffusion in the upper air. This is done on Computing Form P XI. On the contrary, for dealing with the thermal boundary condition at the earth's surface, the procedure is simpler if we use instead the potential temperature at the local and instantaneous surface pressure. Computing Forms P IX, P X are arranged accordingly. It remains an open question as to whether $\sigma$ or $\tau$ is the better when we discuss Stability. As an interim procedure I have used $\tau$ for this purpose on Forms P I, P III taking the standard pressure as $p_G$. In order to explain these values of $\tau$ the formula from which they were computed will now be deduced.

**To compute the potential temperature of air which is clear and which will remain clear at the standard pressure $p_i$.**

According to Hertz's well-known formula which is derived above in Ch. 4/5/1,

$$\sigma = \gamma_p \log_e \theta - b \log_e p + \text{const.} \qquad \text{..............................(12)}$$

and $\gamma_p$ and $b$ are functions of $\mu$ only.

Therefore if air at $\theta$, $p$ changes to $\tau$, $p_i$ while $\sigma$ and $\mu$ remain fixed, $\gamma_p$ and $b$ are also fixed and $\gamma_p \log \tau - b \log p_i = \gamma_p \log \theta - b \log p$ so that, without approximation,

$$\tau = \theta \left(\frac{p_i}{p}\right)^{b/\gamma_p}, \qquad \text{..................................(13)}$$

and, as the moisture in the clear air increases, $b$ and $\gamma_p$, though both increasing, do so in an almost constant ratio: thus

$$\left.\begin{matrix} \mu = \cdot 000 & \cdot 010 & \cdot 020 \\ b/\gamma_p = \cdot 2893 & \cdot 2881 & \cdot 2870 \end{matrix}\right\} \qquad \text{..........................(14)}$$

A formula for $\tau$ will now be derived, which is convenient in the present scheme because we have given $p$ as a function of $h$. On combining the hydrostatic equation $\partial p/\partial h = -g\rho$ with $p = b\rho\theta$ there can be obtained

$$\theta = -\frac{g}{b}\left(\frac{\partial h}{\partial \log p}\right)_{\text{vertically}}, \qquad \text{..............................(15)}$$

which is strictly correct provided $b$ be given its value corresponding to the existing $\mu$. Now on multiplying (15) by $(p_i/p)^{b/\gamma_p}$ and assuming that $b/\gamma_p$ is independent of height, while admitting arbitrary variations in $b$ proportional to those in $\gamma_p$, there results

$$\tau = -\frac{g}{\gamma_p}\, p_i^{b/\gamma_p} \left\{\frac{dh}{d\,(p^{b/\gamma_p})}\right\}_{\text{vertically}} \qquad \text{.......................(16*)}$$

or approximately, with values of the constants corresponding to an ordinary value of $\mu$

$$\tau = -0{\cdot}983 \times 10^{-4} \,.\, p_i^{\cdot 288} \left(\frac{dh}{dp^{\cdot 288}}\right)_{\text{vertically}} \,. \qquad \text{..............(17)}$$

Here $h$ is the height in centimetres. The unit for $p$ is arbitrary provided that the same is used for $p_i$.

A table giving $p^{\cdot 288}$ as a function of $p$ will be found in the *Quarterly Journal of the Royal Meteorological Society* for July 1921.

### CH. 8/2/7. ARRANGEMENT OF LEVELS FOR THE EDDY-FLUXES

The moisture is tabulated at the mean levels of the strata. Its rate of accumulation is also required at these levels, and therefore the fluxes must be tabulated at the intermediate levels where strata meet. At the ground the boundary is somewhat blurred by the presence of vegetation; but if we imagine it as viewed from the neighbouring boundary, 2 kilometres above, it would appear to be sufficiently definite.

### CH. 8/2/8. ESTIMATE OF TURBULENCE AND HETEROGENEITY

The estimate is intended to be based on the instantaneous distribution of entropy, of velocity, and of water-substance. The time taken to establish or destroy a state of turbulence or of patchiness is thus neglected. Common observation of the diurnal variation of gustiness, or of the rising and dissipation of cumuli, show that this time is of the order of an hour, and is therefore unimportant in comparison with the time of passage of a cyclone, with which forecasts are usually concerned.

An alternative plan, more correct and physically more interesting, but not attempted here on account of its probable toilsomeness, would be to take suitable measures of turbulence and of heterogeneity as main variables in addition to the seven chosen in Ch. 4/0, and to trace their time changes step by step. The equations expressing their changes might include Ch. 4/8/1 # 22, Ch. 4/9/7 # 6, Ch. 4/9/8 # 8, but this system is not in itself complete, and I do not know the equations required to complete it.

The coefficient $\overline{\xi}_{G8}$ developed in Ch. 4/8/5 for finding the amount of water moving up through the thick stratum is estimated from the wind measurements set out in the last column of the table on page 84. The coefficient ℓℓ, introduced in Ch. 4/8/4 for finding the upward flux of heat in the thick stratum, is similar to $\overline{\xi}_{G8}$ and is, provisionally, put equal to it.

Observation and theory both suggest—although they do not yet prove—that these two measures $\overline{\xi}_{G8}$ and ℓℓ of turbulence near the ground could be expressed fairly well as functions of three variables, namely of the locality, of $M_{G8}$ the momentum per area of the lowest stratum, and of $\overline{\tau}_{G8}-\tau_A$, where $\overline{\tau}_{G8}$ is the mean potential temperature of the stratum and $\tau_A$ is the temperature of the interface where radiation is converted to

* Compare Exner, *Dynamische Meteorologie* (Teubner), Art. 70.

heat. This functional relationship could probably be made more definite by lowering the height denoted in these symbols by the suffix 8. But however much it is lowered we should still require to base the estimate on the temperature actually at the surface of the soil or vegetation. This temperature is computed below in Ch. 8/2/15, 16 from an equation which is non-linear, and which involves $\bar{\xi}_{G8}$ and ɷ in certain terms. Thus the best way to disentangle these processes appears to be to make first a trial estimate of $\tau_A$, on which to base corresponding values of $\bar{\xi}_{G8}$ and of ɷ. These are then used, together with the radiation, to find a corrected value of $\tau_A$ and thence corrected values of $\bar{\xi}_{G8}$ and ɷ.

On the other hand the radiation cannot be computed until the amount of detached cloud has been estimated, and detached cloud, being an effect of heterogeneity, is naturally grouped together with turbulence. Thus after estimating the amount of detached cloud and of local showers by the aid of statistics (Form P III) we next compute the radiation, and come back to the diffusion produced by turbulence at a later stage (Ch. 8/2/17, 19).

## CH. 8/2/9. RADIATION

From Ch. 8/2/5 we know in each stratum the density of any continuous cloud which there may be. In Ch. 8/2/8 we estimated the amount of detached cloud. The radiation can therefore be traced downwards and upwards. For long-wave radiation I have used the "approximate simplified process" of Ch. 4/7/1, for solar radiation the process and constants of Ch. 4/7/2. The processes have been so fully described there, that it should suffice here to refer to the computing forms P IV, V, VI and to note that the upward long-wave radiation from the interface, although to appear on Form P VI, is not computed until the subsequent Form P X has been filled up according to Ch. 8/2/15.

## CH. 8/2/10. THE EVAPORATION FROM THE SEA

The rate of evaporation depends only on the up-grade of $\mu$ and on the distribution of eddy-motion in the air.

According to Ch. 4/8/5 the rate of evaporation is taken as

$$\frac{\bar{\xi}_{G8}}{g} \frac{(\mu_G - \bar{\mu}_{G8})}{\frac{1}{2}(p_G - p_8)} = \Xi \text{ for brevity,}$$

where $\mu_G$ is the value of $\mu$ which would be in equilibrium with the water at the surface of the sea. It is assumed here that the air actually in contact with the water is practically saturated. Observations made at the height of a ship's deck are no proper test of this assumption.

## CH. 8/2/11. EVAPORATION FROM FOLIAGE

At the mean level of the stratum $(G, 8)$ we know by Ch. 8/2/3 a mean value $\bar{\mu}_{G8}$ or say $\mu_9$ of the water per mass of the atmosphere. In the intercellular spaces of the leaves $\mu$ according to Brown, Escombe and Wilson has, except in case of wilting, its saturated value at the temperature of the leaf. The latter is not a main variable but has to be derived from the temperature of the highest stratum of soil and that of the lowest stratum of air, taking into account the radiation. It will be convenient to defer this problem to Ch. 8/2/16. There is not room for much water to accumulate in the air close to the leaves, so that the rate at which water is coming through all the stomata, above say a hectare of soil, may be put equal to the rate at which it is being carried aloft by eddies across the upper boundary of the vegetation film. There is thus an analogy to the electrical instrument known as the potentiometer in that there is a current through two resistances in series; and $\mu$ plays the part of the potential. The transpiration equation Ch. 4/10/3 # 3 may be written thus in terms of $\mu$

$$\left\{\begin{matrix}\text{Mass of water evaporating per time}\\ \text{from foliage above unit area of land}\end{matrix}\right\} = \kappa\rho\,\{\mu_{\text{leaves}} - \mu_{GL}\} = \Xi, \quad \text{........(1)}$$

where $\mu_{GL}$ is the value of $\mu$ for the air surrounding the leaves and $\mu_{\text{leaves}}$ is the value within the leaves.

For a field of growing barley $\kappa\rho$ is of the order of

$$0{\cdot}25 \times 10^{-3} \text{ grm sec}^{-1} \text{ per horizontal cm}^2. \quad \text{...................(2)}$$

But by Ch. 4/8/5 the upward flux of water between the vegetation film and the middle of the stratum next above is

$$\frac{\bar{\xi}_{LS}\,(\mu_{GL} - \bar{\mu}_{LS})}{g\ \tfrac{1}{2}\,(p_G - p_S)}, \quad \text{..................................(3)}$$

where $\bar{\xi}_{LS}$ may be taken as the same as $\bar{\xi}_{GS}$ given in Ch. 8/2/8 and where $\bar{\mu}_{LS}$ is practically the same as $\bar{\mu}_{GS}$ since the vegetation cannot hold much moisture.

Equating these two values of the same flux,

$$\kappa\rho\,\{\mu_{\text{leaves}} - \mu_{GL}\} = \frac{\bar{\xi}_{GS}\,(\mu_{GL} - \bar{\mu}_{GS})}{g\ \tfrac{1}{2}\,(p_G - p_S)}. \quad \text{......................(4)}$$

By the analogy to the potentiometer or by solving these equations, we see that $\frac{1}{\kappa\rho}$ and $\frac{g\,(p_G - p_S)}{2\bar{\xi}_{GS}}$ play the part of resistances in series. Call them $A$ and $B$ respectively. Then the common flux is

$$\frac{\mu_{\text{leaves}} - \bar{\mu}_{GS}}{A + B} = \Xi, \quad \text{...............................(5)}$$

which is the formula used to determine it on Computing Form P VII. The water per mass in the air surrounding the vegetation is not required if our only object be to make a "lattice-producing" system. But this quantity may be of interest for its own sake. It is determined by

$$\frac{B\,.\,\mu_{\text{leaves}} + A\,.\,\mu_{GS}}{A + B} = \mu_{LG}.$$

A portion of rainfall is caught on vegetation and evaporates there without ever reaching the ground. From what is known about eddy diffusion we should expect that the amount evaporated would be proportional to $\mu_{A\,\text{sat.}} - \mu_9$ the difference between the saturated water-per-mass at the vegetation and the actual water-per-mass at $h_9$, and also to be directly proportional to $2\bar{\xi}_{G9}/\{g\,(p_G - p_8)\}$ which is the eddy-conductance between these two levels. It is for direct observation to show how it is related to rainfall. Two raingauges were kept by the author, one in an open exposure, the other a few hundred metres away in a beech forest. (Compiègne, 1916, Dec. 8 to 23.) The ratio of the daily readings varied less than their difference. On this, admittedly far too scanty evidence, it is suggested that the downcoming water should enter the expression as a factor, thus

$$\left(\begin{matrix}\text{rainfall}\\\text{evaporated}\end{matrix}\right)=\left(\begin{matrix}\text{rainfall}\\\text{received}\end{matrix}\right)\times(\mu_{A\,\text{sat.}}-\mu_9)\times\frac{2\bar{\xi}_{G9}}{g\,(p_G-p_8)}\times n.$$

In this case one-fifth of the whole was evaporated on the average, whence with appropriate values of the other quantities in C.G.S. units

$$n=1{\cdot}2\times10^5 \text{ for beech forest bare of leaves.}$$

### CH. 8/2/12. EVAPORATION FROM BARE SOIL

In the absence of precipitation this may be treated in much the same manner as the evaporation from vegetation, except that there is a term added to the difference of $\mu$. For, in the notation of Ch. 4/10/2 the upward flux of water in the soil is say $F$, where

$$F=\mathfrak{w}\left\{\frac{d\Psi}{dz}-g\right\}+\varkappa\frac{d\mathfrak{F}}{dz},\quad\ldots\ldots(1)$$

which may be written

$$F=\frac{d\mu}{dz}\left\{\mathfrak{w}\frac{d\Psi}{d\mu}+\varkappa\frac{d\mathfrak{F}}{d\mu}\right\}-g\mathfrak{w}.\quad\ldots\ldots(2)$$

The coefficient of $d\mu/dz$ in this is a function of the humidity of the air in the pores of the soil, since the density of this air is known. So for brevity we may write

$$\mathfrak{w}\frac{d\Psi}{d\mu}+\varkappa\frac{d\mathfrak{F}}{d\mu}=f(\mu).\quad\ldots\ldots(3)$$

Now changing to finite differences, make our usual approximation

$$\frac{d\mu}{dz}=\frac{\mu_{11}-\mu_{10}}{z_{11}-0}.\quad\ldots\ldots(4)$$

Then (2) becomes

$$F=\frac{\mu_{11}-\mu_{10}}{z_{11}}.f(\mu)-g\mathfrak{w}.\quad\ldots\ldots(5)$$

But, neglecting accumulation, the flux in the air is the same, so that

$$F=\frac{\mu_G-\bar{\mu}_{G8}}{B},\quad\ldots\ldots(6)$$

where, as in Ch. 8/2/11,

$$B=\frac{g\,(p_G-p_8)}{2\bar{\xi}_{G8}}.\quad\ldots\ldots(7)$$

Let us assume that at the interface $\mu$ is continuous, that is to say

$$\mu_{10}=\mu_G.\quad\ldots\ldots(8)$$

On eliminating $\mu_{10}$ and $\mu_G$ from (5), (6), (8) and solving the resulting equations for $F$, we arrive at an expression for the flux which is suitable for use in computing. Thus at the interface

$$\Xi = F = \frac{\mu_{11} - \bar{\mu}_{G8} - \mathfrak{g}g \,.\, z_{11}/f(\mu)}{z_{11}/f(\mu) + B}. \quad \text{......................(9)}$$

In the electrical analogy $z_{11}/f(\mu)$ is the resistance of half the soil stratum, $B$ is the resistance of half the air stratum and $\mathfrak{g}g \,.\, z_{11}/f(\mu)$ corresponds to the potential difference due to a battery in the circuit. An outside potential difference $\mu_{11} - \bar{\mu}_{G8}$ is also applied.

Provided the soil is not saturated, we do not require to know the temperature $\theta_{10}$ of the interface. The temperature $\theta_{11}$ suffices.

If the soil is **saturated**, as it often is in winter, then $\mu$ in the soil corresponds to the saturated vapour density and scarcely changes with depth, whereas $\Psi$ the "negative pressure in the soil-water" varies. Thus $d\Psi/d\mu$ is nearly infinite and so by (3) is $f(\mu)$ also. Thus (9) reduces to

$$\Xi = \frac{\mu_G - \bar{\mu}_{G8}}{B} \quad \text{..................................(10)}$$

just as for the sea.

### CH. 8/2/13. THE HEAT RENDERED LATENT BY EVAPORATION

This usually concerns the evaporation of liquid water. In that case the heat absorbed is $597 - 0{\cdot}6\,(\theta - 273^\circ)$ calories $\text{grm}^{-1} = 3182 - 2{\cdot}51\,\theta \times 10^7$ ergs $\text{grm}^{-1}$. In Ch. 8/2/6 we have found it convenient to suppose that water carries its latent heat of fusion about with it. But if so, that is supplied by the ground, and received again by the ground when an equal quantity of water falls upon it as rain. Radiation is converted into latent heat of fusion only when ice, snow or hoar frost are actually fused by radiation.

### CH. 8/2/14. THE THERMAL BOUNDARY CONDITION AT THE SEA-SURFACE

From Ch. 4/8/0 it appears evident that the upward flux of entropy can only depend upon (i) the distribution of entropy-per-mass, (ii) the distribution of eddy motion in the air.

Now it is observed* that the temperature of the air in contact with the ocean has a very small daily range $- 1^\circ$ C. or less, and differs very slightly from the slowly changing temperature of the surface water. Thus over the sea we are not so frequently confronted with the principal complication which occurs in respect to the flux of heat on land, namely that the rapid changes of surface temperature produce an upgrade of potential temperature next to the land which has little connection with the mean upgrade in the first kilometre of height. Nevertheless the same difficulty occurs at sea in a lesser degree as is shown by G. I. Taylor's observations in the Report† on the work carried out by S.S. "Scotia" 1913. Apparently the only thorough way out of this

* Hann, *Lehrb. der Meteorologie*, 3 Aufl. pp. 56—70. † Sold by T. Fisher Unwin, London, W.C.

difficulty is to take a thinner lowest stratum of air. To continue however with the thick stratum, we represent the distribution of entropy per mass by

$$\frac{\sigma_G - \bar{\sigma}_{G8}}{\frac{g}{2}(p_G - p_8)}, \quad \text{.......................................(1)}$$

where $\sigma_G$ is the potential temperature of air having the surface pressure $p_G$ and having the temperature of the sea water.

The liveliness of turbulence itself depends on $(\sigma_G - \bar{\sigma}_{G8})$ and on $M_{G8}$ the momentum per area of the stratum. So that, as $(p_G - p_8)$ does not vary much, the flux of entropy might be taken as a function of only two variables $(\sigma_G - \bar{\sigma}_{G8})$ and $M_{G8}$. Observations of the flux, when there are any, might suitably be expressed in this form. However in the meantime let us express the flux as the product of two factors, one ꭙ measuring the liveliness of turbulence and the other being the expression (1) above. So that the entropy rising from unit horizontal area per time is expressed as

$$ꭙ \cdot \frac{\sigma_G - \bar{\sigma}_{G8}}{\frac{g}{2}(p_G - p_8)}. \quad \text{.......................................(2)}$$

Until better values become available, ꭙ may be taken as equal to the mean turbulivity $\bar{\xi}_{G8}$, defined in Ch. 4/8/5 # 9, determined from wind observations, and set out numerically in the last column of the table on p. 84.

#### Ch. 8/2/15. THE THERMAL BOUNDARY CONDITIONS AT THE INTERFACE WHERE BARE SOIL AND ATMOSPHERE MEET

The treatment in Ch. 4/8/4 will now be improved.

It would be convenient if we could believe that the temperature of the soil and of the air were the same where they touch one another. The nearest approach to a test of this question that I have seen was made by T. Bedford Franklin (*Edinburgh Roy. Soc. Proc.* Vol. 39, Part 2, No. 10; 1918—1919). A thermometer just above the soil was compared with one 1·3 cm under the surface. On clear nights the upper thermometer tended to have a minimum about 1° C colder than the minimum of the other, if the soil was firm, or 2° C if the soil was raked loose. But we can hardly regard a thermometer covered by soil as measuring the temperature of the surface. A better test might possibly be made if the surface temperature could be measured by the long-wave radiation which it emits, and the air temperature by a special aspirated thermometer. However in the meantime it appears reasonable to assume that soil and air share a common temperature at the interface. In other words, on crossing the interface, temperature changes continuously. .......................................(1)

In contrast with the temperature, the flux of sensible heat, regarded as a function of height, increases discontinuously at the surface of the soil by a jump equal to the radiation absorbed, plus the heat carried down by precipitation, and minus the heat rendered latent by evaporation. .......................................(2)

This relationship is a sort of equation of continuity.

The surface conditions are seen to be analogous to the conditions in a system of electric conductors at a point where three wires meet. That is to say the ordinary thermal conduction in the ground is analogous to the electric flow in one wire, the eddy conduction in the air is analogous to the flow in the second wire, while the heat brought in or out by radiation, precipitation and evaporation is represented by the third current. On looking along any wire towards the junction, the other two wires appear to be in series with it and in parallel with one another. The meteorological phenomenon is more complicated than its electric analogue, because the "conductances" depend upon the "potential" at the interface.

Now in order that the changes of a temperature may be determinable from the accumulation of heat, it must be the temperature of some substantial body of matter such as a stratum of soil or of air; it cannot be the temperature of a surface. For this reason among others the temperature is not tabulated at the surface, but, just as for moisture in Ch. 8/2/7, at certain heights above and below. On the other hand, the surface* temperature is needed in order to give us the fluxes, which are a necessary part of the scheme of prediction. The surface temperature is also of immediate interest to agriculturalists.

It is not satisfactory to find the surface temperature by interpolation between air temperature at some height above and soil temperature below, as may be seen by considering such facts as the following:—

(i) The surface of dry sand in sunshine on a calm day is sometimes as much as 30° C hotter than the air at head level.

(ii) The grass minimum on a still clear night is sometimes as much as 10° C below both the temperature of the soil at 30 cm depth and the temperature of the air at head level.

On account of the 4th power in Stefan's law, these differences of temperature must make notable differences in the outgoing radiation. The rate of evaporation from bare soil depends on the surface temperature if the soil is wet. Again the turbulence depends much upon the surface temperature when the temperature of the upper air and its velocity are given. For all these reasons we must try to determine the surface temperature exactly.

It will be convenient to have a suffix-notation for conventional strata which shall run on without a break from air to soil. As $h_8 = 2$ km above sea-level and $h_9 =$ roughly halfway between the surface $h_G$ and $h_8$, so let $z_{10}$ denote the surface also, let $z_{11}$ be somewhere in the midst of the top stratum of soil and let $z_{12}$ be the lower surface of this top stratum. The suffixes are then continuous in spite of the fact that $h$ is height above sea-level while $z$ is depth below the surface of the soil. Let $f$ stand for the flux of sensible, as distinct from latent, heat. That is to say let $f$ be the number of ergs of sensible heat rising through one square centimetre per second. Then the following

* The phrase "surface temperature" is used by some writers to denote the temperature in a radiation screen placed at head level. This is such an obvious misuse of the word "surface" that it is not entitled to the respect usually accorded to custom. In the present work "surface" is used in a more geometrical sense.

arrangement of temperatures and fluxes upon the height-scale fits with the schemes proposed for the upper air and for the soil:

$$
\begin{array}{lll}
h_7 & \theta_7 & \\
h_8 \cdots\cdots\cdots\cdots & & f_8 \\
h_9 & \theta_9 & \\
h_G = z_{10} \cdots\cdots & & \begin{cases} f_G \\ f_{10} \end{cases} \\
z_{11} & \theta_{11} & \\
z_{12} \cdots\cdots\cdots\cdots & & f_{12} \\
z_{13} & \theta_{13} &
\end{array}
$$

Here $f_G$ is intended for the value of $f$ just above the surface, and $f_{10}$ for its value just below.

In the air we must work not with $\theta$, but with the potential temperature $\tau$, which the air would have if brought adiabatically to some standard pressure $p_i$. By choosing the surface pressure as the standard, the surface temperature is made to enter in a simple manner into the equations for the fluxes in both the air and the soil.

Similarly in the air, for the reasons explained in Ch. 4/8/0, it is desirable to attend to the flux of (potential temperature × mass) which is connected with the up-grade of potential temperature, thus:

$$\text{Flux of (potential temperature} \times \text{mass)} = \frac{\xi \partial \tau}{g \partial p} \cdots\cdots\cdots\cdots \text{Ch. 4/8/0 \# 17.}$$

By the above choice of surface pressure for the standard, the flux $\xi/g \times \partial\tau/\partial p$ becomes equal, at the surface only, to $1/\gamma_p$ times the flux of heat. So that

$$\left(\frac{\xi}{g}\frac{\partial \tau}{\partial p}\right)_G = \frac{f_G}{\gamma_p}. \cdots\cdots\cdots\cdots\cdots\cdots (3)$$

As we have to replace the differential coefficient $\partial\tau/\partial p$ by the difference ratio $(\theta_G - \tau_9)/\frac{1}{2}(p_G - p_8)$ it is necessary to replace $\xi$ by the compensated value which has been called $\text{ꭍꭍ}$ in Ch. 4/8/4 and which is provisionally taken to be equal to $\xi_{G8}$ of pp. 91 and 84.

Thus

$$f_G = \gamma_p \frac{\text{ꭍꭍ}}{g} \frac{(\theta_G - \tau_9)}{\frac{1}{2}(p_G - p_8)}. \cdots\cdots\cdots\cdots\cdots\cdots (4)$$

To simplify the subsequent algebra introduce $C_{G9}$ defined as:

$$C_{G9} = \gamma_p \frac{2\text{ꭍꭍ}}{g(p_G - p_8)}. \cdots\cdots\cdots\cdots\cdots\cdots (5)$$

Then from (4)

$$f_G = C_{G9}(\theta_G - \tau_9) \cdots\cdots\cdots\cdots\cdots\cdots (6)$$

and $C_{G9}$ may be called a "conductance" from its analogy with electric conductance.

Similarly if $C_{11,10}$ is the ordinary thermal conductance of the soil between $z_{11}$ and $z_{10}$ in a vertical column of unit cross-sectional area, we have:

$$\text{The flux of heat, } f_{10} = (\theta_{11} - \theta_{10})\, C_{11,10}. \cdots\cdots\cdots\cdots (7)$$

There is an error of centering in equations (6) and (7) because the flux is at the surface while the temperature difference is centred above or below. This error appears to be inevitable. It would be reduced if the strata were thinner. But we must leave it and pass on.

In (7)
$$C_{11,10} = \frac{\text{(thermal conductivity)}}{z_{11} - z_{10}}. \quad \text{.......................(8)}$$

The thermal conductivity of soil has been repeatedly measured. For a loam rich in humus, T. B. Franklin* gives for the conductivity 0·004 calories $\text{sec}^{-1}\,\text{cm}^{-1}\,\text{degree}^{-1}$, which is
$$1{\cdot}67 \times 10^5 \text{ ergs sec}^{-1}\,\text{cm}^{-1}\,\text{degree}^{-1}. \quad \text{.......................(9)}$$

Next let us consider the heat taken in and out by radiation, precipitation and evaporation. As our first business will be to find the surface temperature, these quantities should be classified according as they depend on the surface temperature or not. Let Γ be the radiation coming from the sun or the atmosphere and absorbed by the soil. ......................................................................................(10)

Γ has been computed on Forms P IV, V, VI and is thus known. The long wave radiation emitted by the interface may be expressed as ȣ ↄ $\theta_{10}^4$, .................(11)

where ↄ is Stefan's constant $= 5{\cdot}36 \times 10^{-5}$ erg $\text{cm}^{-2}\,\text{sec}^{-1}$ ..............(12)

and ȣ is the emissivity of the solid for long waves. ..............(13)

The effect of precipitation has been mentioned on p. 46 but is not easy to systematize, and with this reference will be omitted for the present, after giving to the ergs per second brought to a horizontal square centimetre by precipitation the symbol $\varpi$.

The rate of evaporation has been denoted by Ξ gram $\text{sec}^{-1}$ per $\text{cm}^2$ of a horizontal surface. ......................................................................................(14)

If the top layer of soil is wet, evaporation takes place there and increases rapidly with the surface temperature. If the top layer of soil is dry, evaporation may still go on underneath but then it is conditioned not by the so-far unknown $\theta_{10}$ but by $\theta_{11}$ which is a main-variable, traced step by step.

Let the latent heat of evaporation be denoted by†
$$ϩ = (3182 - 2{\cdot}51\,\theta) \times 10^7 \text{ erg grm}^{-1}. \quad \text{.........................(15)}$$

The whole flux of heat arriving on a horizontal square centimetre per second by radiation, precipitation and evaporation is therefore:
$$\Gamma - ȣ\,ↄ\,\theta_{10}^4 + \varpi - ϩ\,\Xi. \quad \text{.............................(16)}$$

This, together with the inflow $f_{10}$ coming up from the ground and the inflow $-f_G$ coming down by eddies in the air, must amount to zero, since no heat can accumulate at a surface. But $f_G$ and $f_{10}$ are given by (6) and (7). Therefore
$$0 = C_{G9}(\tau_9 - \theta_G) + C_{11,10}(\theta_{11} - \theta_{10}) + \Gamma - ȣ\,ↄ\,\theta_{10}^4 + \varpi - ϩ\,\Xi. \quad \text{........(17)}$$

This is to be looked upon as an equation to determine the common temperature $\theta_G$ or $\theta_{10}$ at the interface. The best way to solve it is first to assume some trial value

* *Edinburgh Roy. Soc. Proc.* Vol. 39, Part 2, No. 10.

† ϩ is a Coptic letter named "he."

$\theta'$ for the surface temperature. The error of such a guess will seldom exceed 5° C for this is one of the more familiar of meteorological elements. So that if we write

$$\theta_G = \theta' + \text{ⲇ}, \quad \ldots\ldots (18)$$

then ⲇ* is a correction to be determined and $\text{ⲇ}/\theta'$ is small compared with unity.

So that if $\mathrm{F}(\theta_G)$ denote the second member of (17), we have $\mathrm{F}(\theta' + \text{ⲇ}) = 0$, and therefore by Newton's rule or Taylor's theorem

$$-\text{ⲇ} = \frac{\mathrm{F}(\theta')}{d\mathrm{F}(\theta')/d\theta'} \text{ approximately.} \quad \ldots\ldots (19)$$

This written out in full, with dashes to denote values of the various quantities when $\theta_G = \theta'$, reads

$$+\text{ⲇ} = \frac{C'_{G9}(\tau_9 - \theta') + C'_{11,10}(\theta_{11} - \theta') + \Gamma - \text{ȣ}\,\text{ɔ}\,\theta'^4 + \varpi' - \text{ϒ}'\Xi'}{+C'_{G9} + C'_{11,10} - \frac{\partial C'_{G9}}{\partial \theta'}(\tau_9 - \theta') - \frac{\partial C'_{11,10}}{\partial \theta'}(\theta_{11} - \theta') + 4\,\text{ȣ}\,\text{ɔ}\,\theta'^3 - \frac{\partial(\varpi' - \text{ϒ}'\Xi')}{\partial \theta'}} . \quad \ldots\ldots (20)$$

This equation determines ⲇ and so by (18) the surface temperature $\theta_G$. Of course there are simpler less accurate ways of arriving at $\theta_G$. One is to be content with the guess $\theta'$, that is to say to neglect ⲇ altogether. Another simplification would be to neglect various portions of the long equation for ⲇ, but it seems best to draw up Computing Forms P IX, X for the full equation, and to leave the relative importance of the various terms to be decided by further experience†.

Terms such as $\partial C'_{G9}/\partial\theta'$ imply that we have a table, based on observation, giving the conductance $C_{G9}$, which depends on turbulence, as an empirical function of $M_{G8}$, the momentum per area of the stratum, of $(\tau_9 - \theta_G)$, and of the locality. Observation and theory both indicate that such a functional relationship exists, at least approximately. Then taking the fixed actual $M_{G8}$ and locality, we pick out corresponding variations of $C_{G9}$, and of $\theta_G$ around a central value $\theta'$, and we call the ratio of the variations $\partial C'_{G9}/\partial\theta'$.

The best sequence of this portion of the computing-operations is thus seen to be the following. Estimate the detached cloud (Form P III) and compute the solar radiation and the descending radiation from the atmosphere (P IV, V, VI), make a guess $\theta'$ at the surface temperature, and put on Form P III the corresponding measures of turbulence near the ground which we have denoted by $\bar{\xi}_{G8}$ and ϣ. From these find the trial values of the fluxes of evaporation and of heat at the interface. Find also the long-wave radiation which would leave the earth if it had the temperature $\theta'$. Put all these quantities into equation (20) and so determine the temperature correction ⲇ. This is done on Forms P IX, X. Then correct the separate quantities to the true temperature of the surface, and proceed to other parts of the computing.

The partition coefficient, which was discussed in Ch. 4/8/4, does not appear in this revised treatment.

* ⲇ is a Coptic letter pronounced "dalda."

† In the example in Ch. 9 the terms $C'_{11,A} + C'_{A9}$ in the denominator are together ten times greater than the sum of the rest.

## CH. 8/2/16. THE THERMAL BOUNDARY CONDITION FOR VEGETATION

This may be discussed as a variation of the process described in Ch. 8/2/15 for bare soil.

We may neglect any accumulation of heat in the vegetation film itself. Thus, for example, in a field of ripe wheat the mass* of the vegetation is about 2·4 tons per acre, that is 6 tons per hectare, which is 0·06 gram $cm^{-2}$. The mass of the vegetation is therefore of the same order as the mass of the layer of air between the tops of the stalks and the soil; and the sum of the two masses is negligible in comparison with that of the conventional stratum of air which extends up to 2 km.

It is at the height where the foliage is densest that most of the radiation is converted to heat, and most of the water is evaporated. A suffix is needed for this height; let it be $A$. In place of the conductance $C_{11,10}$ of Ch. 8/2/15 we have now two conductances in series, one $C_{11,10}$ as before, the other $C_{10,A}$, so that

$$C_{11,A}=\frac{1}{1/C_{11,10}+1/C_{10,A}}.$$

T. Bedford Franklin (*Edinburgh Roy. Soc. Proc.* Vol. 39, Part 2, No. 10) has called attention to the great resistance offered to heat-flow by moss, by a carpet of dead leaves or by grass, instancing a primrose flowering with its roots and flowers in temperatures differing by 10° C. This thermal resistance may come into either $1/C_{11,10}$ or $1/C_{10,A}$ but not into both. Similarly we may regard $1/C_{AL}$ as in series with $1/C_{L9}$.

For the evaporation from the foliage we have, according to Ch. 8/2/11 # 5, if $\mu(\theta)$ be a contraction for the $\mu$ of air saturated at temperature $\theta$ and pressure $p_G$,

$$\Xi=\frac{\mu(\theta_A)-\mu_9}{A+B}=C_\mu\{\mu(\theta_A)-\mu_9\},$$

where $C_\mu$ is a contraction for $1/(A+B)$ of Ch. 8/2/11.

Otherwise the process is so similar to that for bare soil that it has been possible to set them both out on the same computing forms P IX, X.

## CH. 8/2/17. THE EDDY-FLUX AND ACCUMULATION OF WATER-SUBSTANCE IN THE UPPER LAYERS

This is required at the height where strata meet, and may be typified by the expression, applicable at the height $h_6$,

$$\frac{\xi_6}{g}\cdot\frac{\bar{\mu}_{86}-\bar{\mu}_{64}}{\frac{1}{2}(p_8-p_4)}=\frac{\xi_6}{g}\cdot\frac{\mu_7-\mu_5}{p_7-p_5},$$

which is a mass of water per area and per time.

From the difference between the fluxes at adjacent levels there follows the rate of accumulation of mass of water per area of each stratum. On dividing by the mass of atmosphere per area of the stratum we get the rate of increase of $\mu$. But the mass

* Fream's *Elements of Agriculture*, 10th edn. pp. 320—321. An acre is 4·047 × $10^7$ $cm^2$.

of air is, for example, $(p_6-p_4)/g$. This $g$ cancels with those in the preceding expressions so that it is more convenient to omit $g$ throughout. This may also be seen from the fact that the diffusion equation for $\mu$ is

$$\frac{\bar{D}\mu}{\bar{D}t}=\frac{\partial}{\partial p}\left(\xi\frac{\partial\mu}{\partial p}\right),$$

which does not contain $g$. The computing form is denoted by P XI.

### CH. 8/2/18. THE FLUX AND ACCUMULATION OF ENTROPY BY STIRRING

We have arrived at the flux of entropy-per-mass at the surface of sea, bare soil and vegetation (Ch. 8/2/14, Ch. 8/2/15, Ch. 8/2/16 respectively).

It now remains to continue the process in the upper layers. This is done in a manner so similar to that already described for water in Ch. 8/2/17 that it is only necessary here to say that, although the coefficient $\xi$ is provisionally taken as the same for both $\sigma$ and $\mu$, yet we must expect a discrimination between the two cases as more knowledge is gained.

### CH. 8/2/19. COLLECTING THE VARIOUS GAINS OF ENTROPY-PER-MASS

Those due to eddy diffusion, to emission or absorption of radiation, to precipitation and to molecular-dissipation are brought together on Form P XII and added. The gain of entropy due to radiation is reckoned in the usual way as the gain of energy divided by the temperature.

The entropy-change which would be produced by the change in water-content, if pressure and density were unaltered, is required in connection with the vertical velocity. This entropy-change is $a_\mu \,.\, D\mu/Dt$. These quantities are also collected in Form P XII.

### CH. 8/2/20. THE HORIZONTAL DIVERGENCE OF MOMENTUM PER VOLUME

This, expressed in symbols, is $\frac{\delta M_E}{\delta e}+\frac{\delta M_N}{\delta n}-M_N\frac{\tan\phi}{a}$ and is next computed for each stratum. This expression will be denoted by $\text{div}'_{EN}\,M$; the dash serving to distinguish it from the corresponding expression with differential coefficients in place of difference ratios. The arrangement of the computing has been shown on p. 9. The computing form is P XIII.

### CH. 8/2/21. PRESSURE CHANGE AT THE GROUND

Then $\text{div}'_{EN}\,M$ in each stratum is multiplied by the corresponding value of $g$ and $g\,\text{div}'_{EN}\,M$ is summed for all strata. This sum is equal to the rate of decrease of surface pressure. The pressure changes at higher levels cannot be determined until the vertical momentum has been found, but it has been thought convenient to collect all pressure changes on the same form P XIII.

### CH. 8/2/22. THE UP-GRADE OF THE VERTICAL VELOCITY IN THE STRATOSPHERE

This is found by Ch. 6/6#21 on computing form P XIV. Closely connected with the vertical velocity in the stratosphere is the temperature-change, which has therefore been placed on the same form. It is computed from Ch. 6/7/3 # 8,

### CH. 8/2/23. THE VERTICAL VELOCITY IN THE FOUR LOWER STRATA

This is computed from an equation which is identical in effect with Ch. 5/5 # 9 if the air is clear, but which is more general in so far as expressions in $\alpha_\rho$, $\alpha_p$, $\alpha_\mu$ which are defined with reference to air that is either cloudy or clear, replace those in $\gamma_p$, $\gamma_v$ which are correct only for clear air. This equation is obtained directly from the general form Ch. 5/1 # 12 by using the permissible approximations Ch. 5/2 # 13, 14. It is here employed in the differentiated form as follows

$$\frac{\partial}{\partial p}\left\{\frac{\alpha_\rho}{\alpha_p} g\rho^2 \frac{\partial v_H}{\partial p}\right\} = \text{div}_{EN} v + \frac{\partial}{\partial p}\left\{\frac{\alpha_\rho}{\alpha_p}\rho \,.\, \text{div}_{EN} v\right\} + \frac{\partial}{\partial p}\left\{\frac{1}{\alpha_p}\left(\frac{D\sigma}{Dt} - \alpha_\mu \frac{D\mu}{Dt}\right)\right\}$$
$$-\left\{\left(\frac{\partial v_E}{\partial p}\right)_{\lambda,\phi} \times \left(\frac{\partial p}{\partial e}\right)_{h,\phi} + \left(\frac{\partial v_N}{\partial p}\right)_{\lambda,\phi} \times \left(\frac{\partial p}{\partial n}\right)_{h,\lambda}\right\} \quad \ldots\ldots\ldots(1)$$

Alternatively Ch. 5/2 # 18 might have been used, after reinsertion of $\alpha_\rho$, $\alpha_p$, $\alpha_\mu$, but I doubt if the computations would have been any simpler, especially as the varying thickness of the lowest stratum implies that

$$\int_{h_G}^{h_8} \text{div}_{EN}\, m \,.\, dh \qquad \text{is not equal to} \qquad \text{div}_{EN}\int_{h_G}^{h_8} m \,.\, dh.$$

Whatever process is employed we require it to yield $v_H$ at the heights where strata meet, for those are the heights at which $v_H$ has to be inserted in the equation of continuity of mass. That being so, $\partial v_H/\partial p$ must be tabulated at the middle heights of the strata, and $\partial^2 v_H/\partial p^2$, or other second derivatives, at the same heights as $v_H$.

As explained in Ch. 5/3 the first running sum is made downwards, because $\partial v_H/\partial h$ is known at the top from Ch. 8/2/22; but the second running sum is made upwards, because $v_H$ is known at the ground in terms of the slope and surface wind, thus

$$(v_H)_G = \left(v_E\frac{\partial h}{\partial e} + v_N\frac{\partial h}{\partial n}\right)_G. \quad \ldots\ldots\ldots\ldots\ldots\ldots\ldots\ldots(2)$$

The surface wind is not a main variable and has to be estimated specially, for insertion in (2), from statistics of its relation to $M_{G8}$.

The vertical velocity equation might be called the keystone of the whole system, as so many other equations remain incomplete until the vertical velocity has been inserted.

### CH. 8/2/24. THE PRESSURE CHANGES AT ALL LEVELS

These are found, by way of the equation of continuity of mass, which can now be completed by the introduction of the vertical momentum. It gives first $\partial R/\partial t$ for each stratum. Then forming $g \,.\, \partial R/\partial t$ and making a running sum from above downwards we get $\partial p/\partial t$ at the boundaries of the strata. Computing Form P XIII.

### CH. 8/2/25. THE CHANGES IN THE WATER-CONTENT

For each of the lower strata these are next calculated by means of the equatic Ch. 4/3/ # 8 which brings together the changes due to turbulence and to "advection This equation spreads itself over four coordinate differences in both latitude an longitude. Terms such as $\frac{\delta}{\delta e}\left(M_E \frac{W}{R}\right)$ involve horizontal interpolations, since $M_E$ not given above the same points on the map as is $\mu = W/R$. Form P XVII.

### CH. 8/2/26. CHANGES IN THE SOIL

The changes in the water-content of the soil can now be computed by means c Ch. 4/10/2 # 5 and the change in its temperature by means of Ch. 4/10/3 # 10. Com puting Forms P XVIII and P XIX have been drawn up for this purpose.

### CH. 8/2/27. CHANGES IN THE SEA

The change of temperature of the surface of the sea during our time-step wil next need to be estimated. Its surface temperature may in many places be sufficientl forecasted by mean values for the same date in previous years. If that does no suffice, a more elaborate procedure has been sketched in Ch. 4/10/1.

## CH. 8/3. OPERATIONS CENTERED IN COLUMNS MARKED "M" ON THE CHESSBOARD MAP*

### CH. 8/3/1. EDDY-SHEARING-STRESSES

The surface-shearing-stress is estimated from statistics with reference to the "roughness" of the surface and to the strength of the wind, as represented by $M_E$, $M_N$. The angle by which the vector ($M_E$, $M_N$) is veered from the surface-shearing-stress is also estimated from statistics. The table on p. 84 was prepared for this purpose. Then the east and north components of the stress are calculated by expressions Ch. 4/8/3 # 4, 5, or more easily by a chart ruled with both polar and Cartesian coordinates.

The shearing stresses at the upper levels $h_8$, $h_6$, $h_4$ are computed by equations such as Ch. 4/8/2 # 2, equations which are derived from the diffusion equation Ch. 4/8/0 # 15. But the constant $\xi$ is not necessarily the same for velocity as for water or potential temperature. See the observations on pp. 72 to 76. The computing form is marked M I. The difference between the shearing stresses at the bottom and at the top of each stratum is transferred to the appropriate dynamical equation.

### CH. 8/3/2. DYNAMICS OF THE STRATOSPHERE

The stratosphere has some special terms in its dynamical equations, as set out in Ch. 6/4 # 7, 8. These are computed on Form M II. They spread over six times the smallest coordinate difference in use. We also require $v_E$, $v_N$ at the height $h_2$. To find these, an extrapolation is made to $h_2$ from above, using equations Ch. 6/3 # 7, 8. A

* See the Frontispiece.

second extrapolation is made to $h_2$ from below, assuming that $m_E$, $m_N$ do not vary with height in the stratum $h_4$ to $h_2$. The mean of these two extrapolated values is used in the dynamical equations.

### CH. 8/3/3. DYNAMICS OF LOWER STRATA

Finally $\partial M_E/\partial t$ and $\partial M_N/\partial t$ are determined from the dynamical equations Ch. 4/4 # 11, 12. The terms such as $\frac{\delta}{\delta n}\left(\frac{M_E M_N}{R}\right)$ spread over four times the least coordinate difference in use. They are best expanded in forms such as

$$\frac{\delta}{\delta n}\left(\frac{M_E M_N}{R}\right) = \frac{1}{R}\frac{\delta}{\delta n}(M_E M_N) + M_E M_N \frac{\delta}{\delta n}\left(\frac{1}{R}\right).$$

## CH. 8/4. CONCLUDING REMARKS

The cycle is now complete, for it has been shown that the time-changes of each one of the initially tabulated variables can be computed approximately by means of the stated equations, without bringing in any outside information, except a few statistical data. The system is thus "lattice reproducing." The above applies to points not too near the edge of the region on the map for which the initial values were given. New values after $\delta t$ can only be obtained for a smaller area. Ways of avoiding this loss have been proposed in Ch. 7.

It is curious that of the two very similar equations, one for the conveyance (or "advection") of water-per-mass, the other for the conveyance of entropy-per-mass, given respectively in Ch. 4/3 and Ch. 4/5/2, only the one for water appears explicitly in the calculations for the lower strata. The equation for the conveyance of entropy-per-mass is used in finding the equation for the vertical velocity, and does not arise again.

# CHAPTER IX

## AN EXAMPLE WORKED ON COMPUTING FORMS

### Ch. 9/0. INTRODUCTION

Let us now illustrate and test the proposals of the foregoing chapters by applying them to a definite case supplied by Nature and measured in one of the most complete sets of observations on record. Ch. 9/1 deals with the initial observations, Ch. 9/2 with deductions made from them. The computing forms which are used for this purpose may be regarded as embodying the process and thereby summarizing the whole book. In Ch. 9/3 a large error is investigated.

### Ch. 9/1. INITIAL DISTRIBUTION OBSERVED AT 1910 MAY 20 D 7 H G.M.T.

**The initial pressures** are tabulated at the ground and at exactly 2·0, 4·2, 7·2, 11·8 kilometres above M.S.L. They are read from V. Bjerknes' maps for the instant in question. These maps give the "dynamic height" of the isobaric surfaces, so that various conversions were necessary. V. Bjerknes has provided suitable conversion-tables. In the first place 2·0, 4·2, 7·2, 11·8 km are equivalent to 1·959, 4·113, 7·048, 11·543 "dynamic kilometres" when $g = 980{\cdot}00$ cm sec$^{-2}$ at sea-level. The pressures corresponding to these "dynamic heights" were obtained from the maps with the aid of V. Bjerknes' table "10M." Then a small correction has to be applied for the variation of gravity. This was obtained from tables 2M, 4M and 10M and worked out as follows:

| Kilometres from equator | Corrections in millibars to be SUBTRACTED from pressures at fixed "dynamic heights," to bring them to pressures at fixed heights corresponding to the dynamic height for $g = 980{\cdot}00$ cm sec$^{-2}$ at sea-level | | | | |
|---|---|---|---|---|---|
| | $p = 200$ mb | $p = 400$ mb | $p = 600$ mb | $p = 800$ mb | $p = 1000$ mb |
| 6000 | 0·4 | 0·5 | 0·4 | 0·3 | 0·0 |
| 5800 | 0·4 | 0·5 | 0·4 | 0·3 | 0·0 |
| 5600 | 0·3 | 0·4 | 0·3 | 0·2 | 0·0 |
| 5400 | 0·3 | 0·4 | 0·3 | 0·2 | 0·0 |
| 5200 | 0·2 | 0·3 | 0·2 | 0·1 | 0·0 |
| 5000 | 0·2 | 0·2 | 0·2 | 0·1 | 0·0 |

To find the pressure at the ground, the height of the ground was first assigned by reference to V. Bjerknes' maps of idealized topography*. The pressure at sea-level was next read from V. Bjerknes' map for the particular instant, and was then corrected

* *Dynamic Meteorology and Hydrography*, Plates XXVIII and XXIX.

to the height of the ground by means of the usual tables (Observer's Handbook). In the case of very high land the surface pressure was obtained from the maps of the heights of the 800 mb and 900 mb surfaces, with the aid of table 10M.

At some points there is a large uncertainty as to the appropriate value of $h_G$; for example in Switzerland the uncertainty amounts to several hundred metres. However, as the assigned value of $h_G$ will be used consistently throughout, the resulting errors will be small.

**The initial values of W, the mass-of-water-per-unit-horizontal-area-of-a-conventional-stratum,** were obtained as follows. First the density, $w$, of the water vapour was calculated from the temperature, and relative humidity recorded by the registering balloon at successive heights. Then $w$ was plotted against $h$. From areas on this diagram $W = \int w dh$ was obtained for each conventional stratum. These values of $W$ above the observing stations were plotted on maps, and values above the points marked $P$ on the map shown in this chapter were read off by interpolation. At Vienna and at Trappes observations were lacking and those for a time 24 hours previous were taken as a guide. At Pavia the relative humidity data cease at 7 km and values at greater heights were filled in by reference to statistics.

**The initial momenta-per-unit-area,** $M_E$, $M_N$, were obtained from the data published in *Veröffentlichungen der Internationalen Kommission...Beobachtungen mit bemannten, unbemannten Ballons, u.s.w.*, edited by Hergesell. The first process was to undo the computing already done* by the observing stations, by reconverting the winds to components and the heights to pressures. The component velocities $v_E$, $v_N$ were then plotted against the pressures. The divisions between the conventional strata, at 2·0, 4·2, 7·2 and 11·8 km were then marked off on the pressure scale, and areas on the diagram were measured with an Amsler planimeter so as to obtain $\int v_E dp$, $\int v_N dp$ between the limits of the strata. Then, for instance,

$$M_{EG8} = \int_G^8 \rho v_E dh = -\frac{1}{g}\int_G^8 v_E dp.$$

The mean velocities $\frac{M_E}{R}$, $\frac{M_N}{R}$ were compared with the mean resultant velocities and directions published by V. Bjerknes for his 10 standard sheets, on this occasion, and in general there was good agreement. In the data for Pavia and Strassburg, however, there were some discrepancies for which I could not account.

To extrapolate the velocities upwards to $p=0$, $v_E$ and $v_N$ were plotted against height, and straight lines were fitted to the curves in the stratosphere. In accordance with the theory of Ch. 6/3 the mean velocities $M_{E20}/R_{20}$ and $M_{N20}/R_{20}$ were taken as the velocities on these straight lines at a height of $h_2 + \frac{b\theta_1}{g} = 11{\cdot}8 + 6{\cdot}3 = 18{\cdot}1$ kilometres. The lines could be drawn satisfactorily for Vienna, tolerably well for Hamburg and Strassburg, and with considerable uncertainty for Copenhagen and Zürich. At Zürich

* Probably, but I have no definite information.

the highest observation was neglected. The lines were drawn at Lindenberg by assuming that $\frac{\partial v_E}{\partial h}$, $\frac{\partial v_N}{\partial h}$ were the same at Lindenberg as at Hamburg. This appears probable from the horizontal distribution of temperature in the stratosphere, taken in conjunction with equations Ch. 6/3 # 7, 8.

Having thus obtained $M_E$, $M_N$ for each stratum, they (or, alternatively, $M_E/R$, $M_N/R$) were plotted on maps at the observing stations, and values at the conventional points were estimated by interpolation, or, in some cases, by extrapolation. This process was an uncertain one, especially in the stratosphere, where the data were sparse; and near the ground, where the winds were irregular. It makes one wish that pilot balloon stations could be arranged in rectangular order, alternating with stations for registering balloons, in some such pattern as that formed by the points marked $M$ and $P$ on the frontispiece. In the present example, in spite of all the care taken, the tabulated values of the momenta-per-unit-area remind one of the stories which are "founded on fact." In consequence they give absurd values to

$$\frac{\partial p_G}{\partial t} = -g \operatorname{div}_{EN} \sum_{\text{all}}^{\text{strata}} M.$$

We shall return to this point in Ch. 9/3.

**The initial temperature of the stratosphere.** The observations show that the mean temperature over Central Europe was about 214° A and that the temperature over South Germany, Switzerland and Lombardy was some 13° lower than that over Russia and England. But there were considerable irregular variations with height amounting to $\pm 5°$ C at some stations.

Probably the most exact method* of mapping the temperature in the stratosphere is to use $\frac{\partial\theta}{\partial e}$, $\frac{\partial\theta}{\partial n}$ obtained from the variation of wind with height according to equations Ch. 6/3 # 7, 8. The following table gives the temperature gradients $\frac{\partial\theta}{\partial e}$, $\frac{\partial\theta}{\partial n}$ computed from the observations of wind. They have been used to smooth the distribution of temperature.

| Station | $\frac{\partial v_N}{\partial h}$ (in metres per second per 10 kilometres) | $\frac{\partial v_E}{\partial h}$ | $\frac{\theta \cdot 2\omega \sin\phi}{g}$ (degrees sec / cm) | $\frac{\partial\theta}{\partial e}$ (degrees per 200 km) | $\frac{\partial\theta}{\partial n}$ |
|---|---|---|---|---|---|
| Copenhagen ... | + 22 | − 19 | $2{\cdot}6 \times 10^{-5}$ | + 1·2 | + 1·0 |
| Hamburg ...... | + 10 | − 14 | 2·6 | + 0·5 | + 0·7 |
| Strassburg ... | − 19 | + 22 | 2·4 | − 0·9 | − 1·2 |
| Zürich ......... | − 6 | + 15 | 2·3 | − 0·3 | − 0·7 |
| Vienna ......... | + 3 | + 2 | 2·4 | + 0·1 | − 0·1 |

* Sir Napier Shaw, "Upper-Air Calculus," *J. Scott. Met. Soc.* 1913.

**The temperature of the soil** is required at a series of depths. The depths proposed in Ch. 4/10/0 appear to be unnecessarily thin when their thermal conductance is compared with that of the atmospheric strata, and so the alternate divisions are here omitted, leaving those at $z_{12} = 6{\cdot}39$, $z_{14} = 53{\cdot}6$ cm etc. The temperature is required at intermediate depths which, by following the process of gradation proposed in Ch. 4/10/0, are taken to be $z_{11} = 1{\cdot}72$, $z_{13} = 19{\cdot}1$ cm.

In default of direct observations the temperature at $z_{11}$ may be estimated from the statistics of E. Ebermayer*, for these refer to Bavaria in which the *P* point to be treated is situated. The mean of six stations for May 1868 shows that the air at a

[*continued on p.* 186]

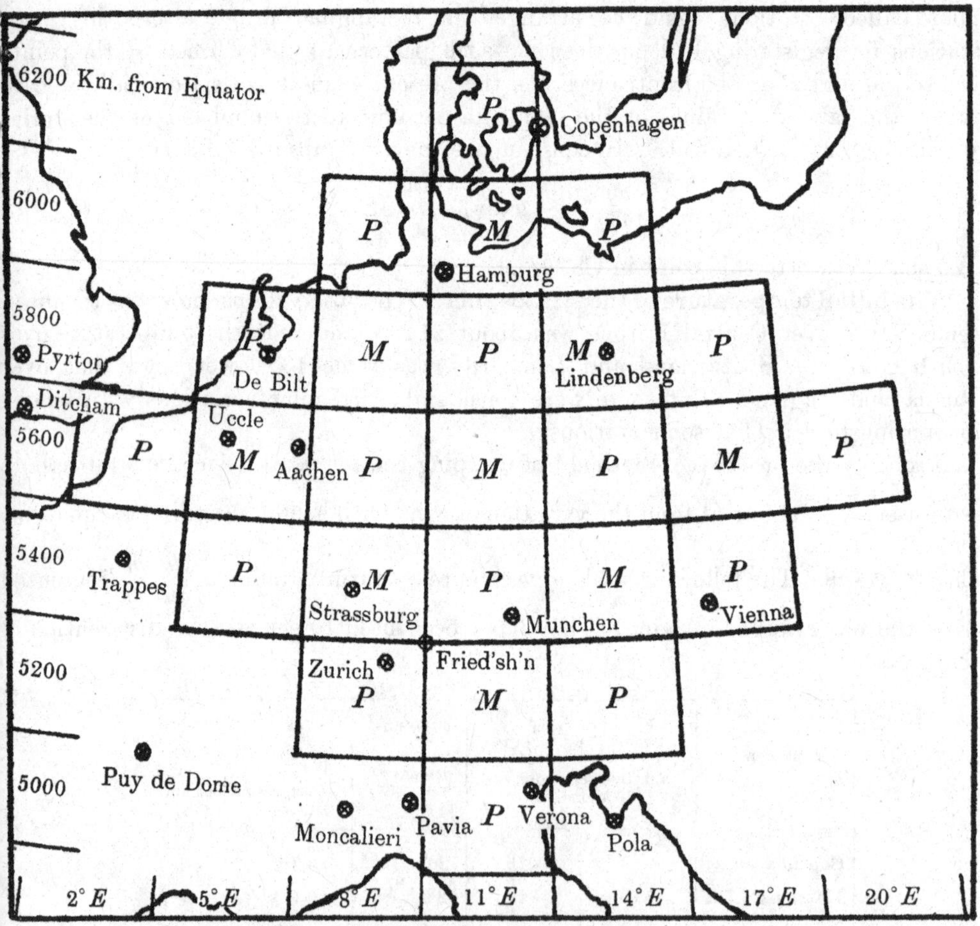


MAP OF POINTS FOR PRESSURE (*P*) AND MOMENTUM (*M*) USED IN THE EXAMPLE of Ch. 9.

Note: These points are placed at the centres of the chequers, and to the centres also the latitude and longitude refer. Each chequer measures 3° from west to east and 200 km from south to north.

* *Die physikalischen Einwirkungen des Waldes*, Berlin, Verlag von Wiegandt, Hempel und Parey, 1873.

*Obtained by interpolating or extrapolating from observations taken at* 1910 *May* $20^d$ $7^h$ *G.M.T.*

| kilometres north from equator | longitude 5° E | longitude 8° E | longitude 11° E | longitude 14° E | longitude 17° E |
|---|---|---|---|---|---|
| 6000 | | $\theta_1$ 214° A | 1000× 1000×<br>$M_{E20}$ −65 $M_{N20}$ +8<br>$M_{E42}$ +127 $M_{N42}$ −104<br>$M_{E64}$ +81 $M_{N64}$ −25<br>$M_{E86}$ −81 $M_{N86}$ zero<br>$M_{EG8}$ −198 $M_{NG8}$ +84<br>$h_G = 0$ | $\theta_1$ 216° A | NOTE: The following stratosphere temperatures have also been used<br>long 11° E, lat 6200 N, 216°<br>,, 2° E, ,, 5600 N, 217°<br>,, 20° E, ,, 5600 N, 216° |
| 5800 | $\theta_1$ 215° A | 1000×<br>$M_{E20}$ −70<br>$M_{E42}$ −62<br>$M_{E64}$ −114<br>$M_{E86}$ −91<br>$M_{EG8}$ −160<br>$h_G$ 15000 | 100×<br>$\theta_1$ 214° A<br>$p_2$ 2047<br>$W_{42}$ 0·0<br>$p_4$ 4090<br>$W_{64}$ 0·2<br>$p_6$ 6086<br>$W_{86}$ 0·4<br>$p_8$ 7983<br>$W_{G8}$ 0·9<br>$p_G$ 9883<br>$h_G$ 20000 | 1000×<br>$M_{E20}$ −160<br>$M_{E42}$ +40<br>$M_{E64}$ −60<br>$M_{E86}$ −60<br>$M_{EG8}$ −219<br>$h_G$ 10000 | $\theta_1$ 216° A |
| 5600 | 1000× 1000×<br>$M_{E20}$ −30 $M_{N20}$ −110<br>$M_{E42}$ −245 $M_{N42}$ +300<br>$M_{E64}$ −223 $M_{N64}$ +158<br>$M_{E86}$ −91 $M_{N86}$ +87<br>$M_{EG8}$ −18 $M_{NG8}$ +15<br>$h_G$ 20000 | 100×<br>$\theta_1$ 212° A<br>$p_2$ 2047<br>$W_{42}$ 0·0<br>$p_4$ 4083<br>$W_{64}$ 0·2<br>$p_6$ 6067<br>$W_{86}$ 0·5<br>$p_8$ 7950<br>$W_{G8}$ 1·2<br>$p_G$ 9834<br>$h_G$ 20000 | 1000× 1000×<br>$M_{E20}$ −56 $M_{N20}$ −18<br>$M_{E42}$ −146 $M_{N42}$ −62<br>$M_{E64}$ −95 $M_{N64}$ +29<br>$M_{E86}$ −52 $M_{N86}$ +58<br>$M_{EG8}$ −110 $M_{NG8}$ +55<br>$h_G$ 40000 | 100×<br>$\theta_1$ 214° A<br>$p_2$ 2049<br>$W_{42}$ 0·0<br>$p_4$ 4091<br>$W_{64}$ 0·1<br>$p_6$ 6087<br>$W_{86}$ 0·4<br>$p_8$ 7979<br>$W_{G8}$ 0·9<br>$p_G$ 9763<br>$h_G$ 30000 | 1000×<br>$M_{E20}$ −100 $M_{N20}$ −32<br>$M_{E42}$ zero $M_{N42}$ −260<br>$M_{E64}$ −55 $M_{N64}$ −135<br>$M_{E86}$ −25 $M_{N86}$ +48<br>$M_{EG8}$ −190 $M_{NG8}$ +160<br>$h_G$ 30000 |
| 5400 | 100×<br>$\theta_1$ 214° A<br>$p_2$ 2030<br>$W_{42}$ 0·0<br>$p_4$ 4049<br>$W_{64}$ 0·2<br>$p_6$ 6044<br>$W_{86}$ 0·7<br>$p_8$ 7928<br>$W_{G8}$ 1·4<br>$p_G$ 9744<br>$h_G$ 20000 | 1000×<br>$M_{E20}$ +27<br>$M_{E42}$ −328<br>$M_{E64}$ −136<br>$M_{E86}$ −33<br>$M_{EG8}$ +48<br>$h_G$ 40000 | 100×<br>$\theta_1$ 212° A<br>$p_2$ 2050<br>$W_{42}$ 0·0<br>$p_4$ 4090<br>$W_{64}$ 0·1<br>$p_6$ 6079<br>$W_{86}$ 0·4<br>$p_8$ 7960<br>$W_{G8}$ 0·9<br>$p_G$ 9626<br>$h_G$ 40000 | 1000×<br>$M_{E20}$ zero<br>$M_{E42}$ −166<br>$M_{E64}$ −95<br>$M_{E86}$ −19<br>$M_{EG8}$ −65<br>$h_G$ 40000 | 100×<br>$\theta_1$ 214° A<br>$p_2$ 2044<br>$W_{42}$ zero<br>$p_4$ 4082<br>$W_{64}$ 0·1<br>$p_6$ 6068<br>$W_{86}$ 0·4<br>$p_8$ 7978<br>$W_{G8}$ 0·9<br>$p_G$ 9882<br>$h_G$ 20000 |
| 5200 | | 100×<br>$\theta_1$ 214° A<br>$p_2$ 2039<br>$W_{42}$ 0·0<br>$p_4$ 4062<br>$W_{64}$ 0·2<br>$p_6$ 6050<br>$W_{86}$ 0·6<br>$p_8$ 7949<br>$W_{G8}$ 0·9<br>$p_G$ 8746<br>$h_G$ 120000 | 1000× 1000×<br>$M_{E20}$ −50 $M_{N20}$ +80<br>$M_{E42}$ −280 $M_{N42}$ +41<br>$M_{E64}$ −175 $M_{N64}$ +150<br>$M_{E86}$ −105 $M_{N86}$ +80<br>$M_{EG8}$ −155 $M_{NG8}$ +40<br>$h_G$ 180000 | 100×<br>$\theta_1$ 214° A<br>$p_2$ 2043<br>$W_{42}$ 0·0<br>$p_4$ 4075<br>$W_{64}$ 0·1<br>$p_6$ 6065<br>$W_{86}$ 0·4<br>$p_8$ 7967<br>$W_{G8}$ 1·0<br>$p_G$ 8458<br>$h_G$ 150000 | |
| 5000 km north of equator | | | 100×<br>$\theta_1$ 213° A<br>$p_2$ 2034<br>$W_{42}$ 0·0<br>$p_4$ 4034<br>$W_{64}$ 0·2<br>$p_6$ 6031<br>$W_{86}$ 0·6<br>$p_8$ 7957<br>$W_{G8}$ 1·4<br>$p_G$ 9972<br>$h_G$ 10000 | | |

All the quantities in any one square refer to the same latitude and longitude, namely to those stated in the margin. The notation is defined in Ch. 12. A numerical value, when multiplied by the integral power of ten, if any, standing at the head of its column, is then expressed in C.G.S. units. To save space the equality symbols are omitted, thus 1000× $M_{E20}-65$ stands for $M_{E20} = -65000$ grm cm$^{-1}$ sec$^{-1}$.

height of 1·6 m had a temperature exceeding that of the soil at $z_{11}$ by 0·8° C in the open or 3·3° C in the forest. Again at Tiflis*, in a latitude only 2° nearer the equator, the difference between soil and air-at-3-metres-height passes through its daily mean value at just about the time of day that concerns us, namely at 8 h L.A.T. On 1910 May 20 d 7 h M.E.Z. the air temperature at the Bavarian stations† is recorded as about 291° A and for the above reasons the earth temperature at $z_{11}$ has been assumed to be 290° A in the open, or 288° in the forest.

**The Table of Initial Distribution** will be found on p. 185.

## CH. 9/2. DEDUCTIONS, MADE FROM THE OBSERVED INITIAL DISTRIBUTION, AND SET OUT ON THE COMPUTING FORMS

The process described in Ch. 8 has been followed so as to obtain $\partial/\partial t$ of each one of the initially tabulated quantities.

The **units** employed on the computing forms follow the same rule as elsewhere in this book. Any numerical value, after being multiplied by the integral power of ten, if any, standing at the head of its column, is in centimetre-gram-second units; unless the contrary is stated. Temperatures are in degrees centigrade absolute. Energy, whether by itself, or as involved in specific heat and entropy, is expressed, not in calories, but in ergs. Some numerical quantities are total amounts during the time step $\delta t$ of $2{\cdot}16 \times 10^4$ seconds, for example the energies of radiation. Other quantities are reckoned as totals for unit area of the individual strata. Others again are reckoned per second and per gram.

The arithmetical accuracy is as follows. All computations were worked twice and compared and corrected. The last digit is often unreliable, but is retained to prevent the accumulation of arithmetical errors. Multiplications were mostly worked by a 25 centim slide rule. But in calculating $\partial P/\partial e$, $\partial P/\partial n$ it was found to be essential to use five-figure logarithms. The chief features of the arithmetic are displayed on the computing forms, and it will suffice here to call attention to a few of the more interesting points, and to confess to certain imperfections and errors.

The forms are divided into two groups marked $P$ or $M$ according as the point on the map to which they refer is one where pressures ($P$) or momenta ($M$) are tabulated.

Only two points on the map have been taken as centres for the equations, namely the "$P$" point at 11° east, 5400 km north, and the "$M$" point 200 kilometres due north of it. These two points are really not enough, because the vertical velocity is required over the "$M$" points, but is only found directly over the "$P$" points. At an "$M$" point the vertical velocity should be taken as the mean of that at the four surrounding "$P$" points. However here the vertical velocity from the single "$P$" point has been inserted for illustration.

The results for radiation on forms P IV, P V, P VI have quite a plausible appearance. Much fuller exemplification of P VI or its equivalent will be found in a paper

* Hann, *Lehrb. der Meteorologie*, 3 Aufl. p. 64.

† *Deutsches Meteorologisches Jahrbuch für* 1910, Bayern, München in Kommission bei A. Buchholz.

by W. H. Dines*. The fact that some clouds were present was not noticed until after these forms had been filled in and for that reason a cloudiness of 1·4 tenths has been neglected.

The potential temperature at pressure $p_G$ increases considerably aloft, as shown in the last column of Form P I. By contrast the vertical changes of entropy-per-mass on Form P XII are small and irregular, because the upgrade of water-per-mass compensates that of potential temperature.

The rate of rise of surface pressure, $\partial p_G/\partial t$, is found on Form P XIII as 145 millibars in 6 hours, whereas observations show that the barometer was nearly steady. This glaring error is examined in detail below in Ch. 9/3, and is traced to errors in the representation of the initial winds.

The vertical velocity (Form P XVI) is nowhere more than $\pm 3$ cm sec$^{-1}$; and, as it is probably exaggerated by the errors in the winds, the real value must have been very small—quite beyond the possibility of observation†. The small values given on Form P XVI are found to have quite notable effects when inserted in the equation of continuity, in the equation for the transport of water and in the dynamical equations.

It is interesting to notice on Form P XIII that the vertical velocity is such as to smooth the pressure changes, making them increase regularly from above downwards.

Coming now to the horizontal velocities in the stratosphere, it is seen on Form M II that there is a misfit at $h_2$ between $m_E$ as extrapolated from below and from above. The misfit may well be due to unnatural initial values of $\partial\theta/\partial e$, $\partial\theta/\partial n$ in the stratosphere. This opinion is strengthened when we consider the dynamical equations in the stratosphere, as set out on forms M III and M IV, because the terms $\delta P_{20}/\delta e$ and $\delta P_{20}/\delta n$ are seen to be unnaturally large; and these two terms depend, on this occasion, mainly on $\partial\theta/\partial e$ and $\partial\theta/\partial n$ since $\partial p_2/\partial e$ and $\partial p_2/\partial n$ are small. In fact it appears necessary to have the temperature in the stratosphere tabulated to 0°·1 C. If it cannot be observed with this accuracy, the observations will have to undergo a preliminary smoothing, before being used as initial conditions. Near the base‡ of the stratosphere $\partial v/\partial h$ and $\partial^2 v/\partial h^2$ are often so large, that their numerical treatment is sure to be difficult.

In the dynamical equations in the lower strata the terms $\partial P/\partial e$, $\partial P/\partial n$, $-2\omega \sin\phi\, M_N$, $+2\omega\sin\phi\, M_E$ are usually the largest, but the "curvature of the path" has a considerable effect and so also has that much neglected term, the rate of change along the path of the square of momentum. In the lowest strata the terms $\partial P/\partial e$, $\partial P/\partial n$ are very large owing to the slope of the ground, and are for the greater part balanced by $p\dfrac{\partial h_G}{\partial e}$, $p\dfrac{\partial h_G}{\partial n}$. Special care has to be taken in computing these, as described in Ch. 4/4 # 14.

* *Q. J. R. Met. Soc.* April 1920.

† Of course there may have been larger vertical currents locally; the statement in the text refers to the mean over a chequer on the map.

‡ *Vide* G. M. B. Dobson, *Q. J. R. Met. Soc.* Jan. 1920.

COMPUTING FORM P I. Pressure, Temperature, Density, Water and Continuous Cloud

Longitude = *11° East* Latitude = *5400 km North* Instant *1910 May 20^d 7^h G.M.T.*

| REF.:— | | p. 185 | Helmert | previous | | Ch. 8/2/2 | | p. 185 | | ρ and p | | | | | |
|---|---|---|---|---|---|---|---|---|---|---|---|---|---|---|---|
| Height $h$ | $\delta h$ | $p$ | $g$ | $R$ | $\rho=\frac{R}{\delta h}$ | $P$ | $p=\frac{P}{\delta h}$ | $W$ | $\mu=\frac{W}{R}$ | $\theta$ | Density of saturated vapour $w_s$ | $W_s=w_s\delta h$ | Precipitated = $W-W_s-0{\cdot}4$ | $Q$ in continuous cloud | Potential temperature at surface pressure |
| $10^5\times$ | $10^5\times$ | $10^2\times$ | | | $10^{-3}\times$ | $10^3\times$ | $10^2\times$ | | $10^{-3}\times$ | °A | $10^{-6}\times$ | | | | |
| $h_0$ | | *0* | | | | | | | | | | | | | |
| | | | *975?* | *210·3* | | *1278* | | zero | zero | *212* | | | | | By Ch. 8/2/6 # 13 |
| $h_2=11{\cdot}8$ | | *2050* | | | | | | | | | | | | | |
| | 4·6 | | *978·2* | *209·0* | *0·454* | *1362* | *2965* | *0·0* | *0·0* | *227·5* | | *0·0* | *0·0* | *0·0* | *320·8* |
| $h_4=7{\cdot}2$ | | *4090* | | | | | | | | | | | | | |
| | 3·0 | | *979·2* | *203·3* | *0·677* | *1501* | *5003* | *0·1* | *0·5* | *257·7* | *1·3* | *0·4* | *0·0* | *0·0* | *312·0* |
| $h_6=4{\cdot}2$ | | *6079* | | | | | | | | | | | | | |
| | 2·2 | | *980·0* | *192·1* | *0·875* | *1538* | *6990* | *0·4* | *2·1* | *279·0* | *6·0* | *1·6* | *0·0* | *0·0* | *305·8* |
| $h_8=2{\cdot}0$ | | *7960* | | | | | | | | | | | | | |
| | *1·6* | | *980·6* | *169·9* | *1·061* | **1402* | *8770* | *0·9* | *5·3* | *287·5* | *12·3* | *2·0* | *0·0* | *0·0* | *295·3* |
| $h_G=0{\cdot}4$ | | *9626* | | | | | | | | | | | | | |

* But see Ch. 4/4 # 14.

## COMPUTING FORM P II. Gas constant. Thermal capacities. Entropy derivatives

Longitude $= 11°$ *East* Latitude $= 5400$ *km North* Instant *1910 May* $20^d\ 7^h$ *G.M.T.*

| REF.:— | Ch. 4/5/1 | Ch. 4/5/1 | Ch. 4/5/1 | | Ch. 4/5/1 | Ch. 4/5/1 | Ch. 8/2/13 | Ch. 4/5/1 | |
|---|---|---|---|---|---|---|---|---|---|
| Height | Gas constant $b$ ergs grm$^{-1}$ deg$^{-1}$ | Thermal capacities at constant density $\gamma_\rho$ ergs per grm per degree | Thermal capacities at constant pressure $\gamma_p$ ergs per grm per degree | | $a_p = \left(\frac{\partial\sigma}{\partial p}\right)_{\rho,\mu,\text{const}}$ | $a_\rho = \left(\frac{\partial\sigma}{\partial\rho}\right)_{p,\mu,\text{const}}$ | $a_\mu = \left(\frac{\partial\sigma}{\partial\mu}\right)_{p,\rho,\text{const}}$ | $\beta = -\frac{a_\rho}{a_p}$ | $-\frac{a_\rho}{a_p}\rho = \rho\beta$ |
| | $10^6 \times$ | $10^6 \times$ | $10^6 \times$ | | | $10^{10} \times$ | $10^6 \times$ | $10^6 \times$ | $10^4 \times$ |
| $h_0$ | | | | | | | | | |
| | *2·870* | *7·05* | *9·92* | | *+ 68·7** | | | *858* | *14·42** |
| $h_2$ | | | | | | | | | |
| | *2·870* | *7·05* | *9·92* | | *23·8* | *−2·182* | | *919* | *41·7* |
| $h_4$ | | | | | | | | | |
| | *2·871* | *7·05* | *9·93* | | *14·10* | *−1·464* | *100 approx.* | *1038* | *70·4* |
| $h_6$ | | | | | | | | | |
| | *2·874* | *7·07* | *9·94* | | *10·09* | *−1·135* | *100 approx.* | *1126* | *98·4* |
| $h_8$ | | | | | | | | | |
| | *2·879* | *7·09* | *9·97* | | *8·08* | *−0·938* | *100 approx.* | *1160* | *123·3* |
| $h_G$ | | | | | | | | | |
| | *2·884* | *7·12* | *10·00* | | | | | | |

* Taking $p$ as $\frac{1}{2}p_2$.

## Computing Form P III. Stability, Turbulence, Heterogeneity, Detached Cloud

Longitude *11° E* Latitude *5400 km N* Time, from *1910 May 20ᵈ 4ʰ G.M.T.* to *10ʰ*

| Ref.:— | | | | | Form P I | | | | | | | |
|---|---|---|---|---|---|---|---|---|---|---|---|---|
| Height $h$ $10^5\times$ | Mean $\delta h$ $10^5\times$ | Velocities $v_E$ | $v_N$ | $\left(\frac{\partial v_E}{\partial h}\right)^2+\left(\frac{\partial v_N}{\partial h}\right)^2$ | $\frac{\delta_H\tau}{\tau}$ when $p_G$ is standard | $\frac{g}{\tau}\frac{\partial\tau}{\partial h}$ * | $\mu$ | | $\xi$ | $\bar{\xi}_{GS}$ | [illegible] | Fraction of area of stratum clouded |
| | | $10^2\times$ | $10^2\times$ | $10^{-4}\times$ | | $10^{-4}\times$ | $10^{-3}\times$ | | | | | |
| 11·8 | | | | | | | | | *0* | | | |
| | | *−11·0* | *−0·5* | | | | *0·0* | | | | | |
| 7·2 | 3·8 | | | *0·03* | *·028* | *0·7* | | | *50* | | | |
| | | *−6·1* | *+4·4* | | | | *0·5* | | | | | |
| 4·2 | 2·6 | | | *0·02* | *·021* | *0·8* | | | *1000* | | | |
| | | *−2·7* | *+3·6* | | | | *2·1* | | | | | |
| 2·0 | 1·9 | | | *0·01* | *·034* | *1·8* | | | *10⁵* | | | |
| | | *−4·2* | *+2·8* | | | | *5·3* | | | Interim values | | |
| *0·4* | | | | | | | | | | *10⁵* | *10⁵* | |
| | | | | | | | | | | for trial surface temperature of $\theta_A = 290°$, for revised estimates see Forms P IX, P X | | *Mean observed cloudiness at 6ʰ G.M.T. about 1·4 tenths. Distribution unknown. Neglected.* |

* Formulae relating to the stability of dry air. The stability of cloud remains to be investigated.

COMPUTING FORM P IV. For Solar Radiation in the grouped ranges of wave-lengths known as *BANDS*

Longitude *11° East*, Latitude *5400 km North*; Time from *1910 May 20ᵈ 4ʰ* to *10ʰ G.M.T.* — Interval *6 hours*; Mean cos $\zeta = 0{\cdot}535$, Solar constant $= 1{\cdot}16 \times 10^{11}$ ergs $cm^{-2}$ $day^{-1}$ — Incident radiation during interval in whole spectrum *1550*

Unit for radiation during interval of time is throughout $10^7$ ergs $cm^{-2}$ = 1 joule $cm^{-2}$

| LEVEL | DIRECT BEAM: Fractions of initial for stratum: Absorbed | DIRECT BEAM: Fractions of initial for stratum: Scattered | DIRECT BEAM: Flux | LOSSES from Direct beam: Amount absorbed | LOSSES from Direct beam: Amount scattered | DIFFUSE DOWNWARDS: Flux | DIFFUSE DOWNWARDS: Fraction absorbed | DIFFUSE DOWNWARDS: Amount absorbed | DIFFUSE UPWARDS: Flux | DIFFUSE UPWARDS: Fraction absorbed | DIFFUSE UPWARDS: Amount absorbed | TOTAL AMOUNT ABSORBED |
|---|---|---|---|---|---|---|---|---|---|---|---|---|
| $h_0$ | | | *230* | | | *0* | | | *12* | | | |
| | *0·067* | *0·026* | | *15* | *6* | | *0·062* | *0* | | *0·062* | *1* | *16* |
| $h_2$ | | | *209* | | | *3* | | | *10* | | | |
| | *0·067* | *0·026* | | *14* | *5* | | *0·061* | *0* | | *0·061* | *0* | *14* |
| $h_4$ | | | *190* | | | *5* | | | *7* | | | |
| | *0·130* | *0·024* | | *25* | *5* | | *0·122* | *1* | | *0 122* | *0* | *26* |
| $h_6$ | | | *160* | | | *7* | | | *5* | | | |
| | *0·288* | *0·020* | | *46* | *3* | | *0·271* | *2* | | *0·271* | *1* | *49* |
| $h_8$ | | | *111* | | | *7* | | | *5* | | | |
| | *0·500* | *0·015* | | *55* | *2* | | *0·476* | *3* | | *0·476* | *3* | *61* |
| $h_L$ | | | *54* | | | *5* | | | *7* | | | |
| | Absorbed *0·88** | Reflected *0·12** | | *48* | Reflected *6* | | *0·88** | *4* | | *0·88** | *0* | *52* |
| $h_G$ | | | | | | | | | | | | |
| | | | | | | | | | | | ABSORBED, SUM | *218* |
| | | | | | | | | | | | TO SPACE | *12* |
| | | | | | | | | | | | SUM = INCIDENT | *230* |

* Arbitrary reflectivity assumed for illustration.

For masses per area of stratum, water per area of stratum, and amount of continuous and detached cloud, see Forms P I and P III.

# Computing Form P v. For Solar Radiation in the grouped ranges of wave-lengths known as *REMAINDER*

Longitude *11° East* — Latitude *5400 km North*

Time from *1910 May 20^d 4^h* to *10^h G.M.T.* — Interval *6 hours*

Mean cos $\zeta = 0{\cdot}535$ — Solar constant $= 1{\cdot}16 \times 10^{11}$ ergs cm$^{-2}$ day$^{-1}$ — Incident radiation during interval in whole spectrum *1550*

Unit for radiation during interval of time is throughout $10^7$ ergs cm$^{-2}$ = 1 joule cm$^{-2}$

For masses per area of stratum, water per area of stratum and amount of continuous and detached cloud see Forms P I, P III.

| LEVEL | DIRECT BEAM: Fractions of initial for stratum: Absorbed | DIRECT BEAM: Fractions of initial for stratum: Scattered | DIRECT BEAM: Flux | LOSSES from Direct beam: Amount absorbed | LOSSES from Direct beam: Amount scattered | DIFFUSE DOWNWARDS: Flux | DIFFUSE DOWNWARDS: Fraction absorbed | DIFFUSE DOWNWARDS: Amount absorbed | DIFFUSE UPWARDS: Flux | DIFFUSE UPWARDS: Fraction absorbed | DIFFUSE UPWARDS: Amount absorbed | TOTAL AMOUNT ABSORBED |
|---|---|---|---|---|---|---|---|---|---|---|---|---|
| $h_0$ | | | *1320* | | | *0* | | | *257* | | | |
| | *0·009* | *0·059* | | *12* | *78* | | *0·008* | *0* | | *0·008* | *2* | *14* |
| $h_2$ | | | *1230* | | | *39* | | | *220* | | | |
| | *0·009* | *0·059* | | *11* | *72* | | *0·008* | *0* | | *0·008* | *1* | *12* |
| $h_4$ | | | *1147* | | | *75* | | | *185* | | | |
| | *0·010* | *0·060* | | *11* | *69* | | *0·009* | *1* | | *0·009* | *1* | *13* |
| $h_6$ | | | *1067* | | | *109* | | | *152* | | | |
| | *0·013* | *0·064* | | *14* | *68* | | *0·012* | *1* | | *0·012* | *2* | *17* |
| $h_8$ | | | *985* | | | *142* | | | *120* | | | |
| | *0·018* | *0·071* | | *18* | *70* | | *0·017* | *2* | | *0·017* | *1* | *21* |
| $h_L$ | | | *897* | | | *175* | | | *86* | | | |
| | *0·92** | Reflected *0·08** | | *825* | Reflected *72* | | *0·92** | *161* | | *0·92** | *0* | *986* |
| $h_G$ | | | | | | | | | | | | |

* Arbitrary reflectivity assumed for illustration.

| | |
|---|---|
| ABSORBED, SUM | *1063* |
| TO SPACE | *257* |
| SUM = INCIDENT | *1320* |

# COMPUTING FORM P VI. For Radiation due to atmospheric and terrestrial temperature

The equations are typified by Ch. 4/7/1 # 18, namely $\Sigma\mathcal{E}_8 = (1-\bar{\eta})\,\Sigma\mathcal{E}_6 + \bar{\eta}$ Ↄ $\theta^4$ for descending radiation

The unit for radiation during the interval of time is throughout $10^7$ ergs $cm^{-2}$ = 1 joule $cm^{-2}$

| Longitude *11° East*<br>Latitude *5400 km North* | Time from *1910 May 20ᵈ 4ʰ* to *10ʰ G.M.T.* } Interval *6 hours* | Ↄ = 5·36 × $10^{-5}$ ergs $cm^{-2}$ $sec^{-1}$<br>$\Lambda = 1{\cdot}48\,(1+40\mu)\times 10^{-3}$ $cm^2\,grm^{-1}$ |
|---|---|---|

Leave this until Form P x has been filled in (applies to the Upward and Total gain columns)

| Height | Fraction transmitted $=1-\bar{\eta}$ $=e^{-\Lambda \sec 55°.R}$ if clear | Radiation emitted to either side $=\bar{\eta}$Ↄ$\theta^4$ | Downward: Amount transmitted by stratum | Downward: Flux $\Sigma\mathcal{E}$ | Downward: Amount gained by stratum | Upward: Amount transmitted by stratum | Upward: Flux $\Sigma\mathcal{E}$ | Upward: Amount gained by stratum | Total gain by stratum |
|---|---|---|---|---|---|---|---|---|---|
| $h_0$ | | | | *0* | | | *398* | | |
| | *0·581* | *97* | *0* | | *− 97* | *301* | | *+121* | *+ 24* |
| $h_2$ | | | | *97* | | | *519* | | |
| | *0·583* | *129* | *57* | | *− 89* | *390* | | *+150* | *+ 61* |
| $h_4$ | | | | *186* | | | *669* | | |
| | *0·586* | *212* | *109* | | *−135* | *457* | | *+111* | *− 24* |
| $h_6$ | | | | *321* | | | *780* | | |
| | *0·584* | *292* | *187* | | *−158* | *488* | | *+ 56* | *−102* |
| $h_8$ | | | | *479* | | | *836* | | |
| | *0·588* | *326* | *282* | | *−129* | *510* | | *+ 33* | *− 96* |
| $h_L$ | | | | *608* | | | *869* | | |
| | *zero* | *869** | *0* | | *−261* | *0* | | *0* | *−261* |
| | | | | *869* | | | *869* | | |

* At *294°·3 abs* see Form P x.

Loss to space = sum = *− 398*

For *R*, $\theta$ and $\mu$ see Computing Form P I
For continuous cloud see Computing Form P I
For detached cloud see Computing Form P III

Long. *11° E* Lat. *5400 km N* Time from *1910 May 20ᵈ 4ʰ* to *10ʰ G.M.T.*

Note: the process is applicable to bare soil when the modifications in column $Y$ are made

| | Character of surface......... | | forest | other crops | water | $Y$ | bare soil |
|---|---|---|---|---|---|---|---|
| Trial values at $\theta_A = \theta'$ | Trial value of surface temperature, as on Forms P IX, P X, | $= \theta'°$ Abs | | *290* | | | |
| | "Resistance" of half air stratum $= g\,(p_G - p_8)/(2\bar{\xi}_{G8}) = B$ | cm² grm⁻¹ sec | | *780* | | $B$ | |
| | "Resistance" of stomata $= (\kappa\rho)^{-1} = A$ | cm² grm⁻¹ sec | | *4000* | — | $z_{11}/f(\mu)$ | |
| | Sum of two preceding = whole "resistance" $= 1/C'_\mu$ | cm² grm⁻¹ sec | | *4780* | | whole resistance | |
| | $\mu' = \mu$ for air saturated at $\theta'$ and $p_G$. A pure number | | | $12{\cdot}7 \times 10^{-3}$ | | $\mu_{11} = F_{11}/\rho_G$ | |
| | Difference $\mu' - \mu_9$. For $\mu_9$ see Form P I | | | $5{\cdot}3 \times 10^{-3}$ | | $\mu_{11} - \mu_9$ | |
| | Flux $= (\mu' - \mu_9)\,C'_\mu$ | | | $1{\cdot}11 \times 10^{-6}$ | | $z_{11}$ Ш$g/f(\mu)$ | |
| | For evaporation from wet vegetation add a portion of simultaneous rainfall. This portion is provisionally taken to be $n\dfrac{(\mu' - \mu_G)}{B}$ (rainfall per second) where *$n = 1{\cdot}2 \times 10^5$ for beech forest in winter* | | | *0* | | sum of last two; sum ÷ whole resistance | |
| | Total rate of evaporation $= \Xi'$ | grm cm⁻² sec⁻¹ | | $1{\cdot}11 \times 10^{-6}$ | | | |
| Here complete Forms P IX, P X to find true surface temperature $\theta_A$ | | | | *294·3* | | | |
| Correction to $(\mu' - \mu_9)\,C'_\mu$ that is $(\theta_A - \theta')\left\{C'_\mu \dfrac{d\mu'}{d\theta'} + \mu' \dfrac{dC'_\mu}{d\theta'}\right\}$ For differential coefficients see Form P X | | | | $1{\cdot}04 \times 10^{-6}$ | | | |
| Correction to rainfall evaporated | | | | | | | |
| True rate of evaporation $= \Xi$ | | grm cm⁻² sec⁻¹ | | $2{\cdot}15 \times 10^{-6}$ | | | |
| Rate of evaporation × interval of time $= \Xi\delta t$ | | grm cm⁻² | | *0·0464* | | | |
| Fractions of coordinate chequer having the several characters | | | *0·25* | *0·70* | *0·00* | | *0·05* |
| Mean evaporation over the whole chequer in $\delta t$ | | grm cm⁻¹ | | *say 0·05* | | | |

## COMPUTING FORM P VIII. Fluxes of Heat at the interface

Long. *11° E.* Lat. *5400 km N.* Time from *1910 May 20$^d$ 4$^h$* to *10$^h$ G.M.T.*

Note: Water and bare soil are treated by the same process as is the vegetation, but terms marked with a heavy **O** vanish.

| | | Character of surface......... | forest | other crops | bare soil | water |
|---|---|---|---|---|---|---|
| Height $(h_A - h_G)$ at which radiation converted | | cms | *1000* | *30* | — | — |
| Part of depth of top soil stratum $= z_{11}$ | | cms | *1·7* | *1·7* | *1·7* | — |
| Thermal conductivity of soil $= k$ | | erg cm$^{-2}$ sec$^{-1}$ degree$^{-1}$ $10^5 \times$ | *1·6* | *1·6* | *1·6* | very large |
| Trial value of $\theta_A$ the temperature of the surface $= \theta'$ | | °Abs | | *290* | | |
| Trial Resistances above $h_A$ | $1/C'_{L9} = g\,(p_6 - p_8)/(2\mathfrak{R}\mathfrak{R}'\gamma_p)$ | cm$^2$ grm$^{-1}$ sec $10^{-4} \times$ | *0·780* | *0·780* | *0·780* | |
| | $1/C'_{AL} =$ | ,, ,, ,, | *0·005* † | *0·010* † | **0·00** | **0** |
| | Sum of previous pair $= 1/C'_{A,9}$ | ,, ,, ,, | *0·785* | *0·785* | *0·780* | |
| Trial Resistances below $h_A$ | $1/C'_{11,10} = z_{11}/k$ | ,, ,, ,, | *0·106* | *0·106* | *0·106* | |
| | $1/C'_{10,A} =$ | ,, ,, ,, | *1·5* * | *0·05* † | **0** | |
| | Sum of previous pair $= 1/C'_{11,A}$ | ,, ,, ,, | *1·61* | *0·156* | *0·106* | |
| Here fill up Forms P IX, P X to find corrected $\theta_A$, which is | | | | *294·3* | | |
| Corrected value $C_{A9}$ of $C'_{A9}$ | For $\partial/\partial\theta'$ see Form P X | $10^4 \times$ | | *1·70* | | |
| Corrected value $C_{11A}$ of $C'_{11A}$ | | $10^4 \times$ | | *7·27* | | |
| Temperature at $z_{11}$ in soil $= \theta_{11}$ | | °Abs | *288* | *290* | | |
| $\tau_9 = \theta_9$ plus 9·9° C per km of $\frac{1}{2}(h_8 - h_G)$ if air clear | | °Abs | *295·4* | *295·4* | | |
| Trial jump in Flux of sensible heat $= \Gamma - \mho \supset \theta'^4 - \varpi'\Xi' + \varpi'$. See form P IX | | $10^4 \times$ | | *34·6* | | |
| Correction to preceding $= \lambda \cdot \frac{\partial}{\partial\theta'}(-\mho \supset \theta'^4 - \varpi'\Xi' + \varpi')$ | | $10^4 \times$ | | *4·7* | | |
| Corrected jump, $f_{10} - f_G$ | | erg cm$^{-2}$ sec$^{-1}$ $10^4 \times$ | | *29·9* | | |
| Flux of sensible heat upward just above $h_A$, $= f_G = C_{A9}(\theta_A - \tau_9)$ | | $10^4 \times$ | | *−1·7* | | |
| ,, ,, ,, below ,, , $= f_{10} = C_{11,A}(\theta_{11} - \theta_A)$ | | $10^4 \times$ | | *−31·3* | | |
| Difference of preceding pair. This should agree with $f_{10} - f_G$ | | $10^4 \times$ | | *29·3* | | |
| Mean $f_G$ for whole chequer | | erg cm$^{-2}$ sec$^{-1}$ $10^4 \times$ | *say −1·7* | | | |

* Layer of dead leaves on ground. Guess. † Guesses inserted for illustration.

Long. *11° E* Lat. *5400 km N* Time from *1910 May 20^d 4^h* to *10^h G.M.T.*

| Character of surface...... | | forest | other vegetation | | bare soil |
|---|---|---|---|---|---|
| Trial value of $\theta_A$ used on Forms P VII, P VIII is $\theta' =$ | | | *290* | | |
| $C'_{A9}$ = "conductance" of upper portion corresponding to $\theta'$ | $10^4 \times$ | | *1·27* | | |
| $C'_{11,A} =$ ,, ,, lower ,, ,, ,, ,, | $10^4 \times$ | | *6·41* | | |
| sum $= C'_{A9} + C'_{11,A}$ | $10^4 \times$ | | *7·68* | | |
| $C'_{A9}(\tau_9 - \theta')$ since $\tau_9 - \theta' = 5°{\cdot}4\ A$ | $10^4 \times$ | | *6·86* | | |
| $C'_{11,A}(\theta_{11} - \theta')$ since $\theta_{11} - \theta' =$ for forest, *0* for other crops, for bare soil | $10^4 \times$ | | *0·00* | | |
| sum $= C'_{11,A}(\theta_{11} - \theta') + C'_{A9}(\tau_9 - \theta')$ | $10^4 \times$ | | *6·9* | | |
| $\Gamma$ = sum of radiation absorbed from solar "bands," solar "remainder" and from the atmosphere. Forms P IV, P V, P VI erg $cm^{-2}$ $sec^{-1}$ | $10^4 \times$ | | *76·3* | | |
| − ℧ Ↄ $\theta'^4$ = minus the long wave radiation emitted, where Ↄ $= 5{\cdot}36 \times 10^{-5}$ erg $cm^{-2}$ $sec^{-1}$ $degree^{-4}$ and ℧ = emissivity = for forest, *1·0* for other crops, and for bare soil | $10^4 \times$ | | *−37·9* | | |
| − ɤ′ Ξ′ = minus the heat rendered latent if $\theta_A = \theta'$, where ɤ′ $= (3182 - 2{\cdot}51\theta') \times 10^7$ erg $grm^{-1}$ provided no ice is being fused, and where Ξ′ is taken from Form P VII | $10^4 \times$ | | *−3·8* | | |
| Heat brought to surface by precipitation $\varpi$ erg $cm^{-2}$ $sec^{-1}$ | $10^4 \times$ | | *0* | | |
| Numerator of equation Ch. 8/2/15 # 20 is the sum of the preceding numbers up to the thick line. (For denominator see Form P X.) | $10^4 \times$ | | *+41·5* | | |

COMPUTING FORM P X. For Temperature of Radiating Surface. Part II, Denominator of Ch. 8/2/15 # 20

Long. *11° E* Lat. *5400 km N* Time from *1910 May 20ᵈ 4ʰ* to *10ʰ G.M.T.*

Note the several terms here are $-\partial/\partial\theta'$ of the corresponding terms on Form P IX

| Character of surface...... | | forest | other vegetation | | bare soil |
|---|---|---|---|---|---|
| $\partial C''_{A9}/\partial\theta'$ (*the numbers are guesses*) | $10^4\times$ | | $+0{\cdot}1$ | | |
| $\partial C''_{11,A}/\partial\theta'$ " " " | $10^4\times$ | | $+0{\cdot}2$ | | |
| $-(\tau_9-\theta')\,.\,\partial C''_{A9}/\partial\theta'$ | $10^4\times$ | | $-0{\cdot}54$ | | |
| $-(\theta_{11}-\theta')\,.\,\partial C''_{11,A}/\partial\theta'$ | $10^4\times$ | | $0{\cdot}000$ | | |
| $C''_{11,A}+C''_{A9}$ | $10^4\times$ | | $7{\cdot}68$ | | |
| $4\,ϒ\,Ↄ\,\theta'^3$ (for ϒ and Ↄ see previous form P IX) | $10^4\times$ | | $0{\cdot}523$ | | |
| $\dfrac{\partial(ϒ'\Xi')}{\partial\theta'}$. This must be worked differently for vegetation and for bare soil | | — | — | — | — |
| For Vegetation see Form P VII: $\dfrac{\partial C'_\mu}{\partial\theta'}=5\times10^{-6}$ (*guess*). $\therefore\ \dfrac{\partial C'_\mu}{\partial\theta'}\,.\,ϖ'\,.\,(\mu'-\mu_9)=$ | $10^4\times$ | | $0{\cdot}091$ | | — |
| $\dfrac{\partial ϖ'}{\partial\theta'}=-2{\cdot}51\times10^7$. $\therefore\ \dfrac{\partial ϖ'}{\partial\theta'}\,.\,C'_\mu\,.\,(\mu'-\mu_9)=$ | $10^4\times$ | | $-0{\cdot}004$ | | — |
| $\dfrac{\partial\mu'}{\partial\theta'}=0{\cdot}85\times10^{-3}$. $\therefore\ \dfrac{\partial\mu'}{\partial\theta'}\,.\,C'_\mu\,.\,ϖ'=$ | $10^4\times$ | | $0{\cdot}463$ | | — |
| For bare soil see Ch. 8/2/12 | $10^4\times$ | | | | |
| $-\partial ϖ/\partial\theta'$ | $10^4\times$ | | | | |
| Denominator = algebraic sum of all numbers up to thick line | $=10^4\times$ | | $8{\cdot}213$ | | |
| ƛ = "numerator" from Form P IX divided by denominator | °Abs | | $5{\cdot}0$ | | |
| Corrected surface temperature $=\theta'+$ ƛ | = °Abs | | $295{\cdot}0$ | | |
| Second approximation | °Abs | | $294{\cdot}3$ | | |

COMPUTING FORM P XI. Diffusion produced by eddies. See Ch. 4/8, Ch. 8/2/13

Longitude *11° East* Latitude *5400 km North* Time from *1910 May $20^d$ $4^h$* to *$10^h$ G.M.T.*

$$\frac{\bar{D}\mu}{\bar{D}t} = \frac{\partial}{\partial p}\left(\xi \frac{\partial \mu}{\partial p}\right) \qquad \frac{\bar{D}\sigma}{\bar{D}t} = \frac{\partial}{\partial p}\left(\xi \frac{\partial \sigma}{\partial p}\right) + \text{terms expressing irreversible increase}$$

| REF.:— | | Form P I | Form P I | Form P I | | | | | | | | | |
|---|---|---|---|---|---|---|---|---|---|---|---|---|---|
| Height | | $\delta_H p$ | $\delta_H p$ interpolated | $\mu$ | $\frac{\delta_H \mu}{\delta_H p}$ | $\xi$ for $\mu$ | $\delta t \,.\, \xi \frac{\delta_H \mu}{\delta_H p}$ | $\delta_H$ (previous) $\div \delta_H p = \frac{D\mu}{Dt} . \delta t$ | $\delta_H \sigma$ | $\xi$ for $\sigma$ | $\delta t \,.\, \xi \frac{\delta_H \sigma}{\delta_H p}$ | $\delta_H$ (previous) $\div \delta_H p$ = reversible $\frac{D\sigma}{Dt} . \delta t$ | Irreversible part of $\frac{D\sigma}{Dt} . \delta t$ See Ch. 8/2/13 |
| | $10^5 \times$ | $10^2 \times$ | $10^2 \times$ | $10^{-3} \times$ | $10^{-3} \times$ | | | $10^{-3} \times$ | $10^6 \times$ | | $10^6 \times$ | $10^4 \times$ | $10^4 \times$ |
| | | −2050 | | | | | | | | | | | *Omitted pending investigation* |
| $h_2$ | 11·8 | | −2045 | | | 0 | 0 | | 2·44 | 0 | 0 | | |
| | | −2040 | | 0·0 | | | | | | | | −·0004 | |
| $h_4$ | 7·2 | | −2015 | | 0·25 | 50 | 0·003 | | 0·16† | 50 | −0·86 | | |
| | | −1989 | | 0·5 | | | | 0·001 | | | | ·003 | |
| $h_6$ | 4·2 | | −1935 | | 0·83 | 1000 | 0·179 | | −0·05‡ | 1000 | 5·6 | | |
| | | −1881 | | 2·1 | | | | 0·207 | | | | −·326 | |
| $h_8$ | 2·0 | | −1774 | | 1·81 | $10^5$ | 39·1 | | 0·05‡ | $10^5$ | −608· | | |
| | | −1666 | | 5·3 | | | | 0·059 | | | | 2·47 | |
| $h_G$ | 0·4 | | | | | | 49·0* | | | | 3500§ | | |

* This quantity is $g$ times the evaporation in gram $cm^{-2}$ during $\delta t$. It is obtained from the last row of Form P VII.

† $d\sigma = \gamma_p d_H \log_e \theta - b d_H \log_e p + a_\mu d\mu$ (see Ch. 8/2/13). ‡ By the *Hertz* diagram.

§ From Forms P VII, P VIII. This figure is $g$ times the entropy leaving 1 $cm^2$ during $\delta t$. The water is reckoned as carrying both latent heats with it. But to fit with *Hertz* the specific heat of ice is *neglected.*

COMPUTING FORM P XII. Summary of gains of entropy and of water, both per mass of atmosphere during $\delta t$

Longitude *11° East* Latitude *5400 km North* Time from *1910 May 20ᵈ 4ʰ* to *10ʰ G.M.T.* Interval *6 hours*

| | Gains of Energy per mass by Radiation: Solar in Bands | Solar in Remainder | Atmospheric and Terrestrial | Gain of Energy by Precipitation | Total gain of Energy per mass by Radiation and Precipitation | Temperature | Gain of $\sigma$ by Radiation and Precipitation | Gains of $\sigma$ by Stirring: Reversible Part | Irreversible Part | Total gain of $\sigma = \frac{D\sigma}{Dt} \cdot \delta t$ | Gains of Water per mass: by Stirring | by Precipitation | total $= \frac{D\mu}{Dt} \cdot \delta t$ | Partial gain of $\sigma$ due to gain of $\mu$ $= \alpha_\mu \frac{D\mu}{Dt} \delta t$ . | $\left\{\frac{D\sigma}{Dt} - \alpha_\mu \frac{D\mu}{Dt}\right\} \delta t$ |
|---|---|---|---|---|---|---|---|---|---|---|---|---|---|---|---|
| REF. | P IV | P V | P VI | | | | | P XI | | | P XI | | | $\alpha_\mu$, P II | |
| | Unit for Energy: $10^4$ erg gram$^{-1}$ | | | | | °A | $10^4$ erg gram$^{-1}$ degree$^{-1}$ | | | | $10^{-3}\times$ | $10^{-3}\times$ | $10^{-3}\times$ | $10^4\times$ | $10^4\times$ |
| $h_0$ | | | | | | | | | | | | | | | |
| | *76* | *67* | *114* | 0 | *257* | *212* | *1·21* | *0·00* | *Omitted pending further theoretical investigation* | *1·21* | *0·00* | *0* | *0·00* | *0·0* | *1·21* |
| $h_2$ | | | | | | | | | | | | | | | |
| | *67* | *57* | *292* | 0 | *416* | *227·5* | *1·83* | *0·00* | | *1·83* | *0·00* | *0* | *0·00* | *0·0* | *1·83* |
| $h_4$ | | | | | | | | | | | | | | | |
| | *127* | *64* | *−118* | 0 | *73* | *257·7* | *0·28* | *0·00* | | *0·28* | *0·00* | *0* | *0·00* | *0·0* | *0·28* |
| $h_6$ | | | | | | | | | | | | | | | |
| | *255* | *89* | *−531* | 0 | *−187* | *279·0* | *−0·67* | *−0·33* | | *−1·00* | *0·21* | *0* | *0·21* | *2·1* | *−3·10* |
| $h_8$ | | | | | | | | | | | | | | | |
| | *359* | *124* | *−565* | 0 | *−82* | *287·5* | *−0·29* | *2·47* | | *2·18* | *0·06* | *0* | *0·06* | *0·6* | *1·56* |
| $h_G$ | | | | | | | | | | | | | | | |

## COMPUTING FORM P XIII. Divergence of horizontal momentum-per-area. Increase of pressure

The equation is typified by: $-\frac{\partial R_{86}}{\partial t} = \frac{\partial M_{E86}}{\partial e} + \frac{\partial M_{N86}}{\partial n} - M_{N86}\frac{\tan\phi}{a} + m_{H6} - m_{H8}^{*} + \frac{2}{a}M_{H86}$. (See Ch. 4/2 #5.)

* In the equation for the lowest stratum the corresponding term $-m_{G8}$ does not appear

Longitude *11° East* $\delta e = 441 \times 10^5$ Latitude *5400 km North* $\delta n = 400 \times 10^5$ Instant *1910 May 20ᵈ 7ʰ G.M.T.* $a^{-1}.\tan\phi = 1{\cdot}78 \times 10^{-9}$ Interval, $\delta t$ *6 hours* $a = 6{\cdot}36 \times 10^8$

| REF.:— | | | | previous 3 columns | previous column | | Form P XVI | Form P XVI | equation above | previous column | previous column | previous column |
|---|---|---|---|---|---|---|---|---|---|---|---|---|
| $h$ | $\frac{\delta M_E}{\delta e}$ | $\frac{\delta M_N}{\delta n}$ | $-\frac{M_N \tan\phi}{a}$ | $\mathrm{div}'_{EN}M$ | $-g\delta t\,\mathrm{div}'_{EN}M$ | Leave the subsequent columns to be filled up after the vertical velocity has been computed on Form P XVI | $m_H$ | $\frac{2M_H}{a}$ | $-\frac{\partial R}{\partial t}$ | $+\frac{\partial R}{\partial t}\delta t$ | $g\frac{\partial R}{\partial t}\delta t$ | $\frac{\partial p}{\partial t}\delta t$ |
| | $10^{-5}\times$ | $10^{-5}\times$ | $10^{-5}\times$ | $10^{-5}\times$ | $100\times$ | | $10^{-5}\times$ | $10^{-5}\times$ | $10^{-5}\times$ | | $100\times$ | $100\times$ |
| $h_0$ | | | | | | | 0 | | | | | 0 |
| | −61 | −245 | −6 | −312 | 656 | | | | −229 | 49·5 | 483 | |
| $h_2$ | | | | | | | −83 | | | | | 483 |
| | 367 | −257 | 2 | 112 | −236 | | | 0·06 | −136 | 29·4 | 287 | |
| $h_4$ | | | | | | | 165 | | | | | 770 |
| | 93 | −303 | −16 | −226 | 478 | | | 0·11 | −124 | 26·8 | 262 | |
| $h_6$ | | | | | | | 63 | | | | | 1032 |
| | 32 | −55 | −12 | −35 | 74 | | | 0·07 | −110 | 23·8 | 233 | |
| $h_8$ | | | | | | | 138 | | | | | 1265 |
| | −256 | 38 | −8 | −226 | 479 | | | 0·03 | −88 | 19·0 | 186 | |
| $h_G$ | | | | | SUM = 1451 $= \frac{\partial p_G}{\partial t}\delta t$ | | | | | | | 1451 |
| | | | | | | | | | | | | check by $\Sigma - g\delta t\,\mathrm{div}'_{EN}M$ |

NOTE: $\mathrm{div}'_{EN}M$ is a contraction for $\frac{\delta M_E}{\delta e} + \frac{\delta M_N}{\delta n} - M_N\frac{\tan\phi}{a}$

# COMPUTING FORM P XIV. Stratosphere. Vertical Velocity by Ch. 6/6 # 21 and Temperature Change by Ch. 6/7/3 # 8

Longitude *11° East* Latitude *5400 km North* Instant *1910 May 20ᵈ 7ʰ G.M.T.* Interval $\delta t = 6$ *hours*

$\delta e = 4{\cdot}41 \times 10^{7}$ for $\delta\lambda = 6°$; $2\omega \sin\phi = 1{\cdot}092 \times 10^{-4}$; $\dfrac{\tan\phi}{a} = 1{\cdot}78 \times 10^{-9}$; $\dfrac{\cot\phi}{a} = 1{\cdot}38 \times 10^{-9}$

$g = 975$; $b = 2{\cdot}870 \times 10^{6}$; $\gamma_v = 7{\cdot}05 \times 10^{6}$; $\gamma_p = 9{\cdot}92 \times 10^{6}$; $p_2 = 2{\cdot}050 \times 10^{5}$; $\theta = 212°$

| | | | by cross multiplication | by direct multiplication | | |
|---|---|---|---|---|---|---|
| $\dfrac{M_E}{R} = +65$ | $\dfrac{\delta\theta}{\delta e} = zero$ | $\dfrac{\delta p_2}{\delta e} = +158 \times 10^{-7}$ | $\dfrac{\delta\theta}{\delta e} \cdot \dfrac{\delta p_2}{\delta n} = 0$ | $\dfrac{M_E}{R}\dfrac{\delta\theta}{\delta e} = 0$ | $\dfrac{M_E}{R}\dfrac{\delta p_2}{\delta e} = 1{\cdot}032 \times 10^{-3}$ | $\dfrac{b\theta}{g} = 6{\cdot}24 \times 10^{5}$ |
| $\dfrac{M_N}{R} = +148$ | $\dfrac{\delta\theta}{\delta n} = +1{\cdot}25 \times 10^{-8}$ | $\dfrac{\delta p_2}{\delta n} = +162 \times 10^{-7}$ | $-\dfrac{\delta\theta}{\delta n} \cdot \dfrac{\delta p_2}{\delta e} = -198 \times 10^{-15}$ | $\dfrac{M_N}{R}\dfrac{\delta\theta}{\delta n} = +185 \times 10^{-8}$ | $\dfrac{M_N}{R}\dfrac{\delta p_2}{\delta n} = 2{\cdot}40 \times 10^{-3}$ | $R\theta\dfrac{D\sigma}{Dt} = +115$ |
| | | | sum $= -198 \times 10^{-15}$ | sum $= +185 \times 10^{-8}$ | $\dfrac{\partial p_2}{\partial t} = 2250 \times 10^{-3}$ | |
| | | | | | sum $= 2253 \times 10^{-3}$ | |

$$\frac{1}{1{\cdot}405\,\theta p_2}\left\{\frac{\delta\theta}{\delta e}\cdot\frac{\delta p_2}{\delta n} - \frac{\delta\theta}{\delta n}\cdot\frac{\delta p_2}{\delta e}\right\} = -0{\cdot}323 \times 10^{-20}$$

$$-\frac{1}{\theta}\frac{\delta\theta}{\delta e}\left(\frac{\cot\phi}{a}\right) = zero$$

$$\text{sum} = -0{\cdot}32 \times 10^{-20}$$

$$-\frac{b\theta}{2\omega\sin\phi} \times \text{sum} = +0{\cdot}18 \times 10^{-7}$$

From Form P xv: $0{\cdot}289\,\mathrm{div}'_{EN}\left(\dfrac{M}{R}\right) = -43{\cdot}0 \times 10^{-7}$

$$\text{sum of last two} = -42{\cdot}8 \times 10^{-7} = \frac{\Upsilon}{g}$$

$$-\frac{0{\cdot}289g}{2\omega\sin\phi}\left(\frac{\cot\phi}{a}\right)\frac{1}{\theta}\cdot\frac{\delta\theta}{\delta e} = zero = \frac{\Phi}{g}$$

Up-grade of vertical velocity =

$$\frac{\partial v_H}{\partial h} = -\frac{\Upsilon}{g} + \frac{D\sigma}{\gamma_p Dt} - \frac{\Phi}{g}(h - h_2) = +43{\cdot}3 \times 10^{-7} + (h - h_2) \times 0$$

by Ch. 6/6 # 21

$$M_{H20} = R\left[v_{H2} + \frac{b\theta}{g}\left\{-\frac{\Upsilon}{g} + \frac{1}{\gamma_p}\frac{D\sigma}{Dt}\right\} - \frac{\Phi}{g}\left(\frac{b\theta}{g}\right)^2\right] = +62{\cdot}2$$

$$-\frac{M_E}{R}\frac{\delta\theta}{\delta e} - \frac{M_N}{R}\frac{\delta\theta}{\delta n} = +0{\cdot}0019 \times 10^{-3}$$

$$\frac{b\theta}{\gamma_v \cdot p_2}\left\{\frac{\partial p_2}{\partial t} + \frac{M_E}{R}\frac{\partial p_2}{\partial e} + \frac{M_N}{R}\frac{\partial p_2}{\partial n}\right\} = +0{\cdot}95 \times 10^{-3}$$

$$-\frac{g}{\gamma_v}\frac{M_H}{R} = -0{\cdot}041 \times 10^{-3}$$

$$+\frac{\theta}{\gamma_v}\cdot\frac{D\sigma}{Dt} = +0{\cdot}00008 \times 10^{-3}$$

Temperature change by Ch. 6/7/3 # 8 $\Big\}$ $\dfrac{\partial\theta}{\partial t} = \text{sum} = +0{\cdot}91 \times 10^{-3}$

Alternative, involving more assumptions,

$$\frac{\partial\theta}{\partial t} = \theta\frac{\partial v_H}{\partial h} = +0{\cdot}92 \times 10^{-3} + (h - h_2) \times 0$$

See Ch. 6/7/2 # 14

Computing Form P xv. For Vertical Velocity in general, by equation Ch. 8/2/23 # 1. Preliminary

Longitude *11° East* Latitude *5400 km North* Instant *1910 May 20ᵈ 7ʰ G.M.T.*

Ref.:—

| Height | $\delta p$ | $\frac{\delta v_E}{\delta e}$ | $v_E$ 220·5 km to West | $v_E$ 220·5 km to East | $v_E$ mean* at centre | $\left(\frac{\delta v_E}{\delta p}\right)_{\lambda,\phi}$ | $\left(\frac{\delta p}{\delta e}\right)_{h,\phi}$ | $\frac{\delta v_N}{\delta n}$ | $v_N$ 200 km to South | $v_N$ 200 km to North | $v_N$ mean* at centre | $-\frac{\tan\phi . v_N}{a}$ | $\left(\frac{\delta v_N}{\delta p}\right)_{\lambda,\phi}$ | $\left(\frac{\delta p}{\delta n}\right)_{h,\lambda}$ |
|---|---|---|---|---|---|---|---|---|---|---|---|---|---|---|
| | $10^2\times$ | $10^{-7}\times$ | | | | $10^{-5}\times$ | $10^{-6}\times$ | $10^{-7}\times$ | | | | $10^{-7}\times$ | $10^{-5}\times$ | $10^{-6}\times$ |
| $h_0$ | | | | | | | | | | | | | | |
| | 2050 | − 29 | + 129 | 0 | + 65 | | | −117 | +382 | − 86 | +148 | −3 | | |
| $h_2$ | | | | | | −613 | +16 | | | | | | − 97 | +16 |
| | 2040 | +180 | −1583 | −792 | −1188 | | | −124 | +199 | −298 | − 50 | +1 | | |
| $h_4$ | | | | | | +307 | +37 | | | | | | +244 | +70 |
| | 1989 | + 46 | − 669 | −468 | − 569 | | | −149 | +738 | +143 | +441 | −8 | | |
| $h_6$ | | | | | | +224 | +27 | | | | | | − 43 | +69 |
| | 1881 | + 17 | − 172 | − 98 | − 135 | | | − 28 | +412 | +301 | +357 | −6 | | |
| $h_8$ | | | | | | + 52 | +57 | | | | | | − 57 | +33 |
| | 1666 | −143 | + 271 | −357 | − 43 | | | + 1 | +209 | +303 | +256 | −5 | | |
| $h_G$ | | | | | | | | | | | | | | |

* It might be better to use the mean of four contiguous values instead of the mean of two only.

## COMPUTING FORM P XVI. For Vertical Velocity. Conclusion

The equation is Ch. 8/2/23#1, which reads $\frac{\partial}{\partial p}\left\{\frac{\alpha_\rho}{\alpha_p} g\rho^2 \frac{\partial v_H}{\partial p}\right\} = \text{div}_{EN} v + \frac{\partial}{\partial p}\left\{\frac{\alpha_\rho}{\alpha_p}\rho \,.\, \text{div}_{EN} v\right\} - \left\{\left(\frac{\partial v_E}{\partial p}\right)_{\lambda,\phi} \cdot \left(\frac{\partial p}{\partial e}\right)_{h,\phi} + \left(\frac{\partial v_N}{\partial p}\right)_{\lambda,\phi} \cdot \left(\frac{\partial p}{\partial n}\right)_{h,\lambda}\right\} + \frac{\partial}{\partial p}\left\{\frac{1}{\alpha_p}\left(\frac{D\sigma}{Dt} - \alpha_\mu \frac{D\mu}{Dt}\right)\right\}$

Longitude *11° East* Latitude *5400 km North* Instant *1910 May 20ᵈ 7ʰ G.M.T.*

| REF.:— | | P XV | | previous column | | | P II P XII | previous column | equation above | previous column | previous column | previous column | Form P I | | | |
|---|---|---|---|---|---|---|---|---|---|---|---|---|---|---|---|---|
| $h$ | $\delta_H p$ For interpolated $\delta_H p$ see Form P XI | $\text{div}'_{EN} v$ | $\frac{\alpha_\rho}{\alpha_p}\rho \,.\, \text{div}'_{EN} v$ | $\frac{\delta}{\delta p}\left(\frac{\alpha_\rho}{\alpha_p}\rho \,.\, \text{div}'_{EN} v\right)$ | $\text{div}'_{EN} v$ interpolated at boundaries | $\left(\frac{\delta v_E}{\delta p}\right)_{\lambda,\phi} \cdot \left(\frac{\delta p}{\delta e}\right)_{h,\phi} + \left(\frac{\delta v_N}{\delta p}\right)_{\lambda,\phi} \cdot \left(\frac{\delta p}{\delta n}\right)_{h,\lambda}$ | $\frac{1}{\alpha_p}\left(\frac{D\sigma}{Dt} - \alpha_\mu \frac{D\mu}{Dt}\right)$ | $\frac{\delta}{\delta p}\left\{\frac{1}{\alpha_p}\left(\frac{D\sigma}{Dt} - \alpha_\mu \frac{D\mu}{Dt}\right)\right\}$ | $\frac{\partial}{\partial p}\left\{\frac{\alpha_\rho}{\alpha_p} g\rho^2 \frac{\partial v_H}{\partial p}\right\}$ | $\frac{\alpha_\rho}{\alpha_p} g\rho^2 \frac{\partial v_H}{\partial p} \equiv -\frac{\alpha_\rho}{\alpha_p}\rho \frac{\partial v_H}{\partial h}$ | $\frac{\partial v_H}{\partial h}$ | $v_H$ | $\rho$ | $m_H$ | $M_H$ | $h$ |
| | $10^2 \times$ | $10^{-7} \times$ | $10^{-2} \times$ | $10^{-7} \times$ | $10^{-7} \times$ | $10^{-7} \times$ | $10^{-4} \times$ | $10^{-7} \times$ | $10^{-7} \times$ | $10^{-2} \times$ | $10^{-5} \times$ | | $10^{-6} \times$ | $10^{-6} \times$ | | |
| $h_0$ | | | | | | | | | | | | | | | | $h_0$ |
| | *2050* | *−149* | *215* | | | | *82* | | | *62** | *0·433†* | | | | | |
| $h_2$ | | | | *−221* | *−46* | *−1·14* | | *1·3* | *−265* | | | *−2·465* | *337* | *−831* | | $h_2$ |
| | *2040* | *57* | *−237* | | | | *356* | | | *−480* | *−1·152* | | | | *188* | |
| $h_4$ | | | | *506* | *−27* | *2·84* | | *−1·3* | *475* | | | *2·833* | *582* | *1648* | | $h_4$ |
| | *1989* | *−111* | *782* | | | | *92* | | | *477* | *0·678* | | | | *341* | |
| $h_6$ | | | | *−318* | *−64* | *0·31* | | *−7·8* | *−390* | | | *0·800* | *785* | *628* | | $h_6$ |
| | *1881* | *−17* | *167* | | | | *−1422* | | | *−277* | *−0·282* | | | | *221* | |
| $h_8$ | | | | *929* | *−82* | *0·11* | | *13·0* | *860* | | | *1·420* | *975* | *1384* | | $h_8$ |
| | *1666* | *−147* | *1813* | | | | *894* | | | *1248* | *1·012* | | | | *92* | |
| $h_G$ | | | | | | | | | | | | *−0·200‡* | *1143* | *−229* | | $h_G$ |

The equations may be typified by the following, which is modelled on Ch. 4/3 # 8

$$+\frac{\partial W_{86}}{\partial t} = R_{86}\frac{D\mu_{86}}{Dt} - \frac{\partial}{\partial e}\left(M_E\mu\right)_{86} - \frac{\partial}{\partial n}\left(M_N\mu\right)_{86} + \left(M_N\mu\right)_{86}\frac{\tan\phi}{a} + \mu_8 m_{H8}{}^{*} - \mu_6 m_{H6} - \frac{2}{a}\mu_{86} M_{H86}.$$

Longitude *11° East* Latitude *5400 km North* $a^{-1}\tan\phi = 1{\cdot}78 \times 10^{-9}$ Instant *1910 May 20^d 7^h G.M.T.* $\delta t = 6$ *hours*

| REF. | | $h_2$ + | $10^{-8}\times$ | − | $h_4$ + | $10^{-8}\times$ | − | $h_6$ + | $10^{-8}\times$ | − | $h_8$ + | $10^{-8}\times$ | − $h_L$ |
|---|---|---|---|---|---|---|---|---|---|---|---|---|---|
| Form P XI | $R\frac{D\mu}{Dt}$ | | *zero* | | | *zero* | | *187* | | | | | *181* |
| | $-\frac{\delta}{\delta e}(M_E\mu)$ | | *zero* | | | | *163* | | | *123* | *2102* | | |
| | $-\frac{\delta}{\delta n}(M_N\mu)$ | | *zero* | | *228* | | | *193* | | | *348* | | |
| | $+(M_N\mu)\frac{\tan\phi}{a}$ | | *zero* | | *12* | | | *30* | | | *64* | | |
| Form P XVI † | $+\mu m_H$ at lower limit | *33* | | | *69* | | | *470* | | | | no term here* | |
| Form P XVI † | $-\mu m_H$ at upper limit | | *zero* | | | | *33* | | | *69* | | | *470* |
| Form P XVI | $-\frac{2}{a}\mu M_H$ | | *zero* | | | | *0·05* | | | *0·15* | | | *0·15* |
| | sums + and − | *33* | | *0* | *309* | | *196* | *880* | | *192* | *2514* | | *651* |
| | $\partial W/\partial t$ | | $+33\times10^{-8}$ | | | $+113\times10^{-8}$ | | | $+688\times10^{-8}$ | | | $+1863\times10^{-8}$ | |
| | $\partial W/\partial t\,.\,\delta t$ | | $+0{\cdot}007$ *grm cm*$^{-2}$ | | | $+0{\cdot}024$ *grm cm*$^{-2}$ | | | $+0{\cdot}149$ *grm cm*$^{-2}$ | | | $+0{\cdot}402$ *grm cm*$^{-2}$ | |

* Note that in the equation for the lowest stratum the corresponding term $\mu_G m_{HG}$ does not appear.

† At the boundaries $\mu$ was obtained by interpolation on a $p, \mu$ graph.

COMPUTING FORM P XVIII. For water in soil $\frac{\partial w}{\partial t} = \frac{\partial}{\partial z}\left[ \mathfrak{y}(w)\left\{\frac{\partial \Psi(w)}{\partial z} - g\right\} + \mathfrak{x}(w)\frac{\partial F(w,\theta)}{\partial z}\right]$, which is equation Ch. 4/10/2 # 5

Character of Vegetation, if any, covering soil

Longitude Latitude Instant Interval, $\delta t =$

REF. :—

| Depth $z$ cms For explanation see Ch. 4/10/0 | | Mass of water per volume $w$ | Deficit of pressure from atmospheric $\Psi(w)$ | Effective body force $\frac{\delta\Psi}{\delta z} - g$ | Conductivity for percolation $\mathfrak{y}(w)$ | Flux by percolation $\mathfrak{y}\left\{\frac{\delta\Psi}{\delta z} - g\right\}$ | $\frac{\delta\,(\text{flux})}{\delta z} = \frac{\partial w}{\partial t}$ by percolation | Vapour density $F(w,\theta)$ | Down-grade of density $\frac{\delta F}{\delta z}$ | Porosity to vapour $\mathfrak{x}(w)$ | Flux of vapour $\mathfrak{x} \times \frac{\delta F}{\delta z}$ | $\frac{\delta\,(\text{flux})}{\delta z} = \frac{\partial w}{\partial t}$ by evaporation | Form P VII $\frac{\partial w}{\partial t}$ by roots | Total $\frac{\partial w}{\partial t}$ | Total $\frac{\partial w}{\partial t}\delta t$ | $z$ |
|---|---|---|---|---|---|---|---|---|---|---|---|---|---|---|---|---|
| | | * | | | | | | | | | | | | | | |
| 0 | | | | | | rate of precipitation | | | | | † | | | | | |
| | 1·7 | | | | | | | | | | | | | | | 1·7 |
| 6·4 | | | | | | | | | | | | | | | | |
| | 19·1 | | | | | | | | | | | | | | | 19·1 |
| 53·6 | | | | | | | | | | | | | | | | |
| | 147·5 | | | | | | | | | | | | | | | 147·5 |
| | | | | | | | | | | | | | | | | |

* Mass of snow per area.

† Evaporation from surface from Form P VII.

COMPUTING FORM P XIX. For Temperature in Soil. The equation is Ch. 4/10/2 # 10, namely

$$\frac{\partial\theta}{\partial t}=\frac{1}{u}\frac{\partial}{\partial z}\left(k\frac{\partial\theta}{\partial z}\right)+\frac{\text{latent heat of evaporation}}{u}\frac{\partial}{\partial z}\left[\mathbf{x}(w)\frac{\partial F(w,\theta)}{\partial z}\right]+\frac{4{\cdot}2\times10^{7}}{u}\frac{\partial}{\partial z}\left[\theta\,.\,\mathfrak{y}(w)\left\{\frac{\partial\Psi(w)}{\partial z}-g\right\}\right]$$

Character of Vegetation, if any, covering soil — Crops other than Forest

Longitude — Latitude — Instant — Interval, $\delta t=$

| REF.:— | | | | | | | | | Form P XVIII | Form P XVIII | | | | | |
|---|---|---|---|---|---|---|---|---|---|---|---|---|---|---|---|
| Depth $z$ cms (For explanation see Ch. 4/10/0) | | Temperature $\theta$ | Down-grade of temperature $\frac{\delta\theta}{\delta z}$ | Thermal conductivity $k$ | Flux of sensible heat $k\frac{\delta\theta}{\delta z}$ | Accumulation of sensible heat $\frac{\delta}{\delta z}\left(k\frac{\delta\theta}{\delta z}\right)$ | Thermal capacity per volume $u$ | Part of $\frac{\partial\theta}{\partial t}$ due to conduction | Part of $\frac{\partial\theta}{\partial t}$ due to evaporation | $\theta\times4{\cdot}2\times10^{7}\times$ (flux of water by percolation) | $\frac{\delta}{\delta z}$ of foregoing | Part of $\frac{\partial\theta}{\partial t}$ due to percolation | Total $\frac{\partial\theta}{\partial t}$ | Total $\frac{\partial\theta}{\partial t}\delta t$ | $z$ |
| | | | | | $10^{4}\times$ | | | | | | | | | | |
| 0 | | | | $1{\cdot}6\times10^{5}$ | $31{\cdot}3$* | | | | | † | | | | | |
| | 1·7 | 290° | | | | | | | | | | | | | 1·7 |
| 6·4 | | | | | | | | | | | | | | | |
| | 19·1 | | | | | | | | | | | | | | 19·1 |
| 53·6 | | | | | | | | | | | | | | | |
| | 147·5 | | | | | | | | | | | | | | 147·5 |
| | | | | | | | | | | | | | | | |

* This number is $f_{10}$ from Form P VIII.

† Use $\theta$ of precipitation.

## COMPUTING FORM M I. For Stresses due to Eddy Viscosity

For the upper layers the equation is typified by $-\frac{\partial M_{E86}}{\partial t} = \frac{\xi_6}{g}\left(\frac{\partial v_E}{\partial p}\right)_6 - \frac{\xi_8}{g}\left(\frac{\partial v_E}{\partial p}\right)_8$.

See Ch. 4/8/2. The stress at the ground is found by a special process; see below.

Longitude = *11° East* Latitude = *5600 km North* Instant *1910 May 20ᵈ 7ʰ G.M.T.*

| Height | Vertical pressure differences | Vertical pressure differences: interpolated | $\xi$ | $v_E = \frac{M_E}{R}$ | $\frac{\delta v_E}{\delta p}$ | Shearing stress $\frac{\xi}{g}\frac{\delta v_E}{\delta p}$ | $-\frac{\partial M_E}{\partial t}$ due to viscosity alone | $v_N = \frac{M_N}{R}$ | $\frac{\delta v_N}{\delta p}$ | Shearing stress $\frac{\xi}{g}\frac{\delta v_N}{\delta p}$ | $-\frac{\partial M_N}{\partial t}$ due to viscosity alone |
|---|---|---|---|---|---|---|---|---|---|---|---|
| | $10^5 \times$ | $10^5 \times$ | | cm sec$^{-1}$ | $10^{-5} \times$ | $10^{-3} \times$ | | cm sec$^{-1}$ | $10^{-5} \times$ | $10^{-3} \times$ | |
| $h_0$ | | | | | | | | | | | |
| | *2·048* | | | *−266* | | | | *−86* | | | |
| $h_2$ | | *2·04* | *0* | | *−212* | *0* | | | *−103* | *0* | |
| | *2·040* | | | *−698* | | | *0·000* | *−297* | | | *0·000* |
| $h_4$ | | *2·02* | *50* | | *+114* | *+0·06* | | | *+218* | *+0·11* | |
| | *1·992* | | | *−467* | | | *−0·001* | *+142* | | | *−0·001* |
| $h_6$ | | *1·94* | *1000* | | *+103* | *+1·05* | | | *+82* | *+0·85* | |
| | *1·888* | | | *−270* | | | *+0·215* | *+301* | | | *−0·014* |
| $h_8$ | | *1·78* | *100000* | | *−212* | *−214·* | | | *+13* | *+13·* | |
| | *1·67* | | | *−647* | | | *−1·40* | *+324* | | | *+0·153* |
| $h_L$ | | | | | | *+1190·* | | | | *−140·* | |
| $h_G$ | | | | | | | | | | | |

To find, according to Ch. 4/8/3, the stress at ground: $M_{EG8} = -110 \times 10^3$; $M_{NG8} = +55 \times 10^3$; $\mathbf{M}_{G8} = \sqrt{M^2_E + M^2_N} = 123 \times 10^3$; $\overline{\mathfrak{m}} = \frac{\mathbf{M}_{G8}}{h_8 - h_G} = 0{\cdot}77$.

From roughness of the surface, and from thermal stability the resultant stress is *2* times $(\overline{m})^2$.

So resultant stress is *1·2* dynes cm$^{-2}$. Veer of $\overline{\mathfrak{m}}$ from surface-stress estimated to be *20°*.

Component of stress with $\overline{\mathfrak{m}} = -1{\cdot}13 = A$, say. Component of stress to left of $\overline{\mathfrak{m}} = -0{\cdot}41 = B$, say.

Component dragging air eastward $= \frac{1}{M}(AM_E - BM_N)$ | Component dragging air to north $= \frac{1}{M}(AM_N + BM_E)$

Longitude = *11° East* Latitude = *5600 km North* $2\omega \sin\phi = 1{\cdot}124 \times 10^{-4}$ Instant *1910 May 20ᵈ 7ʰ G.M.T.*; $\delta t = $ *6 hours*

$p_2 = 2048 \times 10^2$; $\theta = 213°$; $\rho_2 = 0{\cdot}335 \times 10^{-3}$; $R_{20} = 210$

$\dfrac{\delta\theta}{\delta e} = +0{\cdot}47 \times 10^{-7}$; $\dfrac{\delta\theta}{\delta n} = +0{\cdot}50 \times 10^{-7}$; $g = 975\cdot$ ; $\dfrac{b\theta}{g} = 6{\cdot}27 \times 10^5$

| Eastward velocity | Northward velocity |
|---|---|
| $\dfrac{M_{E20}}{R_{20}} = -267$ *cm sec*$^{-1}$ | $\dfrac{M_{N20}}{R_{20}} = -86$ |

The above are the velocities at the level $h_2 + b\theta/g$

| | |
|---|---|
| $\dfrac{b\theta}{g}\dfrac{\partial v_E}{\partial h} = -\dfrac{b}{2\omega\sin\phi}\dfrac{\partial\theta}{\partial n} = -1280$ | $\dfrac{b\theta}{g}\dfrac{\partial v_N}{\partial h} = +\dfrac{b}{2\omega\sin\phi}\dfrac{\partial\theta}{\partial e} = +1200$ |

So, extrapolating from above by these formulae:—

| | |
|---|---|
| $v_{E2} = +1013$ | $v_{N2} = -1286$ |
| whence $m_{E2} = \rho_2 v_{E2} = +0{\cdot}339$ | $m_{N2} = \rho_2 v_{N2} = -0{\cdot}431$ |

Extrapolating from below, assuming $m$ independent of $h$

| | |
|---|---|
| $\dfrac{M_{E42}}{h_2 - h_4} = m_{E2} = -0{\cdot}317$ | $\dfrac{M_{N42}}{h_2 - h_4} = m_{N2} = -0{\cdot}135$ |

Accept mean of the two extrapolations, as follows:—

| | |
|---|---|
| $m_{E2} = +0{\cdot}011$ | $m_{N2} = -0{\cdot}283$ |

| | |
|---|---|
| $\dfrac{\tan\phi}{a} = 1{\cdot}897 \times 10^{-9}$ | $R_{20}\left[\dfrac{b}{2\omega\sin\phi}\right]^2 = 137 \times 10^{21}$ |

$$\frac{3}{a} R_{20} \frac{b\theta}{g}\left\{\frac{\partial v_H}{\partial h} \text{ at } \left(p_2 + \text{twice } \frac{b\theta}{g}\right)\right\} = (+26{\cdot}9 \times 10^{-7}) = C \text{ say}$$

Here $\dfrac{\partial v_H}{\partial h}$ should be mean of four surrounding values at neighbouring points on the map.

**Plan showing horizontal distribution of various quantities**

| | | |
|---|---|---|
| $\delta e = 3{\cdot}83 \times 10^7$ for $\delta\lambda = 6°$<br>400 km north of 6° west } centre | $\left(\dfrac{\delta\theta}{\delta e}\right)^2 = +0{\cdot}273 \times 10^{-14}$<br>$\dfrac{\delta\theta}{\delta e}\cdot\dfrac{\delta\theta}{\delta n} = +0{\cdot}261 \times 10^{-14}$ | |
| $\delta e = 4{\cdot}25 \times 10^7$ for $\delta\lambda = 6°$<br>$\left(\dfrac{\delta\theta}{\delta n}\right)^2 = +0{\cdot}063 \times 10^{-14}$<br>$\dfrac{\delta\theta}{\delta e}\cdot\dfrac{\delta\theta}{\delta n} = -0{\cdot}294 \times 10^{-14}$ | $\left(\dfrac{\delta\theta}{\delta e}\right)^2 = +0{\cdot}222 \times 10^{-14}$<br>$\left(\dfrac{\delta\theta}{\delta n}\right)^2 = +0{\cdot}250 \times 10^{-14}$<br>$\dfrac{\delta\theta}{\delta e}\cdot\dfrac{\delta\theta}{\delta n} = +0{\cdot}235 \times 10^{-14}$ | $\left(\dfrac{\delta\theta}{\delta n}\right)^2 = +0{\cdot}250 \times 10^{-14}$<br>$\dfrac{\delta\theta}{\delta e}\cdot\dfrac{\delta\theta}{\delta n} = +0{\cdot}235 \times 10^{-14}$ |
| $\delta e = 4{\cdot}58 \times 10^7$ for $\delta\lambda = 6°$ | $\left(\dfrac{\delta\theta}{\delta e}\right)^2 =$ *zero*<br>$\dfrac{\delta\theta}{\delta e}\cdot\dfrac{\delta\theta}{\delta n} =$ *zero* | 400 km south of 6° east } centre |

| | |
|---|---|
| $+\dfrac{\delta}{\delta e}\left(\dfrac{\delta\theta}{\delta n}\right)^2 = +0{\cdot}0220 \times 10^{-21}$ | $-\dfrac{\delta}{\delta e}\left(\dfrac{\delta\theta}{\delta e}\dfrac{\delta\theta}{\delta n}\right) = -0{\cdot}0622 \times 10^{-21}$ |
| $-\dfrac{\delta}{\delta n}\left(\dfrac{\delta\theta}{\delta e}\dfrac{\delta\theta}{\delta n}\right) = -0{\cdot}0326 \times 10^{-21}$ | $+\dfrac{\delta}{\delta n}\left(\dfrac{\delta\theta}{\delta e}\right)^2 = +0{\cdot}0341 \times 10^{-21}$ |
| $+\dfrac{2\tan\phi}{a}\dfrac{\delta\theta}{\delta e}\dfrac{\delta\theta}{\delta n} = +0{\cdot}0089 \times 10^{-21}$ | $+\dfrac{\tan\phi}{a}\left\{\left(\dfrac{\delta\theta}{\delta n}\right)^2 - \left(\dfrac{\delta\theta}{\delta e}\right)^2\right\} = +0{\cdot}0005 \times 10^{-21}$ |
| Sum $= -0{\cdot}0017 \times 10^{-21}$ | Sum $= -0{\cdot}0276 \times 10^{-21}$ |
| $R_{20}\left[\dfrac{b}{2\omega\sin\phi}\right]^2 \times$ sum $= -0{\cdot}23$ | $R_{20}\left[\dfrac{b}{2\omega\sin\phi}\right]^2 \times$ sum $= -3{\cdot}78$ |
| $C\dfrac{b\theta}{g}\dfrac{\partial v_E}{\partial h} = (-0{\cdot}003)$ | $C\dfrac{b\theta}{g}\dfrac{\partial v_N}{\partial h} = (+0{\cdot}003)$ |
| Transfer SUM $= -0{\cdot}23$ | Transfer SUM $= -3{\cdot}78$ |
| as contribution to $-\dfrac{\partial M_{E20}}{\partial t}$ on Form M III | as contribution to $-\dfrac{\partial M_{N20}}{\partial t}$ on Form M IV |

## COMPUTING FORM M III. For the Dynamical Equation for the Eastward Component

$$-\frac{\partial M_{E68}}{\partial t} = \frac{\partial P_{G8}}{\partial e} + \left[p\frac{\partial h}{\partial e}\right]_G^{*} + \frac{\partial}{\partial e}\left(\frac{M_E^2}{R}\right)_{G8} + \frac{\partial}{\partial n}\left(\frac{M_E M_N}{R}\right)_{G8} + \left[m_E v_H\right]_8 \quad \dagger \quad -2\omega\sin\phi M_{NG8} + 2\omega\cos\phi M_{HG8} + \left(\frac{3M_E M_H - 2M_E M_N \tan\phi}{aR}\right)_{G8}$$

* No corresponding term in upper layers. † A term $[m_E v_H]_G$ absent because ground impervious to wind.

Longitude *11° E* Latitude *5600 km N* $2\omega \sin\phi = 1{\cdot}124 \times 10^{-4}$, $2\omega\cos\phi = 0{\cdot}930 \times 10^{-4}$ Instant *1910 May 20ᵈ 7ʰ G.M.T.* $\delta t = 6$ *hours*

| $\frac{\tan\phi}{a} = 1{\cdot}897 \times 10^{-9}$ | $\delta e = 425 \times 10^5$ for $\delta\lambda = 6°$ | $h_0$ | | $h_2$ | | $h_4$ | | $h_6$ | | $h_8$ | $h_G$ |
|---|---|---|---|---|---|---|---|---|---|---|---|
| REF. | Term | + | − | + | − | + | − | + | − | + | − |
| CH. 4/4 #14 | $\delta P/\delta e$ | *31 3* | | *4·9* | | *9·4* | | *12·9* | | | *215·5* |
| CH. 4/4 #14 | $[p\,.\,\delta h/\delta e]_G$ | ——— | | ——— | | ——— | | ——— | | *230·3* | |
| | $\frac{\delta}{\delta e}(M_E^2/R)$ | *0·5* | | | *3·4* | | *2·7* | | *0·5* | *2·6* | |
| | $\frac{\delta}{\delta n}(M_E M_N/R)$ | *0·2* | | | *0·1* | *1·5* | | *0·6* | | | *0·7* |
| Form P XVI | $m_E v_H$ at upper limit | ——— | | | *(0·0)* | | *(0·9)* | | *(0·2)* | | *(0·7)* |
| Form P XVI | $-m_E v_H$ at lower limit | *(0·0)* | | *(0·9)* | | *(0·2)* | | *(0·7)* | | * no term here | |
| | $-2\omega\sin\phi\,.\,M_N$ | *2·0* | | *7·0* | | | *3·3* | | *6·5* | | *6·2* |
| Form P XVI | $+2\omega\cos\phi\,.\,M_H$ | *(0·01)* | | *(0·02)* | | *(0·03)* | | *(0·02)* | | *(0·01)* | |
| Form P XVI | $+3M_E M_H/(aR)$ | | *(0·0)* | | *(0·0)* | | *(0·0008)* | | *(0·0)* | | *(0·0)* |
| | $-2M_E M_N \tan\phi/(aR)$ | | *0·0* | | *0 2* | | *0·1* | | *0·1* | | *0·1* |
| Form M I | viscosity terms | | *0·0* | | *0·0* | | *0·0* | *0·2* | | | *1·4* |
| Form M II | stratosphere, special | | *0·2* | ——— | | ——— | | ——— | | ——— | |
| | sums + and − | *34·0* | *0·2* | *12·8* | *3·7* | *11·1* | *7·0* | *14·4* | *7·3* | *232·9* | *224·6* |
| | $-\partial M_E/\partial t$ | *+33·8* | | *+9·1* | | *+4·1* | | *+7·1* | | *+8·3* | |
| | $+\delta t\,.\,\partial M_E/\partial t$ | $-730 \times 10^3$ | | $-196 \times 10^3$ | | $-89 \times 10^3$ | | $-153 \times 10^3$ | | $-179 \times 10^3$ | |

$$-\frac{\partial M_N}{\partial t} = -g_N R + \frac{\partial P}{\partial n} + \left[p\frac{\partial h}{\partial n}\right]_G + \frac{\partial}{\partial e}\left(\frac{M_N M_E}{R}\right) + \frac{\partial}{\partial n}\left(\frac{M_N^2}{R}\right) + \left[m_N v_H\right]_8 + 2\omega \sin\phi \,.\, M_E + \frac{3M_N M_H}{aR} + \frac{\tan\phi\{M_E^2 - M_N^2\}}{aR}.$$

The above equation relates to the stratum $h_G$ to $h_8$. For upper strata, omit the term $p\frac{\partial h}{\partial n}$, and subtract a term $m_N v_H$ at the lower boundary.

| Longitude *11° E* | Latitude *5600 km N* | $2\omega\sin\phi = 1{\cdot}124\times10^{-4}$, $2\omega\cos\phi = 0{\cdot}930\times10^{-4}$ | | | | Instant *1910 May 20ᵈ 7ʰ G.M.T.* | $\delta t = 6$ *hours* | | | | |
|---|---|---|---|---|---|---|---|---|---|---|---|
| $\frac{\tan\phi}{a} = 1{\cdot}897\times10^{-9}$ | $\delta e = 425\times10^5$ for $\delta\lambda = 6°$ | $h_0$ | | $h_2$ | | $h_4$ | | $h_6$ | | $h_8$ | $h_G$ |
| REF. | Term | + | − | + | − | + | − | + | − | + | − |
| Table in Ch. 4/4 | $-g_N R$ | *2·7* | | *1·5* | | *0·9* | | *0·4* | | *0·1* | |
| Ch. 4/4 # 14 | $\delta P/\delta n$ | *24·1* | | | *1·7* | *2·3* | | *7·7* | | *498·3* | |
| Ch. 4/4 # 14 | $[p\,.\,\delta h/\delta n]_G$ | ——— | | ——— | | ——— | | ——— | | | *487·8* |
| | $\frac{\delta}{\delta e}(M_N M_E/R)$ | | *0·0* | *4·1* | | *2·5* | | *0·4* | | | *2·1* |
| | $\frac{\delta}{\delta n}(M_N^2/R)$ | | *0·4* | *0·5* | | | *1·3* | | *0·4* | *0·3* | |
| Form P XVI | $m_N v_H$ at upper limit | ——— | | *(0·7)* | | | *(0·1)* | *(0·1)* | | *(0·4)* | |
| Form P XVI | $-m_N v_H$ at lower limit | | *(0·7)* | *(0·1)* | | | *(0·1)* | | *(0·4)* | no term here | |
| | $+2\omega\sin\phi\,.\,M_E$ | | *6·3* | | *16·4* | | *10·7* | | *5·8* | | *12·4* |
| Form P XVI | $3M_N M_H/(aR)$ | | *(0·0)* | | *(0·0)* | *(0·0002)* | | *(0·0)* | | *(0·0)* | |
| | $\tan\phi\{M_E^2 - M_N^2\}/(aR)$ | *0·0* | | *0·2* | | *0·1* | | | *0·0* | *0·1* | |
| Form M I | viscosity terms | | *0·0* | | *0·0* | | *0·0* | | *0·0* | *0·2* | |
| Form M II | stratosphere, special | | *3·8* | ——— | | ——— | | ——— | | ——— | |
| | sums + and − | *26·8* | *11·2* | *7·1* | *18·1* | *5·8* | *12·2* | *8·6* | *6·6* | *499·4* | *502·3* |
| | $-\partial M_N/\partial t$ | *+15·6* | | *−11·0* | | *−6·4* | | *+2·0* | | *−2·9* | |
| | $+\delta t\,.\,\partial M_N/\partial t$ | $-337\times10^3$ | | $+238\times10^3$ | | $+138\times10^3$ | | $-43\times10^3$ | | $+63\times10^3$ | |

## Summary of changes deduced from the Initial Distribution

Δ here means the increase of the quantity concerned, during the period of six hours, which was centered at 1910 May 20[d] 7[h] G.M.T.

*At longitude* 11° *East, latitude* 5600 *kilometres North*

From computing forms M III and M IV

| | $10^3\times$ | | $10^3\times$ |
|---|---|---|---|
| $\Delta M_{E20}$ | −730 | $\Delta M_{N20}$ | −337 |
| $\Delta M_{E42}$ | −196 | $\Delta M_{N42}$ | +238 |
| $\Delta M_{E64}$ | − 89 | $\Delta M_{N64}$ | +138 |
| $\Delta M_{E86}$ | −153 | $\Delta M_{N86}$ | − 43 |
| $\Delta M_{EG8}$ | −179 | $\Delta M_{NG8}$ | + 63 |

*At longitude* 11° *East, latitude* 5400 *kilometres North*

From computing forms P XIII, P XIV, P XVII

| | $100\times$ | | |
|---|---|---|---|
| | | $\Delta\theta_1$ | 19°·6 |
| $\Delta p_2$ | 483 | | |
| | | $\Delta W_{42}$ | 0·007 |
| $\Delta p_4$ | 770 | | |
| | | $\Delta W_{64}$ | 0·024 |
| $\Delta p_6$ | 1032 | | |
| | | $\Delta W_{86}$ | 0·149 |
| $\Delta p_8$ | 1265 | | |
| | | $\Delta W_{G8}$ | 0·402 |
| $\Delta p_G$ | 1451 | | |

It is claimed that the above form a fairly correct deduction from a somewhat unnatural initial distribution.

## CH. 9/3. THE CONVERGENCE OF WIND IN THE PRECEDING EXAMPLE

The striking errors in the "forecast," which has been obtained by means of the computing forms, may be traced back to the large apparent convergence of wind. It may be asked whether this spurious convergence arises from the errors of observations with balloons, or from the finite horizontal differences being too large, or thirdly from the process by which the winds at points, arranged in a rectangular pattern, are interpolated between the observing stations. We may examine this question by eliminating the third source of error by computing the convergence in a triangle formed by three observing stations. The formula for a triangle was given by Bennett in an appendix to the *Life History of Surface Air Currents*, by Shaw and Lempfert. Bennett's formula can be expressed for our purposes as follows: Let the triangle formed by the three balloon stations be drawn on a map. The momentum per horizontal area of a conventional stratum at each station is next to be resolved along the perpendicular to the opposite side of the triangle. Each resolved part is to be divided by the length of the corresponding perpendicular and the quotients are to be added. Their sum would be the rate of increase of mass per unit area of the stratum if there were no vertical motion. In selecting three stations to form a triangle I have chosen those at which the extrapolation of momentum in the stratosphere was the best. These were Hamburg, Strassburg and Vienna. They form a triangle which is nearly equilateral but rather large, the sides being roughly 700 kilometres long*. The computations are set out in the adjoining table†, the interesting column of which is the last one headed "sums for all three stations," for the figures in it would be the rates of increase of mass per horizontal area of stratum, if there were no vertical motion. Unfortunately we have no direct observations of vertical motion for this occasion, but thanks to the extrapolation to the upper limit $h_0$ of the atmosphere, we can find the total rate of increase of mass for all strata by adding up the figures in the last column. Their sum is $+305 \times 10^{-5}$ grm cm$^{-2}$ sec$^{-1}$. As the whole mass of the atmosphere is about one kilogram per horizontal square centimetre, this rate of increase implies a barometric rise of 0·003 millibar per second, that is about 60 millibars in six hours. Actually at Bayreuth‡, which lies near the middle of the triangle, the barometric readings were as follows:

| | May 19 | | | May 20 | | |
|---|---|---|---|---|---|---|
| hour ............... | 7 | 14 | 21 | 7 | 14 | 21 |
| 700 mm of Hg + | 25·8 | 25·0 | 25·3 | 25·6 | 24·7 | 25·6 |

Other neighbouring stations observed a similarly constant pressure, and the hourly values for Hohenpeissberg‡ and for Potsdam, which lie outside the triangle on opposite side of it, tell the same tale. Thus there is a marked disagreement between observation and calculation.

* See the map on p. 184. † The second table on p. 213.

‡ *Deutsches Meteorologisches Jahrbuch für* 1909, Bayern.

*Convergence of Winds in a Triangle*

1917 May 20$^d$ 7$^h$

| | HAMBURG | VIENNA | STRASSBURG |
|---|---|---|---|
| Length of perpendicular | $5{\cdot}7 \times 10^7$ cm | $6{\cdot}1 \times 10^7$ cm | $4{\cdot}8 \times 10^7$ cm |
| Azimuth of perpendicular | S (exactly) | W 19° N | N 50° E |

| | Inward component of $M$ along perpendicular divided by length of perpendicular Unit $=10^{-5}$ grm cm$^{-2}$ sec$^{-1}$ | | | Sums for all three stations |
|---|---|---|---|---|
| | Hamburg | Vienna | Strassburg | |
| $h_2$ to $h_0$ | + 33 | − 20 | − 19 | − 6 |
| $h_4$ to $h_2$ | + 74 | + 136 | − 215 | − 5 |
| $h_6$ to $h_4$ | − 14 | + 102 | − 29 | + 59 |
| $h_8$ to $h_6$ | + 7 | + 36 | + 48 | + 91 |
| $h_G$ to $h_8$ | − 142 | + 243 | + 65 | + 166 |
| Sums for all five strata | − 42 | + 497 | − 150 | + 305 |

It is possible to suppose that the marked convergence in the lower strata was actually balanced by a large divergence in the upper part of the stratosphere, a divergence which does not appear in the table and which would have to be explained by casting the blame on the newcomer: the extrapolation in the stratosphere. That explanation would imply an upward current in the middle layers of some 700 metres in 6 hours at a height of 4·2 kilometres. So large an upward speed would probably have produced cloud, whereas the sky remained almost clear*. So we turn to the alternative explanation, which is that stations as far apart as 700 kilometres did not give an adequate representation of the wind in the lower layers. That appears almost certain when one thinks of the irregularities of the surface wind exhibited on the daily weather reports. The wind stations in the proposed rectangular pattern are nearer to one another, being 400 kilometres apart, and that distance might well have to be halved in practice. This being admitted, let us refer again to the convergence of the horizontal winds set out in the last column of the table. It is seen that for the layers above 7·2 kilometres the divergence is quite small and therefore credible. The suggestion is that at these great heights the flow was so smooth and lacking in detail that observations even as far apart as 700 kilometres gave a fair representation of it; and further that the extrapolation for the upper part of the stratosphere was, at least, not strikingly in error.

* An uplift of 1·4 km would have produced general cloud, whereas the mean cloud amount at 16 Bavarian stations was below 2·5/10 until after May 21$^d$ 7$^h$.

# CHAPTER X

## SMOOTHING THE INITIAL DATA

We are not concerned to know all about the weather, nor even to trace the entangled detail of the path of every air-particle. A judicious selection is necessary for our peace of mind. For some such reason it is customary, at stations which report wind by telegraph, to replace the instantaneous velocity by a mean value over about ten minutes. An extension of this process must be contemplated, for there is a good deal of evidence to show that the wind is full of small "secondary cyclones" or other whirls having the most various diameters. The arithmetical process can only take account individually of such whirls as have diameters greater than the distance between the centres of the red chequers in our co-ordinate chessboard, and this length has been taken provisionally as 400 km.

If we smooth out these whirls we shall have to make amends by introducing suitable eddy-diffusivities. So far meteorologists do not appear to have attended to eddy-diffusivities of this kind. We shall refer to them again in Ch. 11/4.

The evidence for the existence of such eddies includes the following:

(i) The impossible rate of accumulation of air deduced in Ch. 9/3 from observations at three stations.

(ii) The irregular wind-arrows shown on the unusually detailed wind-maps prepared by the Norwegian weather service. In Norway many of the irregularities are obviously due to mountains. They none the less come under consideration here.

(iii) The irregularities noticeable on the British Daily Weather Report between the pilot balloon observations at neighbouring stations*.

(iv) The errors of forecasts which have been attributed by G. M. B. Dobson† mainly to small irregular variations in the pressure distribution.

Let us now consider various ways in which the smoothing could be effected.

A. **Space Means.** If instead of one station observing wind at the centre of each red chequer on our chessboard we had a number of stations distributed, preferably regularly, over the area of the chequer, then some sort of mean of their observations would be the proper quantity to choose as an initial datum for the computing. This plan would be technically preferable to the processes described below, but as observations are scarce and stations costly we should explore other ways.

B. **Time Means.** Suppose that there is only one station on each red square of the chessboard, but that the observations of wind at it are made every hour during say $n$ successive hours, and that their mean is taken and used as the initial datum of the computing. If the large eddies are distributed at random, the result will be much the

* See also especially "The variation of wind with place," by Capt. J. Durward, M.A., *London Met. Office Professional Notes*, No. 24.

† "Causes of Errors in Forecasting Pressure Gradients and Wind," *Q. J. R. Met. Soc.* 1921 Oct.

same as if we took a space-mean along the line travelled by a point moving with the mean wind during $n$ hours. We want this line to be equal in length to the distance between the centres of our red co-ordinate chequers. So for an ordinary wind velocity of 10 m/s and a chequer 200 km in the side the observations would need to be continued for $\frac{400\times 10^3}{10\times 3600} = 11$ hours.

The advantage of this scheme over A is that it would require fewer observers. It appears also to be more practical than C, D, E which follow.

C. **Potential Function.** The irregularity in the observations which has forced itself on our attention is the large value of $\text{div}_{EN}\,\mathbf{m}$. It may be possible by slight adjustment of $\mathbf{m}$ to remove large values of $\text{div}_{EN}\,\mathbf{m}$ especially if the latter are scattered at random and if they vary, as is to be anticipated, symmetrically around a mean near to zero. Thus we might introduce a potential function $f_1$, and replace the observed $m_E$, $m_N$ by

$$m_E + \frac{\partial f_1}{\partial e}, \quad m_N + \frac{\partial f_1}{\partial n}.$$

Then $\text{div}_{EN}$ of these new momenta per volume would be

$$\text{div}_{EN}\,\mathbf{m} + \nabla^2{}_{EN} f_1,$$

where $\nabla^2{}_{EN}$ is what $\nabla^2$ becomes when $\partial f_1/\partial h$ is ignored. If then this expression be put equal to zero or to some function of the observed "barometric tendency" we should have a problem to be solved to determine $f_1$ from $\nabla^2{}_{EN} f_1 =$ a given function of position.

Either there is a boundary of the sort contemplated in Ch. 7 in which case we have a "jury" problem*. Or else the region covers the whole globe, as if the boundary had contracted to a point. There must be no discontinuity of $f_1$ at this point and for that reason we have to do, not with a "marching" problem*, but with one more akin to the "jury" variety.

The solution of such problems by analysis is the subject of an extensive literature† and their solution by arithmetic has been illustrated by examples elsewhere‡. In any case it is rather troublesome.

D. **Stream Function.** The method C would leave the distribution of curl $\mathbf{m}$ unaltered by the smoothing. But curl $\mathbf{m}$ is almost certainly irregular and in need of preliminary smoothing if we are to avoid awkward consequences in the application of the dynamical equations. We might imagine a stream function $f_2$ introduced and $m_E$, $m_N$ replaced by

$$m_E + \frac{\partial f_2}{\partial n}, \quad m_N - \frac{\partial f_2}{\partial e}.$$

Proceeding as with $f_1$ we could by the solution of a jury problem find a distribution $f_2$ which would remove or diminish the irregular curl, without affecting the divergence.

* See p. 3 above.

† For example Byerly, *Fourier Series and Spherical Harmonics.* Wiley & Co., New York.

‡ L. F. Richardson, *Phil. Trans.* A, Vol. 210, p. 307 (1910).

E. **Smoothing during the forecast.** While beginning the forecast with the unsmoothed velocities we might temporarily introduce into the dynamical equations terms representing a considerable fictitious viscosity. These would have the effect of smoothing out irregular motions whether waves of compression or whirls. As is well known, the shorter the wave length or the smaller the diameter of the whirl, the more rapidly would the corresponding motion be damped out.

Arithmetical processes, somewhat analogous to that here suggested, have been used* to smooth an arbitrary function of position and to make it approach gradually to $f_n$ where $\nabla^2 f_n = 0$. In the atmospheric case we should aim to remove the fictitious viscosity after it had smoothed the irregularities and if possible before its effect on the larger motions had become noticeable.

* L. F. Richardson, *Phil. Trans.* A, Vol. 210, p. 307 (1910).

# CHAPTER XI

## SOME REMAINING PROBLEMS

### Ch. 11/0. INTRODUCTION

The two great outstanding difficulties are those connected with the completeness necessary in the initial observations and with the elaborateness of the subsequent process of computing. These are discussed in Ch. 11/1 and Ch. 11/2. The scheme of numerical forecasting has developed so far that it is reasonable to expect that when the smoothing of Ch. 10 has been arranged, it may give forecasts agreeing with the actual smoothed weather. When that stage has been attained, the other difficulties will tend to group themselves with questions of the desirability of weather forecasts and of their cost. We need here an estimate of the economic value of a forecast reliable for $n$ days ahead, given as a function of $n$. As with improved methods $n$ is likely to increase, so forecasts will become of more value to agriculturalists. Now the annual value of the world's food crops is at least £1000,000,000, so that a very tiny fractional saving would correspond to a large sum.

### Ch. 11/1. THE PROBLEM OF OBTAINING INITIAL OBSERVATIONS

#### Pattern

When the observations are taken at stations scattered irregularly, an interpolation has to be made to find the initial data at the centres of the chequers of our chessboard. It has been mentioned in Ch. 9/1 that this interpolation was found to be both troublesome and inaccurate, as may be evident from the map on p. 184.

An existing meteorological station in the British Isles has been either an outgrowth from an astronomical or magnetic observatory, or it has adjoined the house of an enthusiast who lived there for reasons unconnected with meteorology, or it has been pushed out to the confines of the islands to grasp as much weather as possible, or it has been placed in charge of coastguards because they are on duty at night, or it has been set on a mountain to test the upper air. Excellent practical reasons all these, but it is remarkable that the properties of the atmosphere, which are expressed by its dynamical equations and its equation of continuity, appear to have had no influence on the selection*.

There would be a great advantage, from the point of view of meteorological science, if observing stations for pressure and for velocity could be arranged alternately in rectangular order in the pattern shown in the frontispiece, modified where necessary by devices such as those proposed in Ch. 7/3, 7/4, 7/5.

#### Wind

Velocities given, as is customary, to 0·1 metre/sec, and at stations 400 km apart, have been found to be nearly but not quite sufficiently accurate (supposing the decimal

* See resolution XXI in the *Report of the International Commission for the Investigation of the Upper Air*, Bergen 1921.

figure correct) to give reasonable values of $\partial p_G/\partial t$ when they are inserted in the equation of continuity of mass. Greater accuracy could, of course, be obtained by using larger theodolites. Further, a pilot balloon observation naturally gives $\int v_E dh$, $\int v_N dh$ and if these integrals were published for certain limits of $h$, the calculation of

$$M_E = \int v_E \rho . dh, \quad M_N = \int v_N \rho . dh$$

could be improved. There remains the problem of gusts and local eddies of larger size which has been discussed in Chapter 10.

The circumstances, in which wind can be observed, are extending. Thus during the war it became customary to observe pilot balloons at night by attaching to the balloon a candle in a small paper lantern. The wind data available at present relate mainly to clear air. But for observing the wind above fog, or low cloud, kite-balloons can be used, or the elaborate method of location by sound*. For the same purpose projectiles have been used at Benson up to heights of 600 m. The projectiles have been spheres, of about the size of a cherry and they have been projected nearly vertically but in a direction slightly inclined towards the wind so that the returning sphere struck earth near to the hut which protected the observer.

## Temperature

The example of Ch. 9 sets out from the records of registering balloons. But these would not serve as a basis for actual forecasts because the balloon is often not found until a week or more after its ascent.

During and since the war temperature observations by aeroplanes have been taken in great number up to about 5 kilometres. Also kite-balloons have been utilized.

Experiments have recently been conducted at Benson Observatory with a view to finding methods of observing temperature, or its equivalent, which should give immediate records and which should be cheaper than aeroplane ascents. The results obtained are described in publications entitled "Lizard Balloons for signalling the ratio of pressure to temperature †," "Cracker balloons for signalling temperature †" and "Sun-flash balloons for continuous signalling‡." It is hoped also to describe some experiments in which the time of flight of a projectile shot upwards served as an indication of the temperature aloft.

## Water in Clouds

Another gap in the existing observational data relates to the amount of water in clouds. Attempts have been made to measure this photometrically. And it has been shown to be possible§ when the cloud particles are all of one known size. But in actual cases we are often without definite information concerning the size of the particles.

* Schereschewsky, *Report Intern. Commiss. Upper Air*, Bergen 1921, p. 22.

† Meteorological Office, London, *Professional Notes*, Nos. 18 and 19.

‡ *Q. J. R. Met. Soc.* 1920 July, p. 293.

§ L. F. Richardson, "Water in Clouds," *Roy. Soc. Lond, Proc.* A, Vol. 96 (1919), p. 19.

## CH. 11/2. THE SPEED AND ORGANIZATION OF COMPUTING

It took me the best part of six weeks to draw up the computing forms and to work out the new distribution in two vertical columns for the first time. My office was a heap of hay in a cold rest billet. With practice the work of an average computer might go perhaps ten times faster. If the time-step were 3 hours, then 32 individuals could just compute two points so as to keep pace with the weather, if we allow nothing for the very great gain in speed which is invariably noticed when a complicated operation is divided up into simpler parts, upon which individuals specialize. If the co-ordinate chequer were 200 km square in plan, there would be 3200 columns on the complete map of the globe. In the tropics the weather is often foreknown, so that we may say 2000 active columns. So that $32 \times 2000 = 64{,}000$ computers would be needed to race the weather for the whole globe. That is a staggering figure. Perhaps in some years' time it may be possible to report a simplification of the process. But in any case, the organization indicated is a central forecast-factory for the whole globe, or for portions extending to boundaries where the weather is steady, with individual computers specializing on the separate equations. Let us hope for their sakes that they are moved on from time to time to new operations.

After so much hard reasoning, may one play with a fantasy? Imagine a large hall like a theatre, except that the circles and galleries go right round through the space usually occupied by the stage. The walls of this chamber are painted to form a map of the globe. The ceiling represents the north polar regions, England is in the gallery, the tropics in the upper circle, Australia on the dress circle and the antarctic in the pit. A myriad computers are at work upon the weather of the part of the map where each sits, but each computer attends only to one equation or part of an equation. The work of each region is coordinated by an official of higher rank. Numerous little "night signs" display the instantaneous values so that neighbouring computers can read them. Each number is thus displayed in three adjacent zones so as to maintain communication to the North and South on the map. From the floor of the pit a tall pillar rises to half the height of the hall. It carries a large pulpit on its top. In this sits the man in charge of the whole theatre; he is surrounded by several assistants and messengers. One of his duties is to maintain a uniform speed of progress in all parts of the globe. In this respect he is like the conductor of an orchestra in which the instruments are slide-rules and calculating machines. But instead of waving a baton he turns a beam of rosy light upon any region that is running ahead of the rest, and a beam of blue light upon those who are behindhand.

Four senior clerks in the central pulpit are collecting the future weather as fast as it is being computed, and despatching it by pneumatic carrier to a quiet room. There it will be coded and telephoned to the radio transmitting station.

Messengers carry piles of used computing forms down to a storehouse in the cellar.

In a neighbouring building there is a research department, where they invent improvements. But there is much experimenting on a small scale before any change is made in the complex routine of the computing theatre. In a basement an enthusiast s observing eddies in the liquid lining of a huge spinning bowl, but so far the arithmetic proves the better way. In another building are all the usual financial, correspondence and administrative offices. Outside are playing fields, houses, mountains and lakes, for it was thought that those who compute the weather should breathe of it freely.

## CH. 11/3. ANALYTICAL TRANSFORMATION OF THE EQUATIONS

It is conceivable that by a change of variables the equations could be much shortened. But as we are always required in the end to arrive at quantities of direct interest to the public, namely wind, rain, temperature and radiation, so it may be that analytical simplicity does not simplify the arithmetic. There is a tale of a philosopher who succeeded in reducing the whole of physics to a single equation $H=0$, but the explanation of the meaning of $H$ occupied twelve fat volumes.

The sort of transformations that suggest themselves are those to $\log p$, $\log \theta$, instead of $p$, $\rho$ or else to stream functions and velocity potentials*.

No use has been made in this book of Mr W. H. Dines' correlations between temperature and pressure, and it is felt that while it would be very rash to assume the correlation to be exactly unity, yet its proved approach towards unity might suggest an economical choice of variables.

Then again experience must decide whether the various transformations, from equations true at any level, to equations true for strata as a whole, are worth the extra trouble they involve. Would it be easier and more exact, for example, to compute with 20 strata using un-transformed equations, rather than with 5 strata and the equations given in the computing forms?

## CH. 11/4. HORIZONTAL DIFFUSION BY LARGE EDDIES

We are led to consider this by the proposal in Ch. 10 that small secondary cyclones should be smoothed out. For any smoothing process requires compensation by the introduction of eddy-diffusion of an appropriate kind.

There is the hypothesis† that we may measure the diffusion of heat, water, momentum, dust, etc., by first measuring the diffusion of mass of air and afterwards considering how much heat, water, momentum or dust is carried by unit mass. This process is easy and the numerical estimates for vertical diffusion show that it gives results at least of the right order. Let us apply it to horizontal diffusion.

The smoke trails from cities have been observed by aviators to be hundreds of miles long. If aviators would also take note of the horizontal breadth of the trail at various distances from the source, and of the speed of the mean wind, it might be possible to extract a measure of the horizontal diffusivity. The formula has been given by the author‡ thus:

* Compare A. E. H. Love, "Notes on the dynamical theory of the tides." *Proc. Lond. Math. Soc.* Ser. 2, Vol. 12, Part 4.

† Due to G. I. Taylor and W. Schmidt. See p. 76 above.

‡ *Phil. Trans.* A, Vol. 221, p. 6. Compare A. Einstein on Brownian movement, *Ann. Phys.* XVII. 1905.

$$\text{Eddy-diffusivity} = \frac{\text{Square of standard deviation of smoke from its middle line}}{\text{Twice time taken by smoke since leaving source}}, \quad \text{.........(1)}$$

provided always that the time is long enough. There might be difficulties owing to the precipitation of the smoke, or to its becoming too faint to be seen.

In these respects manned balloons are better. The Gordon Bennett races furnish suitable data. The balloons simply drifted, the control of the aeronauts being limited to letting out gas or ballast. Even that amount of control rather confuses the data for the present purpose. But as there is no other information this is not to be despised. On 1906 Sept. 30, between $16^{h}$ and $17^{h}$ 20 in the afternoon, sixteen manned balloons started from Paris. The landing places of all of them* and the log of one† are printed in the *Aeronautical Journal*. From these records I estimate, by means of the above formula, that the horizontal eddy-diffusivity‡ was of the order of $2 \times 10^{8}\,\text{cm}^{2}\,\text{sec}^{-1}$. This figure is also the ratio of viscosity to density, and it lies between the observed values for shoemakers' wax and for pitch. The eddy-diffusivity when multiplied by the density of the air, which we may put at $10^{-3}$ grm cm$^{-3}$, gives the eddy viscosity

$$2 \times 10^{5}\,\text{cm}^{-1}\,\text{grm sec}^{-1}.$$

The mean velocity of one balloon on this occasion was about 400 cm sec$^{-1}$.

To explain the geometry let us take rectangular axes $Ox$, $Oy$, $Oh$. Let $Ox$ be drawn horizontally in the direction of the mean wind, so that $x$ increases for a particle moving with this smooth motion. Let the $y$ axis be drawn horizontally to the left, and, in accordance with the right-hand screw rule, let the $h$ axis point upwards. Let $v_X$, $v_Y$, $v_H$ be the corresponding components of the smoothed wind velocity. Then in a viscous liquid free from eddies there would be§ three shearing stresses, $\widehat{yh}$, $\widehat{hx}$, $\widehat{xy}$, connected with three rates of shear by the following equations in which $C_X$, $C_Y$, $C_H$ would be equal to one another and would be called the viscosity:

$$\left.\begin{aligned} \widehat{yh} &= C_X\left\{\left[\frac{\partial v_H}{\partial y}\right] + \frac{\partial v_Y}{\partial h}\right\} \\ \widehat{hx} &= C_Y\left\{\frac{\partial v_X}{\partial h} + \left[\frac{\partial v_H}{\partial h}\right]\right\} \\ \widehat{xy} &= C_H\left\{\frac{\partial v_Y}{\partial x} + \frac{\partial v_X}{\partial y}\right\} \end{aligned}\right\} \quad \text{..................................(2)}$$

In the smoothed motion of the atmosphere the two terms in square brackets are usually negligible. The eddy-viscosity which has usually been measured is either $C_X$ or $C_Y$ or some combination of $C_X$ and $C_Y$ for it has been customary to assume that these viscosities are equal. However in one case the author has found some evidence that $C_Y$ was seven times greater than $C_X$. (See p. 73 above.) It has now been shown that $C_H$ must be taken to be 1000 or more times greater than either $C_X$ or $C_Y$ if we

* *Aer. J.* 1906 Oct. † *Ibid.* 1907 Jan.

‡ The corresponding race of 1921 Sept. gives diffusivity $= 3{\cdot}6 \times 10^{8}$ cm sec$^{-1}$.

§ Lamb, *Hydrodynamics*, 3rd edn. Arts. 30, 311 to 314.

are going to smooth out not only gusts but also "secondary cyclones" as has been proposed in Ch. 10. All eddy-viscosities imply some conventional coordinate element. The element chosen in this book is shaped like a railway ticket as its edge extends about 200 km on the map, while it is only a few kilometres thick. Apparently the flat shape of the element is the cause of the great excess of $C_H$ over $C_X$ or $C_Y$.

Equations (2) are valid if the fluid is isotropic. But, now that we have proved that $C_X$, $C_Y$, $C_H$ to be unequal, the form of the relationship between the six eddy-stresses and the six rates of mean strain becomes again an open question. In Ch. 4/9/5 it has even been suggested that there may be nine independent components of stress. But apart from that possibility, the general theory is to be found ready-made in connection with the elasticity of crystals. May we perhaps liken our coordinate element to a crystal having three unequal axes at right angles to each other? If so equations (2) would still be correct* and the complications would be confined to the direct stresses.

In order to form some idea of the order of magnitude of the eddy-shearing stress $\widehat{xy}$ which would be produced by this enormous viscosity $C_H$, an attempt has been made to estimate the rate of shear $\partial v_Y/\partial x + \partial v_X/\partial y$. For this purpose daily weather maps have been taken, local irregularities have been smoothed out of the isobars, and then the wind has been assumed to be geostrophic. In the well-marked cyclone of 1919 March 27 at $7^h$ the rate of shear on the slope of the depression across the British Isles appears to have been of the order of $4 \times 10^{-5}\,\text{sec}^{-1}$.

Taking the viscosity $C_H$ as $2 \times 10^5\,\text{cm}^{-1}\,\text{grm sec}^{-1}$ we find by multiplication an eddy shearing stress $\widehat{xy}$ of 8 dyne $\text{cm}^{-2}$. This is only some 1 to 10 times greater than the shearing stress on the ground, so that the large viscosity is associated with the small rate-of-mean-shear, and *vice versa*.

## CH. 11/5. A SURVEY OF REFLECTIVITY

We have been led to attribute considerable importance to the fraction of solar radiation which is absorbed at the surface. The mean value of the reflectivity over a large area could perhaps best be observed from aeroplanes. A very light and simple photometer† would serve to compare the brightness of a uniform stratus cloud above the aeroplane with that of the ground beneath.

* Winkelmann, *Handb. der Physik.*

† See, for example, *Roy. Soc. Lond. Proc.* A, Vol. 96 (1919), p. 25.

# CHAPTER XII

## UNITS AND NOTATION

### Ch. 12/1. UNITS

Except where otherwise stated, centimetre-gram-second units are employed. Temperatures are in degrees absolute centigrade. Energy, whether by itself or as involved in entropy or in specific heat, is expressed not in calories but in ergs. A power of ten standing at the beginning of a row or column of figures, and followed by a multiplication sign, is intended to multiply each number in the row or column.

### Ch. 12/2. LIST OF SYMBOLS

The following list shows the meanings which have been used throughout the book. Where the symbol requires an extended definition reference is made to the place where the definition will be found.

Mathematical notation is international, so that a foreigner, who is unable to read the letterpress, may yet grasp the purport of the book if he knows the meanings of the symbols only. So here I should like to explain the symbols in "the second language for all mankind," if there were such a one. Unfortunately there are several rivals, each apparently easier to learn than any national language. Thus there are Esperanto*, Ido†, Esperantido‡, all much alike, and differing considerably from Interlingua§.

A comparative study of these languages is being made by a committee of the International Research Council (at Washington, U.S.A.)¶ and the choice of one language should rest with some supremely authoritative body. Here without expressing any opinion as to which language is best, one namely Ido is selected for illustration. In making the translations I have been guided by my brother Gilbert H. Richardson and by the large "*Dictionnaire Français* = Ido par Beaufort et Couturat‖." Words marked with an asterisk * are not in the dictionary and so are merely suggestions.

* British Esperanto Association, 17 Hart Street, London, W.C. 1.

† International Language (Ido) Society of Great Britain, Hon. Sec. J. W. Baxter, 57 Limes Grove, Lewisham, London, S.E. 13.

‡ "Esperantido," 10 Hôtelgasse, Bern, Switzerland.

§ Headquarters in Turin, Italy.

‖ 1915 Paris, Imprimerie Chaix, 11 Boul. St Michel.

¶ 1701 Massachusetts Avenue, Washington, D.C., U.S.A.

| | ENGLISH | IDO | FURTHER DEFINITION OR REFERENCE |
|---|---|---|---|
| $a$ | Radius of the earth | Radio di la tero | |
| $b$ | Gas constant | Gasala konstanto | Ch. 4/5/1 |
| $c$ | Eddy-viscosity | Viskozeso efektigata da vortici | Ch. 4/8/0 |
| $d, \partial$ | Differentiators | Infinitezima kreskuri di | |
| $e$ | Base of natural logarithms | Bazo di logaritmi naturala | 2·71828 |
| $\partial e$ | $\equiv a \,.\, \cos\phi \,.\, \partial\lambda$ = distance eastwards † | Disto vers esto † | |
| $f$ | Various functions | Diversa funcioni | |
| $g$ | Acceleration of gravity | Acelero efektigata da gravito | |
| $h$ | Height above mean sea-level | Alteso super la meza surfaco dil maro | |
| $i$ | Subscript for arbitrary height | Subskribajo indikanta alteso segun-vola | |
| $j$ | Special coordinate in soil | Specala mezuro di profundeso en la sulo | Ch. 4/10/0 |
| $k$ | Thermal conductivity | Konduktiveso kalorala | Ch. 4/10/2 |
| $l$ | Length. Distance | Longeso. Disto | |
| $m$ | Momentum per volume | Rapideso multiplikata per denseso | |
| $\mathfrak{m}$ | Mass of molecule | Maso di molekulo | p. 37 |
| $n$ | Number | Nombro o numero | |
| $\partial n$ | $\equiv a\partial\phi$ = distance northwards † | Disto vers nordo † | |
| $p$ | Pressure | Preso | |
| $q$ | | | |
| $r$ | Radius. Correlation coefficient | Radio. Korelatala koeficiento | |
| $s$ | Diffusivity of soil for temperature | Difuziveso di la sulo ye temperaturo | Ch. 4/10/2 # 6 |
| $t$ | Time | Tempo | |
| $u$ | Thermal capacity per volume | Kalorala kapaceso po volumino | Ch. 4/10/2 # 7 |
| $v$ | Velocity | Rapideso | |
| $w$ | Mass of water-substance per volume | Maso di aquo-substanco po volumino | $= \mu\rho$ |
| $x$, $y$ | Horizontal rectangular coordinates | Horizontala koordinati ortangula | |
| $z$ | Depth in ground | Profundeso en la sulo | Ch. 4/10/0 |

† Caution: $\frac{\partial^2}{\partial n \partial e}$ is not equal to $\frac{\partial^2}{\partial e \partial n}$.

† Atencez: $\frac{\partial^2}{\partial n \partial e}$ ne egalas $\frac{\partial^2}{\partial e \partial n}$.

| | ENGLISH | IDO | FURTHER DEFINITION OR REFERENCE |
|---|---|---|---|
| *A*, *B*, *C* | Various meanings | Diversa signifiki | |
| *D* | Infinitesimal increase, accompanying the motion, of... | Infinitezima kreskuro, akompananta la movo, di... | |
| *E* | Subscript for eastwards | Subskribajo indikanta ulo direktata vers esto | |
| *E* | Radiant activity in a "parcel" | Radiada energio trairanta "pako" po tempo | pp. 50, 51 |
| *F* | Various functions | Diversa funcioni | |
| *G* | Subscript for ground level | Subskribajo signifikanta ulo ye la surfaco di la sulo | |
| *H* | Subscript for upwards | Subskribajo indikanta vers supre | |
| *I* | Brightness | Grado di brilo | pp. 50, 51 |
| *J* | | | |
| *K* | | | |
| *L* | Subscript for upper surface of vegetation | Subskribajo indikanta ulo ye la supera surfaco di la vegetantaro | |
| *M* | Momentum per area of stratum | Denses-opla rapideso po areo stratala | Ch. 4/2 # 7 |
| *N* | Subscript for northwards | Subskribajo signifikanta vers nordo | |
| *P* | $=\int p\,dh$ across stratum | $=\int p\,dh$ tra strato | Ch. 4/4 # 9 |
| *Q* | Liquid water per area of stratum | Liquida aquo po areo stratala | |
| *R* | Mass per area of stratum | Maso po areo stratala | Ch. 4/2 # 6 |
| *S* | Entropy per area of stratum | Entropio* po areo stratala | |
| *T* | Time | Tempo | |
| *U* | | | |
| $\mathscr{V}$ | Velocity of cloud particles relative to air | Rapideso, relativa al aero, di nubala partikuli | Ch. 4/6 |
| *V* | | | |
| *W* | Water-substance per area of stratum | Juntata maso di vaporo, aquo e glacio po areo di strato | p. 27 |
| *X*, *Y* | Subscripts indicating horizontal rectangular components | Subskribaji indikanta horizontala kompozanti ortangula | |
| *Z* | In theory of stirring | Ye teorio pri vortici | Ch. 4/8/0 |

| | ENGLISH | IDO | FURTHER DEFINITION OR REFERENCE |
|---|---|---|---|
| $\alpha$ $\beta$ | Coefficients relating to entropy | Koeficienti pri entropio* | Ch. 8/2/6 (Ch. 4/5/1) |
| $\gamma_p$ | Thermal capacities per mass | Kalorala kapacesi po maso | |
| $\delta$ | Finite difference operator | Finite-mikra kreskuro di... | |
| $\epsilon$ | Energy per mass | Energio po maso | Ch. 4/5/0 |
| $\zeta$ | Zenith distance | Angulo inter zenito ed ula direciono | |
| $\eta$ | Absorptance of stratum | Fraciono di radiada energio absorbata da strato | Ch. 4/7/1 # 13 |
| $\theta$ | Temperature, absolute | Temperaturo de − 273°·1 C. | |
| $\kappa$ | Molecular diffusivity | Molekulala difuziveso | Ch. 4/9/8 |
| $\lambda$ | Longitude, always eastwards | Longitudo (sempre vers esto) | |
| $\mu$ | Joint mass of vapour, water and ice per mass of atmosphere | Juntata maso di vaporo, aquo e glacio po maso di atmosfero | |
| $\nu$ | Mass of liquid water per mass of atmosphere | Maso di liquida aquo po maso di atmosfero | |
| $\xi$ | Turbulivity* | Specala mezuro di vorticado | Ch. 4/8/0 # 15 |
| $\pi$ | 3·14159... | | |
| $\rho$ | Density | Denseso | |
| $\sigma$ | Entropy per mass of atmosphere | Entropio* po maso di atmosfero | Ch. 4/5/0 Ch. 8/2/6 |
| $\tau$ | Potential temperature | Temperaturo ye ula preso normala se nek kaloro nek aquo esas perdita | Ch. 4/5/0 Ch. 8/2/6 |
| $\upsilon$ | Internal energy per mass of atmosphere | Interna energio po maso di atmosfero | Ch. 4/5/0 |
| $\phi$ | Latitude (reckoned negative in the southern hemisphere) | Latitudo (negativa en la suda mi-sfero) | |
| $\chi$ | In theory of stirring | Ye teorio pri vortici | Ch. 4/8/0 # 6 |
| $\psi$ | Gravity potential (increasing upwards) | Gravitala potencialo (kreskanta ad-supre) | |
| $\omega$ | Angular velocity of earth | Angulala rapideso di la tero | $0{\cdot}729211 \times 10^{-4}\,\text{sec}^{-1}$ |

| | ENGLISH | IDO | FURTHER DEFINITION OR REFERENC |
|---|---|---|---|
| Γ | Radiant energy absorbed at interface per area and per time | Radiada energio absorbata an interfaco po areo e po tempo | Ch. 8/2/15 |
| Δ | Increase of | Kreskuro di | |
| Θ | Eddy-heat per mass | Energio di vortici po maso | p. 77 |
| Λ | | | |
| Π | | | |
| Ξ | Mass of water evaporating from interface per horizontal area and per time | Maso di aquo vaporeskanta de interfaco po horizontala areo e po tempo | Ch. 8/2/11, 12 |
| Σ | Summing operator | Sumigilo | |
| Ψ | Pressure in water in soil | Preso en aquo en sulo | Ch. 4/10/2 |
| Φ | } Relate to vertical velocity in the stratosphere | } Relatas vertikala rapideso en la supra strato | Ch. 6/6 # 23 |
| ϒ | | | Ch. 6/6 # 22 |
| Ω | $= 2\omega \sin \phi$ | $= 2\omega \sin \phi$ | p. 15 |
| Ϝ | Vapour density in soil | Denseso di vaporo en sulo | Ch. 4/10/2 |
| ϭ | Rate of evaporation from leaf | Vaporeskala rapideso de folio | Ch. 4/10/3 # |
| ⲇ (dalda)† | Correction to the estimate of surface temperature | Korektilo al konjekturo di temperaturo interfacala | Ch. 8/2/15 |
| ⲙ (mi) | See flux of heat at the interface | Pri fluado di kaloro ad o de l'interfaco | Ch. 4/8/4 # 7 |
| ϩ (he) | Latent heat of evaporation per mass | Energio vaporigiva mezur-unajo di maso | |
| ϣ (shai) | Conductivity of soil to soil-water | Konduktiveso di sulo por aquo sulala | Ch. 4/10/2 # |
| ϫ (janja) | Porosity of soil to vapour | Porozeso di sulo por vaporo | Ch. 4/10/2 # |
| ⲝ | Partition coefficient of W. Schmidt | Koeficiento di W. Schmidt pri divido | p. 89 |
| Ↄ | Stefan's radiation constant | Konstanto di Stefan pri radiado | Ch. 4/7/1 # |
| ʌ | Absorptivity per density | Absorbiveso po denseso | Ch. 4/7/2 # 5 |
| ϯ | Scatterivity* per density | Dissemiveso po denseso | Ch. 4/7/2 # 5 |
| ɣ | Emissivity of interface for long waves | Emisiveso di interfaco por ondi longa | Ch. 8/2/15 |

† I am indebted to Prof. Flinders Petrie for the names of these Coptic letters.

## CH. 12/3. RELATIONSHIPS BETWEEN CERTAIN SYMBOLS

The symbol in the third column below is equal to $\rho$ times the corresponding symbol in the second column. The symbol in the fourth column is the integral, with respect to height across a conventional stratum, of the symbol in the third column, or if that is absent, of $\rho$ times the symbol in the second column.

| I | II<br>Per mass of atmosphere | III<br>Per volume of atmosphere | IV<br>Per horizontal area of conventional stratum |
|---|---|---|---|
| Mass ... ... ... | 1 | $\rho$ | $R$ |
| Momentum ... ... | $v$ | $m$ | $M$ |
| — ... ... ... | — | $p$ | $P$ |
| Mass of water in all forms jointly ... ... ... | $\mu$ | $w$ | $W$ |
| Mass of condensed water | $\nu$ | | $Q$ |
| Entropy ... ... ... | $\sigma$ | | $S$ |

## CH. 12/4. SUBSCRIPTS FOR HEIGHT

We have to do with many quantities which are functions of height. Any of these, say $\rho$ for illustration, may be limited to one particular height by a subscript. Thus $\rho_8$ means the value of $\rho$ at the dividing surface between two conventional strata, a surface at which the normal pressure is roughly 8 decibars. Similarly $\rho_6$, $\rho_4$, $\rho_2$, $\rho_0$ refer to conventional heights at which the normal pressure is about 6, 4, 2, 0 decibars. The actual heights selected are exactly as follows:

| | | | | |
|---|---|---|---|---|
| Height above M.S.L. ... | 2·0 km | 4·2 | 7·2 | 11·8 |
| Subscripts ... ... | 8 | 6 | 4 | 2 |

For intermediate levels the odd numbers are occasionally used as subscripts. The following letters are also used as subscripts for height:

- $G$ Ground level
- $C$ Base of stratosphere
- $L$ Upper surface of vegetation
- $A$ Height in vegetation at which radiation is converted to heat
- $i$ Arbitrary height.

To make the foregoing notation run on without a break from air to soil, the surface of the soil is denoted by $z_{10}$ and successive depths by $z_{11}$, $z_{12}$, $z_{13}$, etc. (See Ch. 8/2/15.)

The same subscripts are used to denote limits of integration with respect to height, thus

$$\int_G^8 w\,.\,dh \quad \text{is a convenient abbreviation for} \quad \int_{h_G}^{h_8} w\,.\,dh.$$

Again the same subscripts are used in pairs to denote the particular stratum to which *R*, *M*, *P*, *W*, *Q*, *S* refer; thus for example

$$W_{G8} = \int_G^8 w \, . \, dh.$$

## CH. 12/5. VECTOR NOTATION

Scalar quantities are denoted by ordinary letters, vectors by black letters (Clarendon type), their tensors by ordinary letters, and their components by ordinary letters with the suffixes $_E$, $_N$, $_H$ attached. In conformity with the custom in mathematical physics the subscripts $_E$ and $_N$ denote to, not from, the east and north. As this is opposite to the usage in practical meteorology it may sometimes be desirable to emphasize the sense by writing for example $v_{WE}$, $v_{SN}$ for the velocities to the east and to the north respectively. The subscript $_H$ denotes the component directed vertically upwards.

Scalar products are represented by simple juxtaposition, with a dot to separate the factors and parentheses to enclose them, when either or both of these are needed to make the meaning clear. Thus $\mathbf{v}(\nabla\mathbf{v})$ can be distinguished from $(\mathbf{v}\nabla)\,\mathbf{v}$. Vector products are enclosed in square brackets with if necessary a dot to separate the factors. The right-handed screw rule is used. The Laplacian operator

$$\partial^2/\partial x^2 + \partial^2/\partial y^2 + \partial^2/\partial z^2$$

is denoted by $\nabla^2$ not by $\Delta$.

To denote the divergence which a vector would have, if its vertical component became zero, while its horizontal components remained unchanged, the operator $div_{EN}$ is used. So that

$$div_{EN}\mathbf{v} = \frac{\partial v_E}{\partial e} + \frac{\partial v_N}{\partial n} - v_N \frac{\tan\phi}{a}.$$

In describing spatial variation it is very desirable, as Sir Napier Shaw has pointed out, to have a word which makes a clear distinction between vertical and horizontal directions, and for this reason he uses "gradient" for the horizontal, "lapse-rate" for the vertical. This entire change of term gives no reminder that the horizontal and vertical changes are components of the same vector. Again "lapse-rate" is to be reckoned positive when the quantity which varies is greater below than above, and this convention of signs is not always convenient. Lastly, some writers use lapse-rate without specifying the quantity which lapses; one assumes it to be temperature; but having done that, one hesitates to speak of the lapse-rate of anything else. As a way out of these difficulties the following notation is suggested. Let $p$ be any scalar

$\frac{\partial p}{\partial h}$ = up-gradient of $p$; $\frac{\partial p}{\partial e}$ = east-gradient of $p$; $\frac{\partial p}{\partial n}$ = north-gradient of $p$.

$-\frac{\partial p}{\partial h}$ = down-gradient of $p$; $-\frac{\partial p}{\partial e}$ = west-gradient of $p$; $-\frac{\partial p}{\partial n}$ = south-gradient of $p$.

$\sqrt{\left(\frac{\partial p}{\partial e}\right)^2 + \left(\frac{\partial p}{\partial n}\right)^2}$ = level-gradient of $p$; $\sqrt{\left(\frac{\partial p}{\partial e}\right)^2 + \left(\frac{\partial p}{\partial n}\right)^2 + \left(\frac{\partial p}{\partial h}\right)^2}$ = gradient of $p$.

In the foregoing pages "grade" is sometimes used for "gradient," but that was perhaps a mistake.

# INDEX OF PERSONS

*The numbers refer to the pages*

# INDEX OF SUBSIDIARY SUBJECTS

(*For the main topics the reader is referred to the table of contents on p. xi.*)

PRINTED IN ENGLAND BY J. B. PEACE, M.A.,
AT THE CAMBRIDGE UNIVERSITY PRESS

ıted in the United States
3aker & Taylor Publisher Services